Nutrient Cycling and Plant Nutrition in Forest Ecosystems

Special Issue Editors

Scott X. Chang
Xiangyang Sun

MDPI

Special Issue Editors
Scott X. Chang
University of Alberta
Canada

Xiangyang Sun
Beijing Forestry University
China

Editorial Office
MDPI AG
St. Alban-Anlage 66
Basel, Switzerland

This edition is a reprint of the Special Issue published online in the open access journal *Forests* (ISSN 1999-4907) from 2016–2017 (available at: http://www.mdpi.com/journal/forests/special_issues/nutrient_ecosystems).

For citation purposes, cite each article independently as indicated on the article page online and as indicated below:

Author 1; Author 2; Author 3 etc. Article title. *Journal Name*. **Year**. Article number/page range.

ISBN 978-3-03842-384-3 (Pbk)
ISBN 978-3-03842-385-0 (PDF)

Table of Contents

About the Guest Editors

Scott Chang received his BSc from Zhejiang Agricultural University; MSc from the Chinese Academy of Sciences; and PhD from the University of British Columbia. He is currently Professor of Forest Soils and Nutrient Dynamics at the University of Alberta. Scott Chang has held academic positions in New Zealand and the United States prior to taking up his current position. His main research interests are in forest soils, soil nutrient cycling and plant nutrition. Scott Chang served as Chair of the Soil Fertility and Plant Nutrition Commission of the International Union of Soil Science; President of the Association of Chinese Soil & Plant Scientists in North America; Chair for the Forest, Range and Wildland Soils Division of the Soil Science Society of America; and Chair of the Alberta Soil Science Workshop. He has served as an associate/guest editor or editorial board member for *Biology and Fertility of Soils, Pedosphere, Soil Science Society of America Journal, Canadian Journal of Soil Science, Journal of Environmental Quality, Environmental Science and Pollution Research, Forests,* and *Forest Ecology and Management.* He is a fellow of the Soil Science Society of America and the American Society of Agronomy.

Xiangyang Sun received his BSc, MSc and PhD degrees from Beijing Forestry University. He is currently Professor of Forest Soils and Deputy Dean of College of Forestry at Beijing Forestry University. Xiangyang Sun spent time at the University of Toronto and University of Alberta as a visiting scholar and has visited Denmark, Japan, Israel and New Zealand on academic missions. His main research interests are in soil carbon sequestration, greenhouse gas emissions, recycling of forest and agricultural waste material, horticultural growth media development, plant nutrition, and forest soil genesis and classification. He has trained a large number of MSc and PhD students in his capacity as a professor and the leader of the forest soils discipline. Xiangyang Sun has served as an associate editor for *Journal of Beijing Forestry University*, as a guest editor for *Forests*, and as an editorial board member for *Scinetia Silvae Sinicae* (Chinese version). He is a member of the Chinese National Committee for International Geosphere-Biosphere Programme.

Preface to "Nutrient Cycling and Plant Nutrition in Forest Ecosystems"

Understanding nutrient cycling in forest ecosystems is crucial for managing forest resources and mitigating global climate change; one of the applications of understanding nutrient cycling is the management of plant/forest nutrition and forest productivity [1]. Nutrient cycling and plant nutrition have been subjects of research in forest science for a long time but there have been new focuses in dealing with current environmental issues; for example, there has been recent focus on the effect of global change factors (such as drought and warming) on nutrient cycling and plant nutrition in forest ecosystems [2] and the need to better understand the effect of persistent atmospheric pollutants such as nitrogen and sulfur on forest ecosystems when such pollutants eventually return to the soil [3]. Disturbance regimes such as hurricanes that could occur on an annual basis and disturb a large tract of land can be devastating to the ecosystem and the society but little research has been conducted to understand the effect of such disturbances on soil biogeochemistry [4].

The call for submissions to the Special Issue on Nutrient Cycling and Plant Nutrition in Forest Ecosystems was met with an enthusiastic response and we have now finalized the Special Issue with 16 articles. The 16 articles include a review article, a meta-analysis article and 14 research articles. The 16 articles cover a broad range of topics from assessing the influence of sampling methods on the quantification of soil carbon and nitrogen contents and stocks [5] to a study on surface carbon dioxide exchange across a climatic gradient [6]. The Special Issue can be summarized into five themes: 1) disturbance effects on soil properties; 2) nutrient dynamics in forest soils and nutrient management; 3) soil–plant relations and forest ecophysiology; 4) carbon dioxide emissions from forest soils; and 5) methods for studying forest soil properties and productivities.

The first theme of the Special Issue includes a set of four articles that evaluated the effect of disturbance regimes (including harvesting, thinning and hurricane) on stocks of soil carbon and nitrogen and other nutrients. In Menegale et al. [7], the effect of timber harvest intensity (stem-only, whole tree harvest and whole tree harvest plus litter layer removal) on the stocks of available phosphorus and sulfur, total nitrogen, and oxidizable carbon in an Oxisol planted to Eucalyptus (*Eucalyptus grandis*) was studied in Brazil. Twelve years after the treatments were applied, nutrient stocks in the 0–20 cm layer were reduced in all treatments as compared with the pre-experiment values and the reduction increased with increasing harvesting intensity, demonstrating the negative impact of the intensive harvesting regimes. In a similar study on medium-aged Norway spruce (*Picea abies* L. Karst.) and Scots pine (*Pinus sylvestris* L.) stands in Germany [8], whole tree harvesting caused a much greater export of nutrients than whole tree harvesting but with needles retained on site, with the lowest nutrient export in the stem-only harvesting, consistent with Menegale et al. [7] in terms of the harvesting intensity effect on nutrient availability. Lodge et al. [4] considered downed tree trunks (or coarse woody debris) from hurricanes as a critical component of the disturbance effect and showed that coarse woody debris affected root length growth and soil properties in upslope vs downslope positions relative to the coarse woody debris. On this theme, the article by James and Harrison [9] provides a very good summary of the effect of timber harvesting on soil carbon using a meta-analysis approach. They reported that harvesting reduces soil carbon, on average, by 11.2%, with the 95% confidence interval between 14.1 and 8.5%; the magnitude of the loss is dependent on the soil horizon or depth evaluated, and the soil order. The largest loss occurred in the surface organic horizons, or the forest floor.

In the second theme on nutrient dynamics in forest soils and nutrient management, five articles are included. Among the five articles, Huang et al. [10] reported that native and invasive earthworms had different abilities for resource utilization and their subsequent effect on soil carbon and nitrogen dynamics and the colonization of invasive earthworms were also different. The spatial patterns of the distribution of soil nutrient stocks were different among nitrogen, phosphorus and potassium in Moso bamboo (*Phyllostachys heterocycla* (Carr.) Mitford cv. *Pubescens*) forests in southern China, implying a need for differential fertilization rates if those nutrients need to be added to address nutrient deficiencies [11]. Phosphorus is increasingly being recognized as a nutrient that can be limiting for forest productivity; Julich et al. [12] reported that 1) the composition of phosphorus forms in preferential flow pathways was different from that in the soil matrix; and 2) labile organically-bound phosphorus accumulates in preferential flow pathways such as biopores in temperate beech (*Fagus sylvatica* L.) forests with contrasting soil phosphorus contents in Germany. Within this theme, two articles are on fertilizer use efficiency and forest nutrient management. Raymond et al. [1] used enhanced efficiency fertilizer products, as compared with urea, to improve fertilizer nitrogen use efficiency in loblolly pine (*Pinus taeda* L.) plantations in southeastern United States and showed that fertilizer N recovery at the ecosystem level and fertilizer nitrogen use efficiency were greater for enhanced efficiency fertilizer products than for urea, demonstrating a new way to improve fertilizer nitrogen use efficiency. In the last article on this theme, Parent and Coleman [13] provide an excellent overview of nutrient management for grand Fir (*Abies grandis* (Douglas ex D. Don) Lindley) in Inland northwestern United States.

Four articles are published on the third theme: soil–plant relations and forest ecophysiology. In an interesting article by Xu et al. [14], soil and vegetation data from a 30-ha old-growth broad-leaved Korean pine (*Pinuskoraiensis*) forest were used to illustrate that soil properties strongly affect forest community structure and the soil–plant relationship was more strongly expressed with the canopy species. Of the four evergreen Fagaceae species studied in southwestern Japan, Kayama and Yamanaka [15] discovered that the effect of inoculation of the tree species with ectomycorrhizal fungi for planting on a calcareous soil depends on the tree and ectomycorrhizal species. Guo et al. [16] reported that a combination of nitrogen, phosphorus and potassium fertilizer application was most effective in improving the net photosynthesis rate and growth (height and diameter) of *Ginkgo biloba* L. trees in central eastern China. In a unique study by Chen et al. [17], the seasonal dynamics of nutrient resorption and concentration of phenolics in association with leaf senescence of a mangrove shrub species (*Aegiceras corniculatum* (L.) Blanco) was studied. They reported high nitrogen and phosphorus resorption efficiencies and suggested that high total phenolics to nitrogen and total condensed tannin to nitrogen ratios in senescent leaves were likely nutrient conservation strategies for the studied mangrove species.

The fourth theme has a lone article that reports a 3-year dataset on surface net C exchange and ecosystem respiration across different landforms, including upland, peat plateau and collapse scar, in mid-boreal to high subarctic ecoregions in the Mackenzie Valley in northwestern Canada [6]. Such studies are urgently needed in addressing the potential effect of a warmer climate on ecosystem processes in the vulnerable high latitudes. The authors showed that upland forest sites were sources of CO_2, collapsed areas were sinks of CO_2, especially in the high subarctic, and peat plateaus were minor sources of CO_2.

The fifth and last theme on methods for studying forest soil properties and productivities includes two articles. The Metsaranta and Bhatti [18] article explored the possibility of using diameter increment reconstructed annually from tree-ring data to estimate the annual growth increments of the wood volume of five species: jack pine (*Pinus banksiana* Lamb.), lodgepole pine (*Pinus contorta* Dougl.var. *latifolia* Engelm.), black spruce (*Picea mariana* (Mill.) B.S.P.), white spruce (*Picea glauca* (Moench) Voss.), and trembling aspen (*Poplus tremuloides* Michx). They concluded that the success of such estimates depends on the tree species, the diameter class, and the equation used to estimate the volume. In the last article in this Special Issue, Wang et al. [5] compared the effect of two different soil sampling approaches (pedogenetic horizon versus fixed-depth) on the quantification of soil organic carbon and nitrogen content and storage in larch (*Larix gmelinii*) plantations in northeast China. They found that the sampling method effect was dependent on the soil sampling depth, site type and forest stand age. They suggested that future soil assessments should control for methodological differences.

We thank the authors for submitting their manuscripts to this Special Issue and for returning their revisions promptly. We also thank the reviewers for their contribution in the publication of this Special Issue; some of the reviewers graciously accepted to review a manuscript three times to make sure that the final product is of a high quality for publication. We also received strong support from the Journal's Editorial Office—the support substantially reduced the Guest Editors' workload. It was the collective efforts of the authors, the reviewers and the staff in the Office that made the publication of this Special Issue possible.

Scott X. Chang and Xiangyang Sun
Guest Editors

References

1. Raymond, J.E.; Fox, T.R.; Strahm, B.D. Understanding the fate of applied nitrogen in pine plantations of the southeastern United States using ^{15}N enriched fertilizer. *Forests* **2016**, *7*, 270.
2. Cusack, D.F.; Karpman, J.; Ashdown, D.; Cao, Q.; Ciochina, M.; Halterman, S.; Lydon, S.; Neupane, A. Global change effects on humid tropical forests: Evidence for biogeochemical and biodiversity shifts at an ecosystem scale. *Rev. Geophysics* **2016**, *54*, 523-610.
3. Tian, Y.; Haibara, K.; Chang, S.X.; Toda, H.; Fang, S.Z. Acid deposition strongly influenced element fluxes in a forested karst watershed in the upper Yangtze River region, China. *For. Ecol. Manage.* **2013**, *310*, 27-36.
4. Lodge, D.J.; Winter, D.; González, G.; Clum, N. Effects of hurricane-felled tree trunks on soil carbon, nitrogen, microbial biomass, and root length in a wet tropical forest. *Forests* **2016**, *7*, 264.
5. Wang, H.; Wang, W.; Chang, S.X. Sampling method and tree-age affect soil organic C and N contents in larch plantations. *Forests* **2017**, *8*, 28.
6. Startsev, N.; Bhatti, J.S.; Jassal, R.S. Surface CO_2 exchange dynamics across a climatic gradient in McKenzie valley: Effect of landforms, climate and permafrost. *Forests* **2016**, *7*, 279.
7. Menegale, M.L.C.; Rocha, J.H.T.; Harrison, R.; Goncalves, J.L.D.M.; Almeida, R.F.; Piccolo, M.D.C.; Hubner, A.; Junior, J.C.A.; Ferraz, A.D.V.; James, J.N.; Michelsen-Correa, S. Effect of timber harvest intensities and fertilizer application on stocks of soil C, N, P, and S. *Forests* **2016**, *7*, 319.
8. Knust, C.; Schua, K.; Feger, K.H. Estimation of nutrient exports resulting from thinning and intensive biomass extraction in medium-aged spruce and pine stands in Saxony, northeast Germany. *Forests* **2016**, *7*, 302.
9. James, J.; Harrison, R. The effect of harvest on forest soil carbon: A meta-analysis. *Forests* **2016**, *7*, 308.
10. Huang, C.Y.; González, G.; Hendrix, P.F. Resource utilization by native and invasive earthworms and their effects on soil carbon and nitrogen dynamics in Puerto Rican soils. *Forests* **2016**, *7*, 277.
11. Tang, X.; Xia, M.; Guan, F. Spatial distribution of soil nitrogen, phosphorus and potassium stocks in Moso bamboo forests in subtropical China. *Forests* **2016**, *7*, 267.
12. Julich, D.; Julich, S.; Feger, K.H. Phosphorus in preferential flow pathways of forest soils in Germany. *Forests* **2017**, *8*, 19.
13. Parent, D.R.; Coleman, M.D. Grand fir nutrient management in the inland northwestern USA. *Forests* **2016**, *7*, 261.
14. Xu, W.; Hao, M.; Wang, J.; Zhang, C.; Zhao, X.; Gadow, K.V. Soil elements influencing community structure in an old-growth forest in northeastern China. *Forests* **2016**, *7*, 159.
15. Kayama, M.; Yamanaka, T. Growth characteristics of ectomycorrhizal seedlings of *Quercus glauca, Quercus salicina, Quercus myrsinaefolia, and Castanopsis cuspidate* planted in calcareous soil. *Forests* **2016**, *7*, 266.
16. Guo, J.; Wu, Y.; Wang, B.; Lu, Y.; Cao, F.; Wang, J. The effects of fertilization on the growth and physiological characteristics of *Ginkgo biloba* L. *Forests* **2016**, *7*, 293.
17. Chen. H.; Xu, B.; Wei, S.; Zhang, L.; Zhou, H.; Lin, Y. Nutrient resorption and phenolics concentration associated with leaf senescence of the subtropical mangrove *Aegiceras corniculatum*: Implications for nutrient conservation. *Forests* **2016**, *7*, 290.

18. Metsaranta, J.M.; Bhatti, J.S. Evaluation of whole tree growth increment derived from tree-ring series for use in assessments of changes in forest productivity across various spatial scales. *Forests* **2016**, 7, 303.

x

forests

MDPI

Article

Soil Elements Influencing Community Structure in an Old-Growth Forest in Northeastern China

Wei Xu [1,†], **Minhui Hao** [1,†], **Juan Wang** [1,*], **Chunyu Zhang** [1], **Xiuhai Zhao** [1] and
Klaus von Gadow [2,3]

[1]   The key Laboratory for Forest Resources & Ecosystem Processes of Beijing, Beijing Forestry University,
      Beijing 100083, China; xuweistudy@163.com (W.X.); 960954652@qq.com (M.H.); zcy_0520@163.com (C.Z.);
      zhaoxh@bjfu.edu.cn (X.Z.)
[2]   Faculty of Forestry and Forest Ecology, Georg-August-University Göttingen, Büsgenweg 5,
      Göttingen D-37077, Germany; kgadow@gwdg.de
[3]   Department of Forest and Wood Science, University of Stellenbosch, Stellenbosch 7600, South Africa
*   Correspondence: claire_wangj@163.com; Tel./Fax: +86-10-62337605
†   These authors contributed equally to this work.

Academic Editors: Scott X. Chang and Xiangyang Sun
Received: 6 May 2016; Accepted: 26 July 2016; Published: 28 July 2016

Abstract: This study uses detailed soil and vegetation data collected in a 30-ha old-growth broad-leaved Korean pine forest to study the effect of soil properties on tree community structures. Spatial distribution patterns are simulated using a homogeneous Poisson process (HomP) and a homogeneous Thomas process (HomT). The simulated distributions are compared with the observed ones to explore correlations between certain tree species and several soil elements. The HomP model shows that all tested tree species are significantly correlated with at least one principal component in the upper-layer soil elements. The HomT model shows that only 36.4% of tree species are significantly correlated with the principal component of at least one upper-layer soil element. This result shows that the impact of dispersal limitation is greater than impact of environmental heterogeneity on species spatial distributions. The spatial autocorrelation of species induced by the dispersal limitation will largely conceal the plant-soil relationships caused by the heterogeneity of soil elements. An additional analysis shows that the elements in the upper soil layer which have the greatest impact on community niche structure are Pb, total phosphorus (TP), total nitrogen (TN), Cu, Cr, Zn and available nitrogen (AN). The corresponding elements in the lower soil layers are Pb, TP, Cu, organic carbon (OC), Mn, total potassium (TK) and AN. Different species seem to be complementary regarding the demands on the available soil resources. The results of this study show that the tree species in the different growth groups have different habitat preferences. Compared with subcanopy and shrub species, the canopy species have more significant correlations with the soil elements.

Keywords: plant-soil relationships; dispersal limitation; habitat filtering; soil elements

1. Introduction

The impact of species dispersal and habitat filtering on the species distribution at medium (1–100 km^2) and large landscape scales (100–10,000 km^2) can be easily quantified if the soil resources of a particular forest community show a mosaic pattern and exhibit obvious heterogeneity characteristics [1], in which case certain habitat factors may have a significant impact on the species spatial distribution [2–4]. At the local scale (<1 km^2), the pattern induced by a species dispersal limitation is very similar to that induced by the habitat heterogeneity. It is therefore relatively difficult to distinguish between the impact of species dispersal and habitat filtering on a species' spatial distribution. A considerable amount of evidence shows that the species distribution is often closely correlated with certain soil elements, as well as with the soil texture, terrain features, parent material and other factors at a local scale [5]. Such significant plant-soil relationships occur widely in temperate forests, subtropical forests and tropical forests [5–8].

Seed dispersal is an important ecological process affecting community structuring. Dispersal limitation will occur when the seeds cannot arrive at a new suitable location, and available evidence shows that species dispersal limitation is often found in the forest communities [9]. However, whether dispersal limitation will have a significant impact on community spatial patterns is still in dispute [10]. Habitat filtering is another important ecological process affecting community assemblage and species coexistence [11,12]. The habitat filtering effect may significantly impact the growth and survival of seedlings. A species adapted to a particular habitat can survive and coexist with other species, while a species which is not adapted to that habitat may be eliminated [13,14]. In tropical and subtropical forests, certain habitat factors such as topography and soil conditions may cause many plant species to show clustered distributions. The spatial distribution of plant species is often highly correlated with certain habitat factors [15,16]. Habitat heterogeneity isolates those species which are unable to coexist into different habitat patches. For example, different species have different competitive capacities in different habitat patches and respectively occupy the patches where they have the strongest competitive capacity [17]. In undisturbed forests, the resource heterogeneity shows a very low contribution to species diversity, while the resource quantity in heterogeneous habitats is the main driving force impacting on the species diversity [18].

Studies involving the relationship between species distributions and environmental factors often consider the topographic variables or habitat types based on specific topographic variables [8,19]. However, the topographical factors are not suitable to replace the soil property for disclosing the impact of environmental factors on the species distribution [20]. Therefore, it is necessary to adopt direct environmental factors such as soil elements in the study of plant-soil relationships [6]. In addition, different tree species contribute differently to the soil nutrient composition through their mycorrhiza and litter. Bonifacio et al. [21] found that the content of organic matter in the soil of a broad-leaved spruce mixed forest was lower than that in a pure spruce forest. Moreover, the organic matter in the local soil layer of the pure spruce forest was obviously enriched, but all of the soil organic matter in the broad-leaved spruce mixed forest was distributed uniformly.

The importance of the soil resource heterogeneity to the maintenance of community diversity has been widely recognized, but the impact of the soil resource quantity on the species distribution has not yet been evaluated systematically. Temperate forests account for a high proportion of the world's forest resources, and research regarding the relationship between soil resources and species distribution is helpful in revealing the structuring mechanisms of temperate forest communities. This study evaluates the relationship between species spatial distributions and soil resources in an old-growth broad-leaved Korean pine forest community and attempts to answer the following questions: (1) Do the soil elements significantly affect the species spatial distribution in a broad-leaved Korean pine forest? (2) If yes, which soil elements are closely related to the forest community niche structure? In order to determine the actual relationship between the plant and soil elements, we eliminate the disturbance resulting from the species dispersal using specific spatial simulation techniques.

2. Materials and Methods

2.1. Field Data Acquisition

The research area is located within the Jiaohe Forest Experimental Area Administration Bureau of Jilin province, where a temperate continental mountain climate affected by monsoons and an average temperature of 3.7 °C prevails. July is the hottest month, with an average temperature of 21.7 °C. January is the coldest month, with an average temperature of −18.6 °C. The average annual precipitation is 695.9 mm. The soil type is a dark brown forest soil according to the Chinese soil taxonomy [22], with an average depth of 45 cm. In winter, the soil surface freezing time is about

150 days, with a frozen layer depth of 1.0–2.5 m. Because of the strong biological accumulation of the humus layer effect, the soil has high organic carbon content.

In 2010, a 30-ha (500 m × 600 m) research plot was established in a typical old-growth broad-leaved Korean pine forest which is located at 43°57.928′–43°58.214′ N and 127°45.287′–127°45.790′ E, far from urban areas and is virtually unaffected by human disturbance. The altitude of the research plot is 576.03–784.18 m, and the altitudinal difference between the highest and the lowest point is 208 m. All woody plants with a breast height diameter (dbh) exceeding 1 cm were tagged and mapped, their species were identified, and their dbhs, tree heights, crown widths, and heights to live crown were assessed. In this 30-ha observational study, a total of 49,678 woody plants (belonging to 48 species) were recorded. The dominant tree species are *Pinus koraiensis*, *Tilia amurensis*, *Fraxinus mandshurica*, *Ulmus laciniata* and *Acer mono*.

A combination of systematic and random sampling was used to assess soil properties. We randomly selected one direction from the eight directions around each 40 m × 40 m grid intersection and then randomly selected two distances from 2 m, 8 m and 15 m distances to form three combinations (2 m and 8 m, 2 m and 15 m or 8 m and 15 m) in the selected direction (see [23] for a detailed description of this approach). Finally, a total of 540 sample points were obtained in the 30-ha plot. The plant root system in the plots was mainly concentrated in the 0–20 cm soil layer, especially in the upper 0–10 cm layer, where the root system was dominant. Therefore, soil samples were taken from the upper (0–10 cm) and lower soil layer (10–20 cm) at each sample point and 18 soil variables were assessed: available nitrogen (AN), total nitrogen (TN), available potassium (AK), total potassium (TK), available phosphorus (AP), total phosphorus (TP), organic carbon (OC), pH, Cu, Ni, Cd, As, Pb, Zn, Mo, Cr, Mn and Mg. We used ammonium nitrogen-indophenol blue colorimetry to measure the content of available nitrogen; the hydrochloric acid-ammonium fluoride method to assess available phosphorus; the ammonium acetate extraction method to assess available potassium; the Kelvin heating digestion method to assess total potassium, and the high-temperature external-heat potassium dichromate oxidation method to assess organic carbon. Heavy metal elements were extracted using the Mehlich3 general extracting agent. Concentrations were assessed using inductively coupled plasma atomic emission spectrometry (ICP-AES). The above experiments were carried out according to the recommendations of the China Soil Council [24].

2.2. Relationship between Species Distribution and Soil Elements

Principal component analysis (PCA) was used to identify several composite factors representing various original variables by means of the dimensionality reduction method, and to allow these composite factors to retain the greatest possible amount of information for a large number of original variables. These composite factors had to be mutually unrelated. The 18 soil variables used in this paper have a very high correlation, so there is also a serious multicollinearity problem. In order to compress and extract the main information of the soil elements, principal component analysis was used for these soil elements and the pH values.

The Monte Carlo method was used to simulate species spatial distributions. We assume that the heterogeneous spatial distribution of the soil elements causes specific species distribution patterns. To evaluate this assumption, we applied two spatial point process models, the homogeneous Poisson process (HomP) and the homogeneous Thomas process (HomT) for simulating specific spatial distributions. The homogeneous Poisson process (HomP) simulates a completely random spatial species distribution and is used to test the significance of the plant-soil correlation. The homogeneous HomP thus assumes that all individuals of a certain plant species are randomly distributed in the plots. However, most of the tree species are not randomly distributed, thus it is necessary to consider a non-random process [24]. Therefore, we also used a homogeneous Thomas process (HomT) to simulate the autocorrelation of the species spatial distribution, which may be the result of dispersal processes [25–27]. The Poisson process model is usually expressed as Equation (1):

$$\rho\left(\mu\right) = \alpha \cdot \exp\left(z_{1:k}\left(\mu\right)\beta_{1:k}^{T}\right) \tag{1}$$

When $\rho\left(\mu\right)$ is a constant, the point pattern is represented by a homogeneous Poisson process (completely random); $z_{1:k}\left(\mu\right)$ is the function of environment variables where μ is the location of a point; α is the average density of the species distribution and $\beta_{1:k}^{T}$ are the coefficients.

The homogeneous Thomas process is usually expressed by Equation (2):

$$\rho_c\left(\mu\right) = \alpha k\left(\mu - c; \delta\right) \tag{2}$$

where the parameter c defines the location of the parent tree, and $k\left(\mu - c; \delta\right)$ defines the dispersal model of the offspring around the parent trees.

The spatial point process model was used to simulate the distribution pattern of each species in the observational study area [28]. For each species, there are 1000 distribution maps, including 999 simulated maps and one observed species map. The simulated species distribution maps were then matched with the principle components (PCs) distribution maps of the soil elements. The Pearson coefficient between each simulated species distribution and the PC distribution map among the 40 m × 40 m square cells were calculated. Thus, the 95% confidence interval can be calculated based on 999 Pearson coefficients. When the Pearson coefficient between the observed species distribution and the PCs is greater than the upper limit of the 95% confidence interval, then the relationship between the species distribution and the PCs of the soil elements represents a significant positive correlation. The correlation will be negative if the Pearson coefficient between the observed species distribution and the PCs is smaller than the lower limit of the 95% confidence interval. Alternatively, the correlation will be non-significant.

To ensure reasonable sample sizes, the analysis was limited to the 33 most abundant species (Table S1) which are represented by at least 30 individuals in the study area. In accordance with the potential maximum height (hmax), the tree species were assigned to one of three groups: canopy species (hmax $\geqslant$ 15 m), subcanopy species (5 m $\leqslant$ hmax < 15 m) and shrub species (hmax < 5 m). The 33 species include 23 canopy species, 5 subcanopy species and 5 shrub species.

The Canonical Correspondence Analysis (CCA) was used to study the relationship between the species composition and the soil environment. The species composition was expressed by a species-site matrix, and the soil environment was expressed by 18 soil elements.

The ecological niche breadth, which reflects the degree of adaptation to different environments, is reciprocal to the degree of ecological specialization [29]. In this study, we evaluate Levins' ecological niche breadth of tree species represented by at least 30 individuals [30], and test the degree of ecological specialization of each species with regard to the soil elements. The Levins' ecological niche breadth indicator formula is specified in Equation (3):

$$B_i = \left(\sum N_{ij}\right)^{2} / \left(\sum N_{ij}^{2}\right) \tag{3}$$

where B_i is the ecological niche breadth of species i, and N_{ij} is the resource value in the j'th resource grade for species i. In order to evaluate the overall impact of each soil variable on the community niche structure, a boxplot for the ecological niche breadth of all studied species was drawn for each soil variable.

All calculations were performed using the R statistical software [31]. The principal component analysis was implemented using the "vegan", "ade4", "gclus" and "ape" packages.

3. Results

The difference in the species compositions among the 40 m × 40 m cells was significantly correlated to the difference in the average concentrations of soil elements in those cells for both the upper and the lower soil layer. The CCA ordination analysis regarding soil nutrient and community composition shows that most soil variables are significantly correlated with the CCA ordination axes (Figure 1, Table 1). However, some soil elements show similar effects according to the lengths and directions of the soil nutrient vectors. Surprisingly, major soil elements (such as N, P and K) do not show a stronger influence than the metal elements (such as Ni, As and Mg) in both soil layers. The same nutrient shows a consistent direction in both soil layers, but the nutrient vectors in the upper soil layer are longer than those in the lower layer (Figure 1).

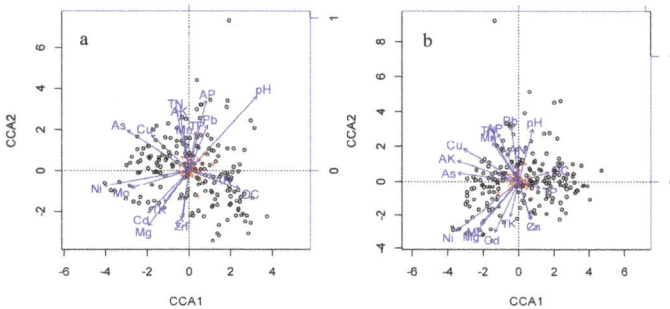

Figure 1. Canonical correspondence analysis (CCA) in two soil layers involving the upper (**a**) and lower layer (**b**). The species abundance and average values of soil nutrients in the 40 m × 40 m plots were used to test the relationship between species composition and soil elements. The species scores and site scores of the first two ordination axes are drawn in the figure. The arrow direction indicates the positive and negative correlation between the soil nutrient and ordination axes, while the length of the arrow represents the degree of correlation between certain soil nutrient variables and the community composition.

Table 1. Pearson's correlations between soil variables and the first two axes of canonical correspondence analysis (CCA).

Soil Variables	Upper Soil Layer		Lower Soil Layer	
	CCA1	CCA2	CCA1	CCA2
available nitrogen	−0.12	−0.13	−0.11	−0.14
total nitrogen	−0.91 ***	0.07	−0.91 ***	0.27 ***
available potassium	0.13	0.05	0.03	−0.05
total potassium	−0.02	−0.41 ***	−0.05	−0.36 ***
available phosphorus	−0.05	−0.36 ***	−0.22 **	−0.44 ***
total phosphorus	−0.05	0.12	0.00	−0.01
organic carbon	0.08	−0.51 ***	−0.07	−0.39 ***
Cu	0.07	−0.26 ***	0.08	0.10
Ni	0.15 *	0.19 *	0.05	0.18 *
Cd	−0.16 *	−0.23 **	−0.19 *	−0.41 ***
As	−0.34 ***	0.02	−0.30 ***	0.03
Pb	−0.10	0.17 *	−0.07	0.15 *
Zn	−0.21 **	−0.31 ***	−0.21 **	−0.29 ***
Mo	0.05	−0.26 ***	0.03	−0.35 ***
Cr	0.03	0.33 ***	0.05	0.28 ***
Mn	−0.16	−0.01	−0.09	−0.03
Mg	0.02	0.31 ***	0.04	0.29 ***
pH	−0.04	−0.23 **	−0.04	−0.36 ***

* indicates $p < 0.05$; ** indicates $p < 0.01$; *** indicates $p < 0.001$.

As soil nutrient variables are significantly correlated with each other (Table S2), we obtained a group of optimum orthogonal unit vectors based on principal component analysis. The first four PC$_S$ (PC1, PC2, PC3 and PC4) are not related to one another and jointly explain 53%–54% of the variation in soil elements in the two soil layers. In the two layers, the first principal component (PC1) accounts for 20.3%–21.9% of the variation in soil nutrient concentrations, while the second (PC2) accounts for 11.3%–14.1%. The third principal component (PC3) accounts for 10.7%–10.9% and the fourth (PC4) for 8.9%–9.1% (Table 2).

Table 2. Soil variable loadings on the four principal components (PCs) in the two soil layers.

Soil Variable	Upper Soil Layer				Lower Soil Layer			
	PC1	PC2	PC3	PC4	PC1	PC2	PC3	PC4
available nitrogen			−0.387	−0.318		−0.445	−0.122	0.111
total nitrogen		−0.130	−0.370	−0.219		−0.419		0.164
available potassium	0.191		−0.306			−0.369	0.140	0.195
total potassium	−0.229		0.296	−0.156				0.260
available phosphorus		−0.379		−0.119		−0.190	0.145	−0.160
total phosphorus							0.142	−0.152
organic carbon	0.318	0.355	−0.203		0.202	−0.267	−0.308	−0.238
Cu		−0.252	−0.345	0.422	−0.103	−0.222	0.442	−0.356
Ni	−0.184	0.248		0.315	−0.259	0.153		−0.264
Cd	−0.389	−0.134	−0.150		−0.415		0.161	0.210
As	−0.270	−0.378	−0.154		−0.293	−0.165	0.318	0.159
Pb	0.113		−0.232	0.493		−0.103	0.393	−0.494
Zn	−0.318	0.401	−0.234	−0.141	−0.348		−0.379	−0.25
Mo	−0.421	−0.216			−0.430		0.127	0.317
Cr	−0.311	0.401	−0.238	−0.157	−0.334	−0.102	−0.380	−0.255
Mn	−0.158		−0.224	0.317	−0.196	−0.296	−0.174	
Mg	−0.292	0.176		0.306	−0.372			
pH	0.196		−0.318	−0.177	0.149	−0.384		
Variance (%)	21.9	11.3	10.9	8.9	20.3	14.1	10.7	9.1
Species ($n \geqslant 5$)	25	25	16	23	24	20	19	17
Species ($n \geqslant 30$)	24	23	14	20	23	20	17	16

PC1 is the most important principal component, reflecting the contribution of the metal elements. In the upper soil layer, PC1 shows a positive correlation with Pb, and a negative one with other elements. In the lower soil layer, PC1 has negative correlations with all metal elements except Pb. The loading values of each soil nutrient variable in PC3 show a large difference. PC3 exhibits a negative correlation with most soil variables in the upper soil layer, but a positive one with most soil variables in the lower soil layer. PC2 and PC4 together represent all the information on various soil elements.

When the Monte Carlo method was used to test the relationship between different principal components and species distributions, most species show a significant correlation with the principal components. The difference in species showing significant correlation with the PCs before and after eliminating rare species ($n < 30$) is minor. Only very few (1–3) rare species show significant correlations with the soil elements (Table 2).

When testing the tree species-soil relationships, four PC axes are used, which replace the individual soil elements to reduce the testing frequency. The HomP model was used to establish the confidence interval and to test whether the species distribution is closely related with the soil elements represented by the PCs. All studied tree species show a significant correlation with at least one principal component of soil elements in the upper soil layer, while 87.9% of the species are significantly related to at least one principal component of elements in the lower soil layer. The HomT model eliminates the effect of the species dispersal limitation, and the results show that only a few tree species exhibit a preference for particular soil elements. A total of 12 tree species are significantly related to at least one principal component of elements in the upper soil layer (accounting for 36.4% of all studied tree species), and 10 tree species are significantly related with at least one principal component of elements in the lower layer (accounting for 30.3% of all studied tree species). The results between the HomT and HomP models are very different. We assume that the spatial autocorrelation of species distributions induced by the dispersal limitation may confound the true plant-soil relationships induced by the spatial heterogeneity of soil elements (Tables 3 and 4).

Table 3. Relationships between species distribution and one of the first four PC axes of soil nutrients in the upper soil layer. HomP represents the homogeneous Poisson process; HomT represents the homogeneous Thomas process.

Tree Species	HomP				HomT			
	PC1	PC2	PC3	PC4	PC1	PC2	PC3	PC4
Betula platyphylla		−	−	−				
Acer mandshuricum	+	−	−				−	
Padus racemosa	−		−	−				
Abies nephrolepis	+	+		+				
Ulmus davidiana var. *japonica*	+		−	+				
Ulmus macrocarpa	+	−	−					
Betula costata	−				−			
Betula dahurica	+							
Pinus koraiensis	−	+	−					
Juglans mandshurica		+	−	−				
Sorbus pohuashanensis	+			+				
Phellodendron amurense	−	−		−				
Tilia mandschurica	+	+				+		
Ulmus laciniata	+	−		−	+			−
Quercus mongolica	+	−			+			
Carpinus cordata	+	+		−				
Acer tegmentosum	−	+	+	−	−		+	
Acer mono	+		−	−				
Abies holophylla		+		−				
Fraxinus mandshurica	+	−			+	−		
Sorbus alnifolia		−				−		
Populus koreana		−						
Tilia amurensis				+				+
Syringa reticulata var. *amurensis*	+	−	−	+		+		+
Acer barbinerve	−	+		+				
Tilia mandschurica		+		+				
Acer ukurunduense	−	+	+	+			+	
Lonicera maackii			+					
Euonymus macropterus		+						
Eleutherococcus senticosus	+		−	+			−	
Euonymus pauciflorus	−	+		+				
Corylus mandshurica	−	−	−	+				
Rhamnus davurica	−							
Significant negative correlation	10	11	11	9	2	2	2	1
Significant positive correlation	14	12	3	11	3	2	2	2
Non-significant correlation	9	10	19	13	28	29	29	30

Symbol "+" indicates significant positive correlation at the 0.05 level; Symbol "−" indicates significant negative correlation at the 0.05 level.

Table 4. Relationships between species distribution and one of the first four PC axes of soil elements in the lower soil layer. HomP represents the homogeneous Poisson process; HomT represents the homogeneous Thomas process.

Plant Species	HomP				HomT			
	PC1	PC2	PC3	PC4	PC1	PC2	PC3	PC4
Betula platyphylla	+	−	+					
Acer mandshuricum	+	−	+	−			+	
Padus racemosa	−	−						
Abies nephrolepis	+			−				
Ulmus davidiana var. japonica	+	−	+	−				
Ulmus macrocarpa	+	−						
Betula costata	−	+			−	+		
Betula dahurica	+							
Pinus koraiensis	−	+				+		
Juglans mandshurica				+				
Sorbus pohuashanensis								
Phellodendron amurense		−	+	+				
Tilia mandschurica	+	+	−	−				
Ulmus laciniata	+			+	+			
Quercus mongolica	+							
Carpinus cordata	+	+	−	−				−
Acer tegmentosum	−	+	−	+		+		+
Acer mono	+							
Abies holophylla	−	+	−			+		
Fraxinus mandshurica	+	−	+	+				
Sorbus alnifolia			+	+				
Populus koreana		+	+				+	
Tilia amurensis								
Syringa reticulata var. amurensis	+	−	+	−		−		
Acer barbinerve	−	+	+					
Tilia mandschurica			−	−				
Acer ukurunduense	−	+	−					
Lonicera maackii			+	−				
Euonymus macropterus								
Eleutherococcus senticosus								
Euonymus pauciflorus	−	+		−		+		−
Corylus mandshurica	−	−	+	+				
Rhamnus davurica	−		−					
Significant negative correlation	10	10	6	9	1	1	1	1
Significant positive correlation	13	10	11	7	1	5	2	1
Non-significant correlation	10	13	16	17	31	27	30	31

Symbol "+" indicates significant positive correlation at the 0.05 level; Symbol "−" indicates significant negative correlation at the 0.05 level.

Significant species-soil relationships show up mainly for the canopy species. A total of 39.1% of canopy species (14 of 23 species) are significantly related to the PCs of soil elements in the upper soil layer, while 34.8% of canopy species (14 of 23 species) are significantly related to the PCs of soil elements in the lower soil layer. Two out of five subcanopy species show significant correlations with the PCs of soil elements in the upper soil layer, and only one out of five subcanopy species shows significant correlations with the PCs in the lower soil layer. Only one out of five shrub species shows significant correlations with PC3 of soil elements in the upper soil layer and with PC2 and PC4 of elements in the lower soil layer (Table 5).

Table 5. Species distribution-PC associations examined by the homogeneous Thomas process in different growth groups.

Soil Layer	Canopy Species	Subcanopy Species	Shrub Species
	23 Species	5 Species	5 Species
Upper			
Number of species related to PC1	5 (3+, 2−)	0 (0+, 0−)	0 (0+, 0−)
Number of species related to PC2	3 (1+, 2−)	1 (1+, 0−)	0 (0+, 0−)
Number of species related to PC3	2 (0+, 2−)	1 (1+, 0−)	1 (1+, 0−)
Number of species related to PC4	3 (3+, 0−)	1 (1+, 0−)	0 (0+, 0−)
Number of species unrelated to PCs	14	3	4
Lower			
Number of species related to PC1	2 (1+, 1−)	0 (0+, 0−)	0 (0+, 0−)
Number of species related to PC2	4 (4+, 0−)	1 (0+, 1−)	1 (1+, 0−)
Number of species related to PC3	3 (2+, 1−)	0 (0+, 0−)	0 (0+, 0−)
Number of species related to PC4	1 (1+, 0−)	0 (0+, 0−)	1 (0+, 1−)
Number of species unrelated to PCs	15	4	4

The first number in parentheses refers to the number of species that showed positive plant-soil relationships; the second number in parentheses is the number of species that showed negative plant-soil relationships.

The soil nutrient variables are highly correlated with each other (Table S2). It is thus difficult to evaluate the relative importance of individual soil elements in a community niche structure. We calculated the ecological niche breadth of all species for individual soil nutrient gradients and conclude that the smaller ecological niche breadth indicates a higher specialization degree of particular focal species. When the ecological niche breadth is relatively small, plant species show a weak adaptability to the variations in soil properties. Thus, the smaller the niche breadth for a given soil variable, the greater is the effect of that soil variable on the niche structure of the community. The soil elements which have the largest impacts on the community niche structure are Pb, TP, TN, Cu, Cr, Zn and AN in the upper soil layer and Pb, TP, Cu, OC, Mn, TK, AN, TN, Zn and Cr in the lower layer (Figure 2). The ecological niche breadths of AK, TK, AP, OC, AN, TN, Ni and pH show significant differences between the two soil layers. These nutritional elements, which are represented differently in the two soil layers (Table S4), are important for plant growth.

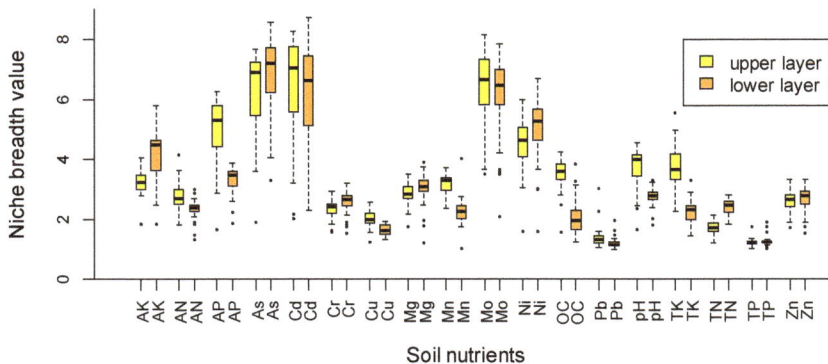

Figure 2. Boxplot of niche breadth values for 18 soil nutrients in two soil layers. The smaller the niche breadth for a given soil variable, the greater is the effect of that soil variable on the niche structure of the community.

4. Discussion

A particular community pattern involves specific ecological processes, and a particular process in turn represents the cause and driving force of many different patterns. Previous studies have shown that the old-growth broad-leaved Korean pine forest exhibits a strong dispersal limitation [32]. The joint effects of environmental heterogeneity and dispersal limitation can more effectively explain the species-area relationships and β diversity in the old-growth broad-leaved Korean pine forest [33]. Our study extracted four principal components to express the spatial variation of soil variables. Each principal component reflects different combinations of the soil characteristics. PC1 indicates the spatial variation of metal elements in soils such as Cu, Ni, Cd, As, Pb, Zn, Mo, Cr, Mn, Mg, while PC2, PC3 and PC4 indicate the spatial variation of the nutrients (nitrogen, phosphorus, potassium, organic carbon) and soil pH.

A species distribution simulation using a spatial point process model can directly evaluate the relationships with the nutrient PCs. The plant-soil relationships tested by the homogeneous Thomas model are less indicative than those tested by the homogeneous Poisson model: they are reduced by 63.6% in the upper soil layer, and by 57.6% in the lower layer. True effects of environmental filtering on the species distribution are reduced after eliminating spatial autocorrelation induced from a dispersal process. Therefore, we can conclude that the dispersal-limitation effect on a particular species distribution is greater than the effect of the environmental heterogeneity.

Previous studies have shown that soil elements strongly affect vegetation patterns in tropical and temperate forest communities [34–37]. Paoli et al. found that the spatial distributions of 18 out of 22 tree species in a tropical forest community were significantly associated with soil elements, especially with P, Mg and Ca [38]. They argued that dispersal and niche processes jointly determine mesoscale beta diversity in a *Bornean Dipterocarp* forest. John et al. [7] detected an effect of soil elements on the spatial distribution of particular tree species in three diverse neotropical forest plots. They found that the spatial distributions of 36%–51% of all the tree species in these plots show strong associations with soil nutrient distributions. B and K in their BCI plot, Ca and Mg in the Yasuni plot and K, P, Fe and N in the La Planada plot showed the strongest effects on community niche structure [7].

In a subtropical broad-leaved forest community, spatial heterogeneity of soil elements was associated with distributions of 88.2% (90 out of 102 species) of tree species after controlling the effects of dispersal limitation. The soil factors most strongly influencing species distributions were TC, TN, TP, K, Mg, Si, soil moisture and bulk density [6]. Our results showed that TP, TN, AN in the upper soil layer and TP, TK, TN, AN in the lower soil layer strongly affect species distributions. Nitrogen, phosphorus and potassium are the essential mineral elements for plant growth, as well as being rather limited soil elements, and their concentration directly affects plant growth and survival [39,40]. The variation coefficient of AN, AP, AK, TN, TP and TK in the broad-leaved Korean pine forest ranges from 0.16 to 0.44, and shows obvious spatial heterogeneity (Table S3). When the quantity of nitrogen, phosphorus and potassium in the habitat cannot meet the plants' demands, they will become important limiting factors for plant growth.

Metal elements show consistent effects on community structure in the two soil layers. Pb, Cu, Cr and Zn in both the upper and lower soil layers strongly affect the community niche structure of the old-growth forest. Previous studies have shown that metal elements have important effects on the growth, development and breeding of plants. Cu is a cofactor influencing the structure and catalytic compounds of proteins and enzymes and is thus essential for the normal growth and biochemical processes. However, excessive Cu will produce a large number of free radicals and malondialdehyde (MDA), causing metabolic disturbances and inhibiting plant growth [41].

Zn is an active cofactor for a large number of enzymes related to the metabolism of DNA transcription, protein, nucleic acid, carbohydrate and lipids [42,43]. Furthermore, Zn is the essential microelement for any living body, and Zn deficiency will cause a series of nutrition problems in plants. Pb is not an essential element and may cause enzyme function disorders, generating active oxygen in plant tissue cells, which will lead to membrane lipid peroxidation, and change the related enzymatic activity of active oxygen metabolism [44,45], which finally affects biochemical processes and the morphological structure of plants [46]. Cr is not an essential element, but may also affect plant growth and development. A high concentration of Cr may inhibit the root cell differentiation of plants, block water absorption, and have a toxic effect, while a low concentration of Cr can promote the growth of roots and root hairs, increase the cortical tissue layers in the roots, and thus facilitate plant growth [47,48]. A low concentration of Cr will also promote the formation of chlorophyll, while a high concentration will inhibit the formation of chlorophyll [49,50]. Mn in the lower soil layer has an especially severe impact on our community niche structure. Mn is an essential element in the PSII oxygen evolution complex [51–53], and its role is to realize water splitting in the reaction center of PSII and to provide electrons to the electron transfer chain as part of thylakoid coupling [54]. Therefore, an Mn deficiency will decrease the photosynthetic capacity of the leaves [55].

Different species groups were formed in accordance with their visible properties. Plant species usually belong to the same growth group under similar environmental conditions, which is the result of adaptive convergence [56]. Plant species in different growth groups have shown different soil habitat preferences in a subtropical forest plot. Only five shrub species were significantly related to the four soil property principal component axes, while most of the canopy species and subcanopy species were significantly related to 1–2 principal component axes [6]. Our results show that the tree species in the different growth groups have different habitat preferences. Compared with subcanopy and shrub species, the canopy species have more significant correlations with the PCs of the soil elements. In addition, different canopy species also show different soil preferences. Canopy species play a dominant role in structuring the forest community. Intense competition will occur when different tree species have similar demands for specific soil resources and some members of a species with weak competitive strength may die. On the other hand, the demands of different tree species on available soil resources may be complementary, which facilitates the full utilization of existing resources and the coexistence of species.

5. Conclusions

This work evaluates plant-soil correlations by comparing simulated the spatial distributions of tree species with observed ones in a large forest observational study. The impact of dispersal limitation is greater than impact of environmental heterogeneity on species spatial distributions. The spatial autocorrelation of species induced by the dispersal limitation will largely conceal the plant-soil relationships caused by the heterogeneity of soil elements. Certain elements in the upper and lower soil layers were identified to influence community niche structure. Tree species in different growth groups have different habitat preferences. Compared with subcanopy and shrub species, the canopy species have more significant correlations with the soil elements.

Supplementary Materials: The following are available online at www.mdpi.com/1999-4907/7/8/159/s1, Table S1: Number of individuals of the 33 most abundant species in the 30-ha study area, Table S2: Pearson's correlations among soil nutrients and pH values, Table S3: Variation coefficients of soil variables within the study area, Table S4: Descriptive statistics of the soil variables.

Acknowledgments: This research is supported by the Fundamental Research Funds for the Central Universities (2015ZCQ-LX-03), the State Key Program of National Natural Science Foundation of China (41330530) and the 12th five-year National Science and Technology plan of China (2012BAC01B03).

Author Contributions: W.X. and J.W. analyzed the data and wrote the manuscript. C.Z. and K.G. modeled and interpreted the results. X.Z. and M.H. performed the laboratory experiments.

Conflicts of Interest: The authors declare no conflict of interest. The founding sponsors had no role in the design of the study; in the collection, analyses, or interpretation of data; in the writing of the manuscript; and in the decision to publish the results.

References

1. Tuomisto, H.; Ruokolainen, K.; Aguilar, M.; Sarmiento, A. Floristic patterns along a 43-km long transect in an Amazonian rain forest. *J. Ecol.* **2003**, *91*, 743–756. [CrossRef]
2. Thessler, S.; Ruokolainen, K.; Tuomisto, H.; Tomppo, E. Mapping gradual landscape-scale floristic changes in Amazonian primary rain forests by combining ordination and remote sensing. *Glob. Ecol. Biogeogr.* **2005**, *14*, 315–325. [CrossRef]
3. Slik, J.W.; Raes, N.; Aiba, S.I.; Brearley, F.Q.; Cannon, C.H.; Meijaard, E.; Nagamasu, H.; Nilus, R.; Paoli, G.; Poulsen, A.D.; et al. Environmental correlates for tropical tree diversity and distribution patterns in Borneo. *Divers. Distrib.* **2009**, *15*, 523–532. [CrossRef]
4. Clark, D.B.; Olivas, P.C.; Oberbauer, S.F.; Clark, D.A.; Ryan, M.G. First direct landscape-scale measurement of tropical rain forest leaf area index, a key driver of global primary productivity. *Ecol. Lett.* **2008**, *11*, 163–172. [CrossRef] [PubMed]
5. Gleason, S.M.; Read, J.; Ares, A.; Metcalfe, D.J. Species-soil associations, disturbance, and nutrient cycling in an Australian tropical rainforest. *Oecologia* **2010**, *162*, 1047–1058. [CrossRef] [PubMed]
6. Zhang, L.; Mi, X.; Shao, H.; Ma, K. Strong plant-soil associations in a heterogeneous subtropical broad-leaved forest. *Plant Soil* **2011**, *347*, 211–220. [CrossRef]
7. John, R.; Dalling, J.W.; Harms, K.E.; Yavitt, J.B.; Stallard, R.F.; Mirabello, M.; Hubbell, S.P.; Valencia, R.; Navarrete, H.; Vallejo, M.; et al. Soil nutrients influence spatial distributions of tropical tree species. *Proc. Natl. Acad. Sci. USA* **2007**, *104*, 864–869. [CrossRef] [PubMed]
8. Zhang, C.; Zhao, Y.; Zhao, X.; Gadow, K.V. Species-Habitat Associations in a Northern Temperate Forest in China. *Silva Fenn.* **2012**, *46*, 501–519. [CrossRef]
9. Hubbell, S.P.; Foster, R.B.; O'Brien, S.T.; Harms, K.E.; Condit, R.; Wechsler, B.; Wright, S.J.; De Lao, S.L. Light-gap disturbances, recruitment limitation, and tree diversity in a Neotropical forest. *Science* **1999**, *283*, 554–557. [CrossRef] [PubMed]
10. Levine, J.M.; Murrell, D.J. The community-level consequences of seed dispersal patterns. *Annu. Rev. Ecol. Evol. Syst.* **2003**, *34*, 549–574. [CrossRef]
11. Beckage, B.; Clark, J.S. Seedling survival and growth of three forest tree species: The role of spatial heterogeneity. *Ecology* **2003**, *84*, 1849–1861. [CrossRef]
12. Lebrija-Trejos, E.; Pérez-García, E.A.; Meave, J.A.; Bongers, F.; Poorter, L. Functional traits and environmental filtering drive community assembly in a species-rich tropical system. *Ecology* **2010**, *91*, 386–398. [CrossRef] [PubMed]
13. Harper, J.L. *Population Biology of Plants*; Academic Press: London, UK, 1977.
14. Tilman, D.; Pacala, S. The maintenance of species richness in plant communities. In *Species Diversity in Ecological Communities*; Ricklefs, R.E., Ed.; University of Chicago Press: Chicago, IL, USA, 1993; pp. 13–25.
15. Webb, C.O.; Peart, D.R. Habitat associations of trees and seedlings in a Bornean rain forest. *J. Ecol.* **2000**, *88*, 464–478. [CrossRef]
16. Webb, C.O. Exploring the phylogenetic structure of ecological communities: An example for rain forest trees. *J. Am. Nat.* **2000**, *156*, 145–155. [CrossRef] [PubMed]
17. Williams, C.B. Patterns in the balance of nature and related problems of quantitative ecology. Academic Press: New York, NY, USA, 1964.
18. Bartels, S.F.; Chen, H.Y. Is understory plant species diversity driven by resource quantity or resource heterogeneity? *Ecology* **2010**, *91*, 1931–1938. [CrossRef] [PubMed]
19. Baldeck, C.A.; Harms, K.E.; Yavitt, J.B.; John, R.; Turner, B.L.; Valencia, R.; Navarrete, H.; Davies, S.J.; Chuyong, G.B.; Kenfack, D.; et al. Soil resources and topography shape local tree community structure in tropical forests. *Proc. R. Soc. Lond. B Biol.* **2013**, *280*, 20122532. [CrossRef] [PubMed]
20. Hall, J.S.; McKenna, J.J.; Ashton, P.M.S.; Gregoire, T.G. Habitat characterizations underestimate the role of edaphic factors controlling the distribution of Entandrophragma. *Ecology* **2004**, *85*, 2171–2183. [CrossRef]
21. Bonifacio, E.; Caimi, A.; Falsone, G.; Trofimov, S.Y.; Zanini, E.; Godbold, D.L. Soil properties under Norway spruce differ in spruce dominated and mixed broadleaf forests of the Southern Taiga. *Plant Soil* **2008**, *308*, 149–159. [CrossRef]
22. Institute of Soil Science, CAS. *Chinese Soil Taxonomy*; Science Press: Beijing, China, 2001.

23. Zhang, C.; Wei, Y.; Zhao, X.; von Gadow, K. Spatial characteristics of tree diameter distributions in a temperate old-growth forest. *PLoS ONE* **2013**, *8*, e58983. [CrossRef] [PubMed]

24. China Soil Council. *Soil Agricultural Chemical Analysis Procedure*; Chinese Agricultural Science Press: Beijing, China, 1999.

25. Harms, K.E.; Condit, R.; Hubbell, S.P.; Foster, R.B. Habitat associations of trees and shrubs in a 50-ha Neotropical forest plot. *J. Ecol.* **2001**, *89*, 947–959. [CrossRef]

26. Waagepetersen, R.P. An estimating function approach to inference for inhomogeneous Neyman-Scott processes. *Biometrics* **2007**, *63*, 252–258. [CrossRef] [PubMed]

27. Shen, G.; Yu, M.; Hu, X.S.; Mi, X.; Ren, H.; Sun, I.F.; Ma, K. Species-area relationships explained by the joint effects of dispersal limitation and habitat heterogeneity. *Ecology* **2009**, *90*, 3033–3041. [CrossRef] [PubMed]

28. Lin, Y.C.; Chang, L.W.; Yang, K.C.; Wang, H.H.; Sun, I.F. Point patterns of tree distribution determined by habitat heterogeneity and dispersal limitation. *Oecologia* **2011**, *165*, 175–184. [CrossRef] [PubMed]

29. Kohn, A.J. Microhabitats, abundance and food of Conus on atoll reefs in the Maldive and Chagos Islands. *Ecology* **1968**, *49*, 1046–1062. [CrossRef]

30. Levins, R. *Evolution in Changing Environments: Some Theoretical Explorations (No. 2)*; Princeton University Press: Princeton, NJ, USA, 1968.

31. Team, R.C. *R: A Language and Environment for Statistical Computing*; R Foundation for Statistical Computing: Vienna, Austria, 2012.

32. Li, B.; Hao, Z.; Bin, Y.; Zhang, J.; Wang, M. Seed rain dynamics reveals strong dispersal limitation, different reproductive strategies and responses to climate in a temperate forest in northeast China. *J. Veg. Sci.* **2012**, *23*, 271–279. [CrossRef]

33. Wang, X.; Wiegand, T.; Hao, Z.; Li, B.; Ye, J.; Lin, F. Species associations in an old-growth temperate forest in north-eastern China. *J. Ecol.* **2010**, *98*, 674–686. [CrossRef]

34. Read, J. Soil and rainforest composition in Tasmania: Correlations of soil characteristics with canopy composition and growth rates in *Nothofagus cunninghamii* associations. *Aust. J. Bot.* **2001**, *49*, 121–135. [CrossRef]

35. Lusk, C.H.; Matus, F. Juvenile tree growth rates and species sorting on fine-scale soil fertility gradients in a Chilean temperate rain forest. *J. Biogeogr.* **2000**, *27*, 1011–1020. [CrossRef]

36. Baltzer, J.L.; Thomas, S.C.; Nilus, R.; Burslem, D.F.R. Edaphic specialization in tropical trees: Physiological correlates and responses to reciprocal transplantation. *Ecology* **2005**, *86*, 3063–3077. [CrossRef]

37. Pyke, C.R.; Condit, R.; Aguilar, S.; Lao, S. Floristic composition across a climatic gradient in a Neotropical lowland forest. *J. Veg. Sci.* **2001**, *12*, 553–566. [CrossRef]

38. Paoli, G.D.; Curran, L.M.; Zak, D.R. Soil nutrients and beta diversity in the Bornean Dipterocarpaceae: Evidence for niche partitioning by tropical rain forest trees. *J. Ecol.* **2006**, *94*, 157–170. [CrossRef]

39. Reich, P.B.; Hobbie, S.E.; Lee, T.; Ellsworth, D.S.; West, J.B.; Tilman, D.; Knops, J.M.; Naeem, S.; Trost, J. Nitrogen limitation constrains sustainability of ecosystem response to CO_2. *Nature* **2006**, *440*, 922–925. [CrossRef] [PubMed]

40. Thuiller, W. On the importance of edaphic variables to predict plant species distributions-limits and prospects. *J. Veg. Sci.* **2013**, *24*, 591–592. [CrossRef] [PubMed]

41. Foyer, C.H.; Lelandais, M.; Kunert, K.J. Photooxidative stress in plant. *Physiol. Plantarum* **1994**, *92*, 708–719. [CrossRef]

42. Broadley, M.R.; White, P.J.; Hammond, J.P.; Zelko, I.; Lux, A. Zinc in plants. *New Phytol.* **2007**, *173*, 677–702. [CrossRef] [PubMed]

43. Marschner, H.; Rimmington, G. Mineral nutrition of higher plants. *Plant Cell Environ.* **1988**, *11*, 147–148.

44. Chaoui, A.; Mazhoudi, S.; Ghorbal, M.H.; El Ferjani, E. Cadmium and zinc induction of lipid peroxidation and effects on antioxidant enzyme activities in bean (*Phaseolus vulgaris* L.). *Plant Sci.* **1997**, *127*, 139–147. [CrossRef]

45. Gallego, S.M.; Benavides, M.P.; Tomaro, M.L. Effect of heavy metal ion excess on sunflower leaves: Evidence for involvement of oxidative stress. *Plant Sci.* **1996**, *121*, 125–159. [CrossRef]

46. Shah, K.; Dubey, R.S. Cadmium elevates level of protein, amino acids and alters activity of proteolytic enzymes in germinating rice seeds. *Acta Physiol. Plant.* **1998**, *20*, 189–196. [CrossRef]

47. Gupta, S.; Srivastava, S.; Saradhi, P.P. Chromium increases photosystem 2 activity in *Brassica juncea*. *J. Biol. Plantarum* **2009**, *53*, 100–104. [CrossRef]

48. Suseela, M.R.; Sinha, S.; Singh, S.; Saxena, R. Accumulation of chromium and scanning electron microscopic studies in *Scirpus lacustris* L. treated with metal and tannery effluent. *Bull. Environ. Contam. Toxicol.* **2002**, *68*, 540–548. [CrossRef]

49. Pandey, V.; Dixit, V.; Shyam, R. Antioxidative responses in relation to growth of mustard (*Brassica juincea* cv. Pusa Jaikisan) plants exposed to hexavalent chromium. *Chemosphere* **2005**, *61*, 40–47. [CrossRef] [PubMed]

50. Shanker, A.K.; Djanaguiraman, M.; Sudhagar, R.; Chandrashekar, C.N.; Pathmanabhan, G. Differential antioxidative response of ascorbate glutathione pathway enzymes and metabolites to chromium speciation stress in green gram (*Vigna radiata* (L.) R. Wilczek. cv. CO 4) roots. *Plant Sci.* **2004**, *166*, 1035–1043. [CrossRef]

51. Spencer, D.; Possingham, J.V. The effect of Mn deficiency on photophosphorylation and the oxygen—Evolving system in spinach chloroplasts. *Biochim. Biophys.* **1961**, *52*, 379–381. [CrossRef]

52. Gerretsen, F.C. Manganese in relation to photosynthesis. II. Redox potentials of illuminated crude chloroplast suspensions. *Plant Soil* **1950**, *11*, 159–193. [CrossRef]

53. Homann, P.H. Studies on the manganese of chloroplast suspensions. *Plant Physiol.* **1967**, *42*, 997–1007. [CrossRef] [PubMed]

54. Yachandra, V.K.; Sauer, K.; Klein, M.P. Manganese cluster in photosynthesis: Where plants oxidize water to dioxygen. *Chem. Rev.* **1966**, *96*, 2927–2950. [CrossRef]

55. Bottrill, D.E.; Possingham, J.V.; Kriedemann, P.E. The effect of nutrient deficiencies on phosynthesis and respiration in spinach. *Plant Soil* **1970**, *32*, 424–438. [CrossRef]

56. Rowe, N.; Speck, T. Plant growth forms: An ecological and evolutionary perspective. *New Phytol.* **2005**, *166*, 61–72. [CrossRef] [PubMed]

forests

MDPI

Review

Grand Fir Nutrient Management in the Inland Northwestern USA

Dennis R. Parent [1] and Mark D. Coleman [2,*]

[1] DRP Forestry, Hayden, ID 83835, USA; drparent@roadrunner.com
[2] Forest, Rangeland and Fire Sciences, University of Idaho; Moscow, ID 83844, USA
* Correspondence: mcoleman@uidaho.edu; Tel.: +1-208-885-7604

Academic Editors: Scott X. Chang and Xiangyang Sun
Received: 31 August 2016; Accepted: 31 October 2016; Published: 4 November 2016

Abstract: Grand fir (*Abies grandis* (Douglas ex D. Don) Lindley) is widely distributed in the moist forests of the Inland Northwest. It has high potential productivity, its growth being nearly equal to western white pine, the most productive species in the region. There are large standing volumes of grand fir in the region. Nutritionally, the species has higher foliage cation concentrations than associated conifers, especially potassium (K) and calcium (Ca). In contrast, it has lower nitrogen (N) foliage concentrations, which creates favorable nutrient balance on N-limited sites. Despite concentration differences, grand fir stores proportionally more nutrients per tree than associated species because of greater crown biomass. Although few fertilization trials have examined grand fir specifically, its response is inferred from its occurrence in many monitored mixed conifer stands. Fertilization trials including grand fir either as a major or minor component show that it has a strong diameter and height growth response ranging from 15% to 50% depending in part on site moisture availability and soil geology. Grand fir tends to have a longer response duration than other inland conifers. When executed concurrently with thinning, fertilization often increases the total response. Late rotation application of N provides solid investment returns in carefully selected stands. Although there are still challenges with the post-fertilization effects on tree mortality, grand fir will continue to be an important species with good economic values and beneficial responses to fertilization and nutrient management.

Keywords: forest nutrition; grand fir; fertilization; Inland Northwest

1. Introduction

Grand fir is widely distributed in the moist forests of the Pacific Northwest (Figure 1). Its range of 39–51 degrees latitude and 114–125 degrees longitude includes southern Canada in the Okanogan and Kootenay Lakes region south through northeastern Washington, northern Idaho, western Montana, and northeastern Oregon [1]. In the Inland Northwest, it occurs on moister sites from 450 to 1800 m elevation on a wide variety of soils derived from basalts, granites, gneiss, and sandstone as long as moisture is adequate [1]. Grand fir does best throughout the region where a deep layer of volcanic ash occurs [2,3]. There are similar occurrences of grand fir throughout western Montana habitat types and northeastern Washington and Oregon plant associations [4,5]. Grand fir is shade tolerant and often takes a subordinate position in the stand in mixture with other inland species.

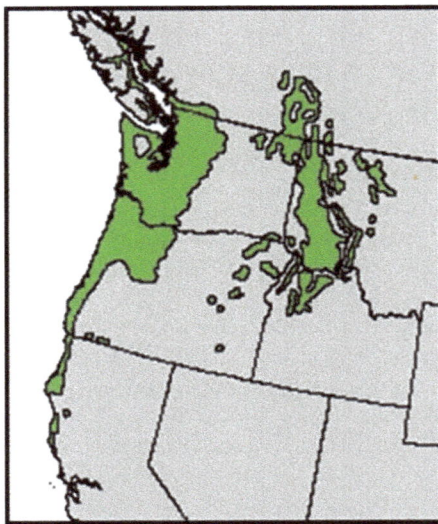

Figure 1. Range map of grand fir in the Pacific Northwest [6].

Grand fir has high potential productivity. Once established, it grows rapidly nearly equal to Douglas-fir (*Pseudotsuga mensiesii* (Mirb.) Franco) on the Pacific Coast and western white pine (*Pinus monticola* Dougl. ex D. Don) in Idaho [1]. As it approaches 50 years old, height is comparable to other associated species in the Inland Northwest including western larch (*Larix occidentalis* Nutt.). At 100-years-old, grand fir may approach or surpass other species in total height and volume yield. Table 1 is a general comparison of grand fir and its associates in the Inland Region using data from various historical authors [7–9]. The last two columns of Table 1 include data for coastal Douglas-fir for an approximate comparison with inland Douglas-fir [10].

Table 1. Grand fir height and total volume yield compared to associated species in the US Inland Northwest. All yields are from yield tables of fully stocked stands.

Stand Age	Grand Fir		Western White Pine		Western Larch		Inland Douglas-Fir		Coastal Douglas-Fir [1]	
	Height (m) [2]	Volume (m³) [3]	Height (m) [2]	Volume (m³) [4]	Height (m) [2]	Volume (m³) [5]	Height (m) [2]	Volume (m³) [6]	Height (m) [7]	Volume (m³) [8]
50	20.6	74.1	21.5	140.3	22.8	37.9	20.0	73.4	23.7	110.0
100	36.3	367.1	42.4	393.3	37.8	141.2	35.4	206.6	33.8	260.1

[1] From McArdle [10], Site Class IV, 100-year site index = 110; [2] From Deitschman [7], Table 2, p. 14; [3] From Stage [9], Table 7, p. 25; [4] From Haig [8], Table 6, p. 20; [5] Stage, op. cit, Table 12, p. 41; [6] Stage, op. cit., p. 9, Table 2; [7] McArdle, op. cit., Table 1, p. 12; [8] Ibid., Table 3, p. 23.

There are large standing volumes of grand fir in the Inland Northwest. It ranks first in total growing stock volume at 26 percent [11]. Douglas-fir ranks second at 23 percent. In Idaho, grand fir also makes up the largest percentage of species harvested, ranking first at 36 percent with Douglas-fir second at 26 percent [12].

Grand fir log and lumber values are lower than Douglas-fir, which is the framing lumber standard in the Inland Region. Grand fir has compensating wood quality advantages over Douglas-fir, such as higher log scale, lighter weight, less shrinkage during drying, and excellent fiber characteristics. Alternative forest products that add value are Christmas trees, floral boughs, and oils obtained from the foliage.

2. Nutrient Characteristics

Grand fir has some unique nutritional characteristics compared to other inland conifers. When grand fir was compared to other conifers in over 160 stands, it had higher foliar base cation levels and also higher variation [13]. Grand fir N foliar concentrations are slightly less than associated conifers (Figure 2a). However, foliar K is higher (Figure 2b) and Ca levels are even higher (Figure 2c). Grand fir also has higher foliar concentrations of sulfur (S) and boron (B) than the other species (not shown). Foliar nutrient data from several studies confirm that N concentrations were similar for all species but other cations were higher in grand fir [14].

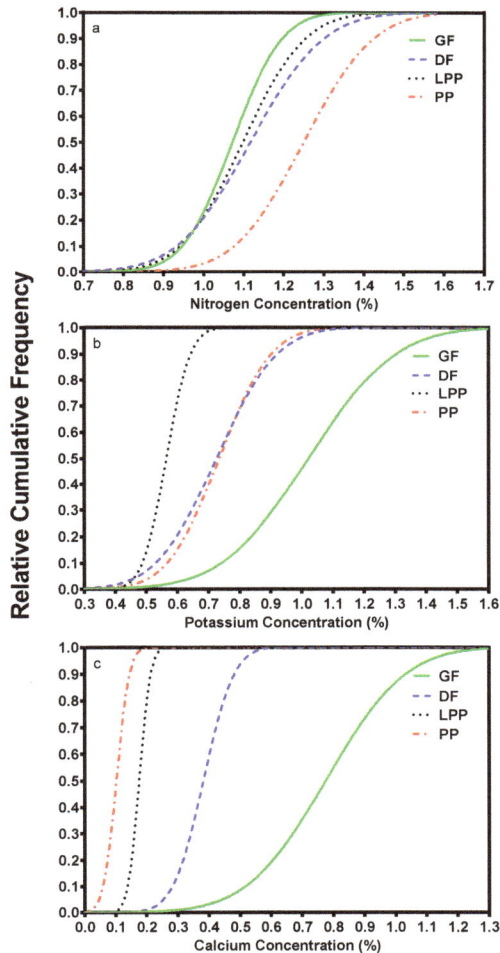

Figure 2. Relative cumulative frequency graphs of foliar N (**a**); K (**b**); and Ca (**c**) concentrations for grand fir (GF), inland Douglas-fir (DF), lodgepole pine (LP), and ponderosa pine (PP). *Y*-axis shows the proportion of trees sampled that are below the nutrient concentration on the horizontal axis. Foliage was collected from over 160 mixed conifer stands that were randomly selected within categories representing geographic region, habitat type, stand density, and rock type to assure that differences in foliar nutrient characteristics reflect species and not sample site differences [13].

Canopy nutrient content shows that grand fir acquires more nutrients than other conifer species (Figure 3). Grand fir holds twice the canopy foliar biomass as ponderosa pine and lodgepole pine and 1.4 times the foliar mass as Douglas-fir for a given diameter class [15], which is in part due to greater needle weight in comparison to other inland species (Figure 4).

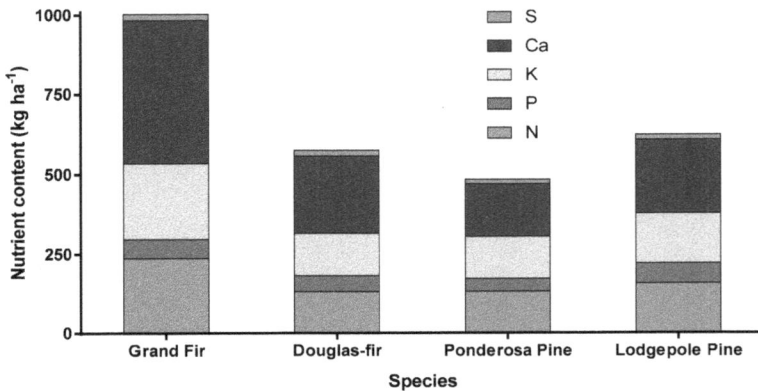

Figure 3. Total nitrogen (N), phosphorus (P), potassium (K), calcium (Ca), and sulfur (S) content of aboveground portion of four inland tree species: grand fir, Douglas-fir, ponderosa pine, and lodgepole pine [16].

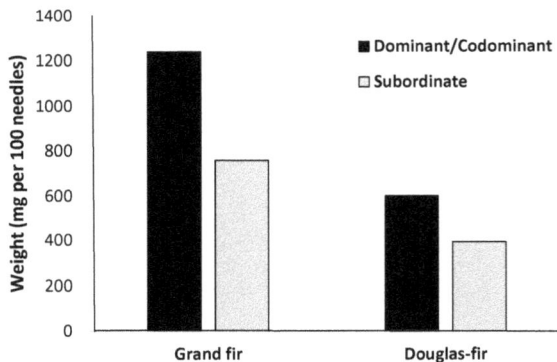

Figure 4. Foliage weight comparison between crown classes of grand fir and Douglas-fir [17].

Shade tolerant tree species (such as grand fir) incorporate cations at higher rates than intolerants (Figure 2b,c). Consequently, in mixed-conifer stands, grand fir may be more competitive for nutrients than less tolerant conifers, particularly when nutrient supply is limited [14]. This indicates that grand fir may have greater nutrient demand than associated species, which would affect its competitive ability in mixed stands and its response to nutrient additions.

We evaluate nutrient status by comparing foliage nutrient concentrations of field-collected test samples to critical foliar nutrient levels determined from greenhouse-grown seedlings. The critical nutrient level is the concentration above which growth does not respond to nutrient additions [14,18]. Table 2 shows the critical foliar nutrient levels for four Inland Northwest tree species. Seedling-based critical levels are used as reference values in mature stands. The critical level of foliar N in grand fir is 11.5 mg·g^{-1}, 18 percent less than Douglas fir. Critical levels of other selected macro- and micro-nutrients are about the same as for other species, suggesting that grand fir may be able to cope with lower N concentrations than other species due to its larger crown biomass, and relatively low N values (Figure 2a). Although critical levels of N are lower, only 20 percent of unfertilized grand fir has adequate levels of N (Figure 2a) indicating that there still may be considerable opportunity for growth increases from N additions. However, there are important limitations in our knowledge of seedling critical levels relate to fertilizer responses in mature trees.

Table 2. Critical nutrient concentrations for inland conifer species: grand fir (GF), Douglas-fir (DF), lodgepole pine (LP), and ponderosa pine (PP) [14].

Foliar Concentration	GF	DF	LP	PP	Ingestad Ratios *
N (mg·g^{-1})	11.5	14.0	12.0	11.0	100
P (mg·g^{-1})	1.5	1.2	1.2	0.8	16
K (mg·g^{-1})	5.8	6.0	5.0	4.8	50
S (mg·g^{-1}))	0.8	1.1	0.9	0.8	7
Cu (μg·g^{-1})	3	2	3	3	0.03
B (μg·g^{-1})	10	10	4.3	20	0.2

* Optimum nutrient ratios, $R = 100$ (M/N), where M is the concentration of the element of interest and N is the concentration of N ([19] except S/N).

Grand fir's shade tolerant ecological status may help it survive and grow in a subordinate position in mixed stands. The lower critical N levels compared to Douglas-fir and lodgepole pine and foliar N levels in untreated stands indicate that grand fir is able to use N conservatively when in low supply. Conversely, when N is available, grand fir may utilize N more aggressively. Higher foliar concentrations of other nutrients such as K and B in untreated grand fir stands suggest that it is commonly in good nutritional balance [14].

Nutrient balance is an important consideration in tree nutrition. Plants require nutrients in specific proportions, which Ingestad [19,20] expressed as optimal nutrient ratios calculated relative to N. Nutrient ratios are consistent among plant taxa due to metabolic requirements, and therefore, are important indicators of nutrient balance (Table 2). As with critical nutrient concentrations, optimal nutrient ratios are based on seedling studies. We assume they are consistent in mature trees because there is important stoichiometric balance among all taxa [21], but we are not aware of studies that consider nutrient balance in mature trees. Grand fir nutrient ratios across three studies found more than 80% of trees measured had K, S, B, and Cu ratios above the Ingestad ratios listed in Table 2. Phosphorous ratios of grand fir averaged near Ingestad's values. Grand fir has relatively high nutrient ratios compared with inland Douglas-fir [22] and perhaps other inland northwest conifers because of the relatively low N concentrations and high concentrations of other nutrients in tree foliage (Figure 2). The resulting good nutritional balance of grand fir further suggests that it may have a competitive advantage.

The previous data show that grand fir has lower N concentrations and N critical levels than Douglas-fir and other inland conifer species. However, the concentrations and ratios of other nutrients are relativity high in grand fir. Grand fir may be better able to utilize available site N than other inland conifer species because of this favorable nutrient balance. Furthermore, grand fir maintains relatively high foliage biomass and therefore stores site nutrients for its own use and out of reach from competitors. These unique features may be partially explained by grand fir's physiology as a shade tolerant species [23]. Perhaps grand fir can tolerate low levels of soil N by sequestering site N in foliage while maintaining a favorable nutrient balance. It is then well-prepared for aggressive growth when opportunity arises.

3. Response to Fertilization

Fertilization is an important tool in nutrient management of conifer stands containing grand fir. The commonly deficient N levels along with subordinate canopy position and shade-tolerance of grand fir suggest that it may have strong foliar and growth response to N additions. However, several complicating factors must be accounted for when considering foliar nutrient and stand growth response to fertilizer. These factors include stand density, the relationships with site quality, and the fact that N fertilizer growth response can be offset by an increase in mortality. We consider each of these factors in the following sections.

Foliar data for grand fir reveals a strong response to N additions, both in young stands at crown closure and mature stands at the end of the rotation (Figure 5). There was also good foliar response to additions of S and B (data not shown). Foliage did not respond positively to K additions (Figure 6) perhaps because all the trees were above critical K levels before fertilization. It is also possible that the amount of added K was insufficient relative to added N, or that the results simply reflect the dilution effect of expanding foliage in response to the N. Limited trials with phosphorous (P) in the Inland Northwest show no response in either foliar nutrient or growth responses [24]. Consequently, P has receive little attention as a limiting nutrient in the Inland Northwest. There is some evidence that mature trees respond to P amendments in more alkaline, N-rich soils where additional research is warranted [25].

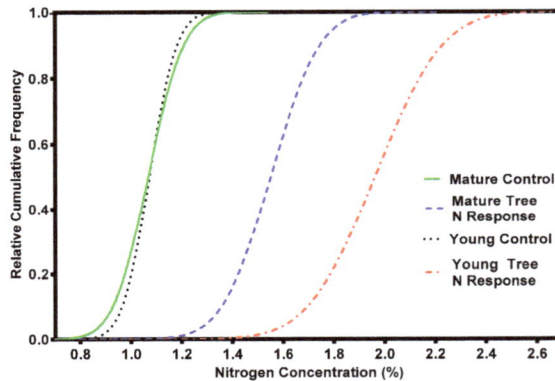

Figure 5. Grand fir foliar N response additions for young (~15 year-old) and mature (~40 year-old) stand ages [16]. (Foliage samples were collected one-year after fertilization with 224–336 kg N ha^{-1} as urea).

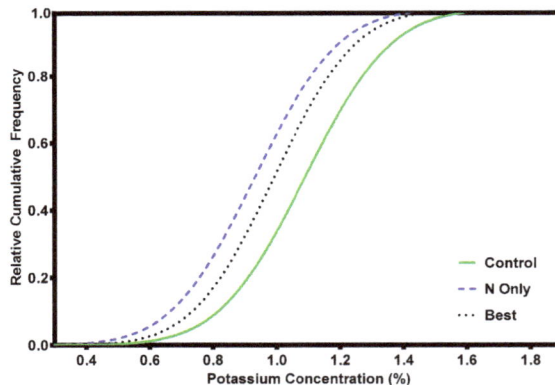

Figure 6. Grand fir foliar K response after fertilization with 224 kg N ha^{-1} (N-only) or a blend of 224 kg N ha^{-1}, 170 kg K ha^{-1} + 73 kg S ha^{-1} + 2.7 kg B ha^{-1} (Best) [16].

Reflecting foliar responses, N fertilization trials generally show a strong response in both tree diameter, and height [26]. Various trials show consistent growth responses to N rates or combinations of N with other nutrients [16,27] (Figure 7). Growth responses over 20 percent are within the realm of economic feasibility. Additional macro- and micronutrients have resulted in a response in some, but not all instances. For example, multi-nutrient trials showed that grand fir responds well to S and B applications but not necessarily K [16,27,28]. Limited responses to K are also observed in Douglas-fir, even though K is suspected of protecting trees from mortality following N-only fertilization [29].

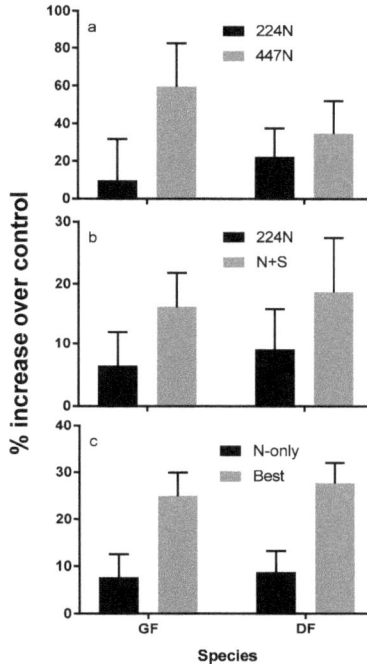

Figure 7. Grand fir (GF) and Douglas-fir (DF) volume increment response in three different fertilization trials at various locations. (**a**) Average annual ten-year response to two levels of N-only fertilization at 224 and 447 kg N ha^{-1} in 30 Douglas-fir dominated (80% basal area) stands in Idaho, Montana, Oregon and Washington including five parent material classes and three vegetation series representing the range of regional productivity [30]; (**b**) Average annual eight-year response to either 224 kg N ha^{-1} or 224 kg N ha^{-1} + 73 kg S ha^{-1}) in seven mixed conifer stands growing on basalt parent materials at 1370 to 1680 m in *Abies grandis* and *Abies lasiocarpa* vegetation series [31]; (**c**) Average annual two-year response to individual-tree treatments of 224 kg N ha^{-1} (N-only) or a blend of 224 kg N ha^{-1}, 190 kg K ha^{-1} + 73 kg S ha^{-1} + 2.7 kg B ha^{-1} (Best) in eight young (5–10 cm average diameter) stands all on metasedimentary parent material [16].

Response trials have shown that the length of the growth response is 4–6 years for unthinned stands [23,32] and up to 14 years in properly thinned stands [32,33]. Thinning concentrates the growth on fewer trees allowing them to take advantage of other site resources and sustain the response for a longer period.

Repeating a fertilization application can have differing effects depending on the site [27]. For example, grand fir showed a strong repeat response with an additional 90 kg N ha^{-1} treatment on the drier sites, but did not respond to higher rates. Grand fir also showed a good repeat response to 89 and 180 kg N ha^{-1} on the wetter sites. However, Douglas fir did not respond to these repeat treatments. Repeat fertilizations of grand fir on moist sites at rates of 90 to 180 kg N ha^{-1} are expected to result in a significant basal area growth response compared to a single fertilization. However, a significant response of grand fir is not likely on drier sites.

Differences in site quality can partly explain the range in fertilizer response. Important factors affecting fertilizer response to site quality include soil parent material, and soil moisture as indicated by climax vegetation series. Soil types modified by bedrock geology can influence growth and response to nutrient additions. One of the earliest investigations involving geologic rock types found no differentiation between responses on three different rock types [34]. More precise experimental designs have since detected parent material effects by using individual tree screening trials. Screening trials include different treatments of replicated trees across a uniform site. Sites with various parent materials then contain replicates of that design.

Rock types divide into three categories when unfertilized controls are compared using N-only fertilization, and a multi-nutrient treatment that included N, K, S, and B (Figure 8). Grand fir does not respond to fertilizer additions on soils derived from some parent material types (schist and basalt), indicating they are not deficient in any of the applied nutrients. Sites with soils from other parent materials respond only to N additions, but not multi-nutrients (granite, carbonate bearing metasedimentary, indicating they are N deficient, but other applied nutrients are not limiting. Heavily weathered quartz metasedimentary rocks have the greatest response to multiple nutrients to achieve optimal growth, indicating more than N is deficient. Productivity of grand fir on quartz-meta parent material is equal to the highest growth observed among the screening trials when treated with multi-nutrient fertilizer.

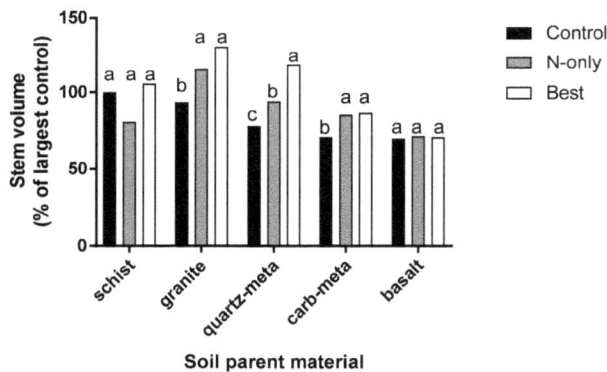

Figure 8. Fertilizer treatment effects on 2-year stem volume by soil parent material. Stem volume is expressed as a percentage of untreated trees growing on schist soils. Fertilizer treatments included 224 kg N ha^{-1} (N-only), or a blend of 224 kg N ha^{-1}, 190 kg K ha^{-1} + 73 kg S ha^{-1} + 2.7 kg B ha^{-1} (Best). Two metasedimentary parent materials include a quartzite (quartz-meta) and carbonate (carb-meta) [24].

Vegetation series as described by habitat type [2] also affect growth response due to soil moisture variations. Habitat types containing more available soil moisture or higher amounts of organic matter would tend to have better response. Soils containing volcanic ash display increased productivity and N response for all conifer species due to the increase in water-holding capacity [3]. Yet, only half of the sites where grand fir is the climax vegetation contain significant ash deposits. Higher elevation vegetation series are associated with greater moisture availability and ash depth, and grand fir is capable of higher productivity on these sites due to increased moisture availability during the growing season. The limited ash deposits and dryer site conditions for the grand fir habitat type suggest that grand fir is tolerant of dryer conditions. However, when grown on sites with abundant moisture grand fir is capable of better growth than on drier sites. The above results demonstrate that there are differences in nutrient availability and fertilization response among soils, bedrock types, and vegetation series but these differences interact with moisture availability and will require focused research effort to discover mechanisms responsible for distinctions among bedrock geology, surficial deposits, and moisture availability.

The development stage of a forest stand explains some of the response variability observed in previous experiments with soils and site types. A good understanding of these stages and their effect on nutrient requirements helps determine when nutrient additions will provide a growth response. Stands of trees have differing nutritional needs depending on where they are in the successional pathway of development [35].

At the seedling and sapling stage, the trees are unrestricted in their quest for nutrients, especially if artificially planted with competition control. Consequently, during the seedling stage of development, sites meet seedling nutrient requirements, and seedlings can explore the soil for available nutrients reducing the need for additions. Furthermore, they can take advantage of the nutrient flush that often occurs after timber harvest. Grand fir plantations are rare because the species abundantly regenerates in association with planted seedlings of higher-value species. Consequently, we are not aware of seedling fertilizer tests specifically for grand fir. Nutrition of natural grand fir seedling development is a potential area of research. Tests using planted seedling of other inland conifer species on multiple sites show few consistent responses to broadcast applied nutrients [36]. During one seedling trial where fertilizer was applied into the planting hole, Douglas-fir and ponderosa pine exhibited increased height and diameter growth after outplanting, but also had higher mortality, especially at greater N application rates [37].

As stands develop through intermediate ages, nutrient demands increase until they reach the self-thinning stage when nutrient demands are greatest [35]. At this stage, the ability to respond to treatments is high because of nutrient limitations and high stand vigor. This stage is not necessarily age-related because stand age at self-thinning can vary depending on the size-density relationship, species, site quality, and stand establishment practices [38,39]. One would expect during self-thinning that nutrient additions would increase growth. Unfortunately, many of the recorded grand fir experiments do not specifically indicate the stage of stand development or measures of density.

Experiments in relatively young (10–30 years old), high-density stands showed strong response to fertilization [26,40–42]. As stands mature, nutrient demands continue and stand density becomes a controlling influence. If artificial fertilization occurs during this time, grand fir should have a strong response in comparison with associated species. Tree spacing control during this period is important to allow sufficient non-nutrient resources to support fertilizer response. As stands reach mature and old growth conditions, nutrient demands depend more on site factors. Grand fir's physiology as a late-successional, shade-tolerant tree may affect its response by allowing it to conserve N when necessary but be more demanding during competitive growth phases such as that which would occur during crown closure [23]. Focused study is required on the ability of grand fir to compete for nutrients during between crown closure and self-thinning stages of stand development.

During the crown closure through self-thinning stages of development, additional growth response is possible when thinning and fertilization are combined as demonstrated from numerous experiments [32,34,43–45]. After thinning a crowded stand, light, water, and temperature become more favorable for growth [46]. Fertilization within a decade after thinning, or in properly spaced stands allows the trees to take advantage of the additional resources [30,47]. Thinning and then nutrient additions are beneficial, especially if relative densities prior to fertilization allow adequate space for increased growth. Research trials show that combined thinning and fertilization are additive compared to the individual treatments [47,48]. In some instances the total growth response of the crop trees is greater than the sum of either fertilization or thinning effects alone [32].

Grand fir exhibits higher mortality after fertilization, especially with N-only amendments. Its mortality is naturally greater than that of Douglas-fir, and N fertilization only increases that trend (Figure 9) [49]. Of the known reasons for mortality, N fertilization increases insect-caused tree mortality more than other causes. This is commonly observed in trees and is attributed to improved foliage nutritional quality [50]. Addition of other macro- and micronutrients tended to reduce mortality but not significantly [16]. There is critical need to learn more about the causes of increase mortality to N-only fertilizer applications.

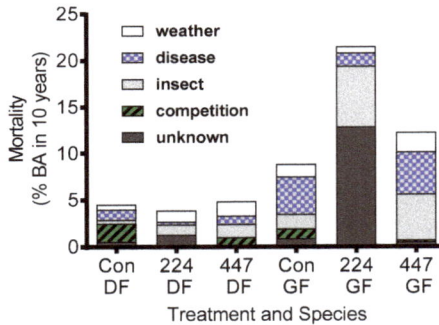

Figure 9. The causes of ten-year basal area mortality of Douglas-fir and grand fir treated with either 224 or 447 kg N ha^{-1} and compared with unfertilized controls. Data are from 30 mature (average 65 year-old) Douglas-fir dominated (80% basal area) stands in Idaho, Montana, Oregon and Washington including five parent material classes and three vegetation series representing the range of regional productivity. All stands were recently thinned or were naturally well spaced. Initial basal area averaged 36 ± 10 m^2 ha^{-1} [30].

Nitrogen fertilization can potentially stimulate pests and disease. Insects such as spruce budworm (*Choristoneura occidentalis* Freeman) reduce current growth, and can diminish fertilizer response [48]. Grand fir is more susceptible to Douglas-fir tussock moth (*Orgyia pseudotsugata* McDunnough) than other inland conifers, but experience confirms that susceptibility tends to decreases with soil volcanic ash content [51] which, as described above, is expected to interact with soil moisture, tree nutrition and fertilizer application. The effects of Armillaria root disease (*Armillaria ostoyae* Romagnesi) can substantially reduce net response to fertilization. Therefore, stand-level responses to fertilization may be lower than expected if root disease is present in the stand. Although there were initial experiments implying that fertilization (especially with K additions) [52] may reduce incidence of root disease, a 10-year study in a diseased stand found no treatment differences due to fertilization [27]. Moore et al. [29] suggested that tree nutrition (K/N ratio) controls production of phenolic defense compounds and resistance to the root rot disease Armillaria in Douglas-fir. If true, grand fir should exhibit a similar relationship because it too is susceptible to Armillaria infection. While the relationship between favorable K/N ratios and phenolic concentrations was confirmed in seedlings [53,54] they could not demonstrate an effect of stand-level N and K fertilization on root disease or mortality after ten years. The potential negative effects of pests and diseases following fertilization are important to recognize [55] and managers should avoid fertilization in stands that display these properties. Increased effort is needed to elucidate the relationship between nutrient balance, defense compounds, and root disease.

Vegetation control using herbicides is expanding in the Inland Northwest. Herbicide application is cheaper, more effective, and protects the soil surface from compaction and displacement by heavy site preparation machinery. Grand fir is generally more tolerant to herbicides than associated species [56] so prudent use presents few difficulties. Specific effects of herbicide application in combination with fertilization are not well understood in the Inland Northwest Region. However, in the southeastern U.S., research confirms the positive combined effects of herbicide treatments and nutrient additions [57].

There is evidence that vegetation control alone may provide additional nutrients to trees by eliminating competition and causing faster breakdown and mineralization of organic matter [55].

In summary, these studies indicate that grand fir growing under the right conditions can respond to N fertilization with volume growth ranging from 15% to 50%. Grand fir generally has a longer response duration than other inland conifers, the greatest response occurring during 3–5 years, tapering off during 5–8 years, with some response lasting through 10–12 years under optimum conditions. The response can be greater after well-timed thinning. Repeated fertilizer applications have the potential to maintain growth response depending on the site and the stage of stand development. It is important to recognize site types and their interaction on fertilizer response. Grand fir is sensitive to higher mortality rates after fertilization so when seeking locations for effective treatments, it is critical to avoid sites with heavily weathered parent material where nutrient imbalance may lead to high pest and disease susceptibility.

4. Economics of Grand Fir Fertilization

Large land-holding companies in the Inland Northwest are interested in the economics of fertilization. As a financial investment, forest fertilization must consider the time value of money. Success is very sensitive to the time between the initial outlay and the final return on investment; thus, fertilization in older stands is considered to be more profitable because the volume gains can be harvested sooner [58] and investment returns increase [59]. Late rotation fertilization has a short payback period with relatively low investment costs relative to other forestry investments [60]. Repeated fertilization can extend the response duration, however a second amendment doubles the cost and the response is not as strong as the first application. These economic principles have led foresters in the region to believe that a one-time fertilization late in the rotation can be a sound forestry investment. A recent Internal Revenue Service (IRS) tax law change that allowed expensing of fertilization costs further increased the investment potential [61].

Since most grand fir in the Inland Northwest grows in mixture with Douglas-fir and other species, fertilization experiments require additional analysis and assumptions to confine the results to just grand fir. Because of this, grand fir response, by itself, is usually inferred using data from mixed stands.

Properly spaced, 60–90 years old mature stands containing grand fir may respond well to N fertilization with responses similar to those obtained for Douglas-fir stands in western Washington [58]. The treatment generally involves application of 150–300 kg ha^{-1} of urea N followed by timber harvest in 8–12 years. This results in a basal area growth response of 20%–80% with comparable volume responses. Addition of multi-nutrients may improve protection against mortality but does not significantly improve individual tree growth response [29], so the added cost does not increase the economic benefit.

Recent operational fertilization treatments in mixed stands including grand fir confirm that investment returns are very good, often surpassing those of other forestry investments, if stands are carefully selected [62]. It is essential to identify stand conditions that will provide an acceptable economic response [63]. Incorporation of a carefully selected fertilization treatment into a general forest management plan can be economically warranted. These growth responses can be used for more sophisticated economic analysis for more accurate projections depending on management requirements. However, such growth responses may require adjustments in coming decades with increased atmospheric carbon dioxide concentrations and accompanying changes in climate.

5. Management Recommendations

We recommend that a grand fir nutrient management and fertilization program consider the following concepts.

1. The greatest response from added nutrients seems to occur just after crown closure when competitive stresses are high.

2. Grand fir is expected to best respond to nutrient additions when it is maintained in a healthy and vigorous condition through stand density management.

3. To conserve site nutrients, soil disturbance should be minimized during timber harvest and stand regeneration activities. Heavy-handed site preparation and slash cleanup may have a negative impact on the site's nutrient resources, especially low-quality sites [14]. Some alternate possibilities to consider include light broadcast burning; leaving small limbs, foliage, and larger material scattered across the site; and the use of herbicides.

4. Carefully analyze timber stands to determine tree and site conditions, species mix, stand density, soil parent material, and harvest plans before making fertilization decisions to assure effective results. Consult the latest fertilization research results to guide grand fir nutrient management.

6. Conclusions

Grand fir is ubiquitous in Inland Northwest forests growing in mixture with other species, mainly Douglas-fir. It often contains higher crown nutrient content than associate species [14]. Grand fir has good growth response to N fertilizer and possibly K. The effect of nutrition, especially K, on root disease potential remains unconfirmed. Good investment returns are possible from fertilization treatments after careful site evaluation. With good management, and a solid understanding of nutrient dynamics, grand fir will remain an important species with solid economic values and a positive response to fertilizer under the right conditions in Inland Northwest forests.

Acknowledgments: Funding for analysis of data included in this report and preparation of the manuscript was supplied by the Intermountain Forestry Cooperative, formerly Intermountain Forest Tree Nutrition Cooperative (IFTNC). We wish to thank the following people who assisted with the preparation and review of this manuscript: Mark J. Kimsey, Research Assistant Professor, IFTNC/University of Idaho, Terry M. Shaw Research Scientist, IFTNC/University of Idaho.

Author Contributions: Dennis Parent and Mark D. Coleman preformed equally in drafting, editing and revising the manuscript.

Conflicts of Interest: The authors declare no conflict of interest.

References

1. Foiles, M.W.; Graham, R.T.; Olson, D.F. Grand fir. In *Silvics of Forest Trees of the United States, Conifers*; Burns, R.M., Honkala, B.H., Eds.; USFS: Washington, DC, USA, 1990; Volume 1, pp. 52–59.
2. Cooper, S.V.; Neiman, K.E.; Roberts, D.W. *Forest Habitat Types of Northern Idaho: A Second Approximation*; USFS, Intermountain Research Station: Ogden, UT, USA, 1991.
3. Garrison-Johnston, M.T.; Mika, P.G.; Miller, D.L.; Cannon, P.; Johnson, L.R. Ash cap influences on site productivity and fertilizer response in forests of the Inland Northwest. In *Volcanic-ash-derived Forest Soils of the Inland Northwest: Properties and implications for management and restoration*; USFS, RMRS: Coeur d'Alene, ID, USA, 2007; pp. 137–163.
4. Pfister, R.D.; Kovalchik, B.L.; Arno, S.F.; Presby, R.C. *Forest Habitat Types of Montana*; USFS, INT Station: Ogden, UT, USA, 1977.
5. Williams, C.K.; Kelley, B.F.; Smith, B.; Lillybridge, T.R. *Forested Plant Associations of the Colville National Forest*; Pacific Northwest Research Station: Portland, OR, USA, 1995.
6. USGS; Little, E.L., Jr. Digital representations of "Atlas of United States Trees". Available online: http://esp.cr.usgs.gov/data/atlas/little/abiegran.pdf (accessed on 31 August 2016).
7. Deitschman, G.H.; Green, A.W. *Relations between Western White Pine Site Index and Tree Height of Several Associated Species*; USFS: Ogden, UT, USA, 1965.
8. Haig, I.T. *Second-growth Yield, Stand, and Volume Tables for the Western White Pine Type*; USFS, Northern Rocky Mountain Forest Experiment Station: Washington, DC, USA, 1932.
9. Stage, A.R.; Renner, D.L.; Chapman, R.C. *Selected Yield Tables for Plantations and Natural Stands in the Inland Northwest forests*; USFS, INT Research Station: Ogden, UT, USA, 1988.
10. McArdle, R.E.; Meyer, W.H.; Bruce, D. *The Yield of Douglas-Fir in the Pacific Northwest*; USFS: Washington, DC, USA, 1961.

11. USFS. Forest Inventory Data Online (FIDO). Available online: http://apps.fs.fed.us/fia/fido/index.html (accessed on 2 Jaunary 2016).
12. Simmons, E.A.; Hayes, S.W.; Morgan, T.A.; Keegan, C.E., III; Witt, C. *Idaho's Forest Products Industry and Timber Harvest 2011 with Trends through 2013*; USFS: Fort Collins, CO, USA, 2013.
13. Moore, J.A.; Mika, P.G.; Shaw, T.M.; Garrison-Johnston, M. Foliar nutrient characteristics of four conifer species in the interior Northwest United States. *West. J. Appl. For.* **2004**, *19*, 13–24.
14. Garrison, M.T.; Moore, J.A. Nutrient Management: A Summary and Review. Available online: http://digital. lib.uidaho.edu/cdm/singleitem/collection/iftnc/id/3357/rec/40 (accessed on 29 September 2016).
15. Brown, J.K. *Weight and Density of Crowns of Rocky Mountain Conifers*; USFS, INT Forest & Range Experiment Station: Ogden, UT, USA, 1978.
16. Mika, P.G. Grand Fir: Nutrient Ecology and Response to Fertilization. Available online: http://digital.lib. uidaho.edu/cdm/singleitem/collection/iftnc/id/3446/rec/6 (accessed on 29 September 2016).
17. Garrison-Johnston, M. Nutrient Cycling in a North Idaho Conifer Stand. Ph.D. Thesis, University of Idaho, Moscow, ID, USA, 2003.
18. Walker, R.B.; Gessel, S.P. *Mineral Deficiencies of Coastal Northwest Conifers*; College of Forest Resources, University of Washington: Seattle, WA, USA, 1991.
19. Ingestad, T. Mineral nutrient requirements of *Pinus silvestris* and *Picea abies* seedlings. *Physiol. Plant.* **1979**, *45*, 373–380. [CrossRef]
20. Ingestad, T. A definition of optimum nutrient requirements in birch seedlings. II. *Physiol. Plant.* **1971**, *24*, 118–125. [CrossRef]
21. Elser, J.J.; Sterner, R.W.; Gorokhova, E.; Fagan, W.F.; Markow, T.A.; Cotner, J.B.; Harrison, J.F.; Hobbie, S.E.; Odell, G.M.; Weider, L.W. Biological stoichiometry from genes to ecosystems. *Ecol. Lett.* **2000**, *3*, 540–550. [CrossRef]
22. Coleman, M.D.; Shaw, T.M.; Kimsey, M.J.; Moore, J.A. Nutrition of Douglas-fir in the Inland Northwest. *Soil Sci. Soc. Am. J.* **2014**, *78*, 11–22. [CrossRef]
23. Graham, R.T.; Tonn, J.R. *Ten-Year Results of Fertilizing Grand Fir, Western Hemlock, Western Larch, and Douglas-Fir with Nitrogen in Northern Idaho*; USFS, INT Station: Ogden, UT, USA, 1985.
24. Shaw, T.M.; Kimsey, M. Screening trial data. Intermountain Forestry Cooperative, University of Idaho: Moscow, ID, USA, Unpublished data. 2016.
25. Mainwaring, D.B.; Maguire, D.A.; Perakis, S.S. Three-year growth response of young Douglas-fir to nitrogen, calcium, phosphorus, and blended fertilizers in Oregon and Washington. *For. Ecol. Manag.* **2014**, *327*, 178–188. [CrossRef]
26. Loewenstein, H.; Pitkin, F.H. *Growth Responses and Nutrient Relations of Fertilized and Unfertilized Grand Fir*; University of Idaho, Forestry, Wildlife, and Range Experiment Station: Moscow, ID, USA, 1971.
27. Shaw, T.; Coleman, M.; Kimsey, M.; Mika, P. *Forest Health and Nutrition Ten-year Growth, Mortality and Foliar Nutrition*; Intermountain Forest Tree Nutrition Cooperative: Moscow, ID, USA, 2014.
28. Garrison, M.T.; Moore, J.A.; Shaw, T.M.; Mika, P.G. Foliar nutrient and tree growth response of mixed-conifer stands to three fertilization treatments in northeast Oregon and north central Washington. *For. Ecol. Manag.* **2000**, *132*, 183–198. [CrossRef]
29. Moore, J.A.; Mika, P.G.; Schwandt, J.W.; Shaw, T.M. Nutrition and forest health. In *Interior Cedar-Hemlock-White Pine Forests: Ecology and Management*; Washington State University: Spokane, WA, USA, 1994; pp. 173–176.
30. Moore, J.A.; Mika, P.G.; Vanderploeg, J.L. Nitrogen fertilizer response of Rocky Mountain Douglas-fir by geographic area across the Inland Northwest. *West. J. Appl. For.* **1991**, *6*, 94–98.
31. Garrison, M.T.; Moore, J.A.; Mika, P.G.; Shaw, T.M. Six-Year Growth Response of the Umatilla and Four-Year Growth Response of the Okanogan Mixed Conifer Stands to N, N + K and N + S Fertilization. Available online: http://digital.lib.uidaho.edu/cdm/ref/collection/iftnc/id/3358 (accessed on 3 October 2016).
32. Shafii, B.; Moore, J.A.; Olson, J.R. Effects of nitrogen fertilization on growth of grand fir and Douglas-fir stands in northern Idaho. *West. J. Appl. For.* **1989**, *4*, 54–57.
33. Scanlin, D.C.; Lowenstein, H. *Forest Fertilization in the Intermountain Region-six Year Results*; University of Idaho, Forestry, Wildlife and Range Experiment Station: Moscow, ID, USA, 1980.
34. Scanlin, D.C.; Loewenstein, H.; Pitkin, F.H. *Two-year Response of North Idaho Stands of Douglas-Fir and Grand Fir to Urea Fertilizer and Thinning*; University of Idaho: Moscow, ID, USA, 1976.

35. Kimmins, J.P. *Forest Ecology: A Foundation for Sustainable Forest Management and Environmental Ethics in Forestry*, 3rd ed.; Prentice Hall: Upper Saddle River, NJ, USA, 2004.

36. Shaw, T.M.; Johnson, L.R. Two-Year Growth Response of Douglas-fir and Ponderosa Pine Seedlings to Boron and Multi-Nutrient Fertilization in Northeast Oregon. Available online: http://digital.lib.uidaho.edu/cdm/ref/collection/iftnc/id/4860 (accessed on 4 October 2016).

37. Fan, Z.; Moore, J.A.; Osborne, H.L. Three-year response of ponderosa pine seedlings to controlled-release fertilizer applied at planting. *West. J. Appl. For.* **2002**, *17*, 154–164.

38. Oliver, C.D.; O'Hara, K.L.; McFadden, G.; Nagame, L. Concepts of thinning regimes. In *Douglas-Fir: Stand Management for the Future*; Oliver, C.D., Hanley, D.P., Johnson, J.A., Eds.; Institute of Forest Resources: Seattle, WA, USA, 1986; pp. 246–257.

39. Weiskittel, A.; Gould, P.; Temesgen, H. Sources of variation in the self-thinning boundary line for three species with varying levels of shade tolerance. *For. Sci.* **2009**, *55*, 84–93.

40. Loewenstein, H.; Pitkin, F.H. Response of grand fir and western white pine to fertilizer applications. *Northwest Sci.* **1963**, *37*, 23–30.

41. Graham, R.T.; Tonn, J.R. *Response of Grand Fir, Western Hemlock, Western White Pine, Western Larch, and Douglas-Fir to Nitrogen Fertilizer in Northern Idaho*; USFS, INT: Ogden, UT, USA, 1979.

42. Powers, R.F. Response of California true fir to fertilization. In *Forest Fertilization Conference*; Gessel, S.P., Ed.; University of Washington: Union, WA, USA, 1979; pp. 95–101.

43. Olson, J. *Four-Year Fertilization and Thinning Response Analysis of Douglas-fir and Grand Fir Stands in Northern Idaho*; Potlatch Corporation: Lewiston, ID, USA, 1979.

44. Olson, J. *Response of Intermountain Grand Fir and Douglas-Fir Stand Types to Nitrogen Fertilization and Thinning*; Potlatch Corporation: Lewiston, ID, USA, 1981.

45. Olson, J.; Hatch, C.R. Volume response of intermountain grand fir stand types to nitrogen fertilizer and thinning treatments. In *The Biology and Management of True Fir in the Pacific Northwest*; University of Washington: Seattle, WA, USA, 1981; pp. 247–250.

46. Chase, C.W.; Kimsey, M.J.; Shaw, T.M.; Coleman, M.D. The response of light, water, and nutrient availability to pre-commercial thinning in dry inland Douglas-fir forests. *For. Ecol. Manag.* **2016**, *363*, 98–109. [CrossRef]

47. Scanlin, D.C.; Loewenstein, H. Response of inland Douglas-fir and grand-fir to thinning and nitrogen fertilization in northern Idaho. In *Forest Fertilization*; Gessel, S.P., Kenady, R.M., Atkinson, W.A., Eds.; University of Washington: Alderbrook Inn, Union, WA, USA, 1979; pp. 82–88.

48. Cochran, P.H. *Response of Thinned White Fir Stands to Fertilization with Nitrogen Plus Sulfur*; USFS, PNW Research Station: Portland, OR, USA, 1991.

49. Shen, G.; Moore, J.A.; Hatch, C.R. The effect of nitrogen fertilization, rock type, and habitat type on individual tree mortality. *For. Sci.* **2001**, *47*, 203–213.

50. Herms, D.A. Effects of fertilization on insect resistance of woody ornamental plants: Reassessing an entrenched paradigm. *Environ. Entomol.* **2002**, *31*, 923–933. [CrossRef]

51. Stoszek, K.J.; Mika, P.J. Douglas-fir tussock moth outbreaks related to stand conditions and pest management. In *1980 SAF Convention: Land Use Allocation*; Society of American Foresters: Bethesda, MD, USA, 1981; pp. 134–137.

52. Mika, P.G.; Moore, J.A. Foliar potassium status explains Douglas-fir response to nitrogen fertilization in the inland Northwest, USA. *Water Air Soil Pollut.* **1990**, *54*, 477–491. [CrossRef]

53. Schwandt, J.W. Fertilization effects on root disease in two Douglas-fir stands in northern Idaho. In *Proceedings of the Fiftieth Western International Forest Disease Work Conference*; USFS: Powell River, BC, Canada, 2002; pp. 1–6.

54. Shaw, T.M.; Moore, J.A.; Marshall, J.D. Root chemistry of Douglas-fir seedlings grown under different nitrogen and potassium regimes. *Can. Jour. For. Res.* **1998**, *28*, 1566–1573. [CrossRef]

55. Mandzak, J.M.; Moore, J.A. The role of nutrition in the health of the inland western forest. *J. Sustain. For.* **1994**, *2*, 191–210. [CrossRef]

56. Miller, D.L.; Gravelle, P.J. *Species Selection Guidelines for Planting, Natural Regeneration and Crop Tree Selection on Potlatch Land in Northern Idaho*; Potlatch Corporation: Lewiston, ID, USA, 2009.

57. Will, R.E.; Munger, G.T.; Zhang, Y.; Borders, B.E. Effects of annual fertilization and complete competition control on current annual increment, foliar development, and growth efficiency of different aged *Pinus taeda* stands. *Can. J. For. Res.* **2002**, *32*, 1728–1740. [CrossRef]

58. Miller, R.E.; Webster, S.R. Fertilizer response in mature stands of Douglas-fir. In *Forest Fertilization*; University of Washington: Union, WA, USA, 1979; pp. 126–132.

59. Miller, R.E.; Fight, R.D. *Fertilizing Douglas-fir Forests*; USFS, PNW Research Station: Portland, OR, USA, 1979.

60. Olson, J.R. Experience with nitrogen fertilization in northern Idaho. In *Forest Fertilization: Sustaining and Improving Nutrition and Growth in Western Forests*; Chappell, H.N., Weetman, G.F., Miller, R.E., Eds.; University of Washington: Seattle, WA, USA, 1992; pp. 232–236.

61. National Timber Tax Website. Timber Fertilization. Available online: http://www.timbertax.org/research/revenuerulings/bda/04-62/ (accessed on 14 January 2015).

62. Shaw, T.M.; Kimsey, M.; Parent, D.R. *Growth Response of an Operationally Fertilized Stand at Brickel Creek, North Idaho*; IFTNC, University of Idaho: Moscow, ID, USA, 2010.

63. Olson, J. The fertilization decision: Using what we have learned to evaluate a nitrogen fertilization program. In *IFTNC Fertilization Workshop*; IFTNC: Moscow, ID, USA, 1988; pp. 1–15.

forests

MDPI

Article

Effects of Hurricane-Felled Tree Trunks on Soil Carbon, Nitrogen, Microbial Biomass, and Root Length in a Wet Tropical Forest

D. Jean Lodge [1],*, Dirk Winter [2], Grizelle González [3] and Naomi Clum [4]

[1] United States Department of Agriculture, Forest Service, Northern Research Station, Luquillo 00773-1377, Puerto Rico

[2] Department of Psychiatry, Columbia University, 180 Ft. Washington Ave., New York, NY 10032, USA; dcw4@caa.columbia.edu

[3] United States Department of Agriculture, Forest Service, International Institute of Tropical Forestry, Jardín Botánico Sur, 1201 Ceiba St.-Río Piedras 00926, Puerto Rico; ggonzalez@fs.fed.us

[4] Carmel Valley Middle School, 3800 Mykonos Lane, San Diego, CA 92007, USA; naomi.clum@sduhsd.net

* Correspondence: dlodge@fs.fed.us or dlodgester@gmail.com or djlodge@caribe.net; Tel.: +1-787-889-7445; Fax: +1-787-889-2080

Academic Editors: Scott X. Chang and Xiangyang Sun
Received: 29 August 2016; Accepted: 31 October 2016; Published: 4 November 2016

Abstract: Decaying coarse woody debris can affect the underlying soil either by augmenting nutrients that can be exploited by tree roots, or by diminishing nutrient availability through stimulation of microbial nutrient immobilization. We analyzed C, N, microbial biomass C and root length in closely paired soil samples taken under versus 20–50 cm away from large trunks of two species felled by Hugo (1989) and Georges (1998) three times during wet and dry seasons over the two years following the study conducted by Georges. Soil microbial biomass, % C and % N were significantly higher under than away from logs felled by both hurricanes (i.e., 1989 and 1998), at all sampling times and at both depths (0–10 and 10–20 cm). Frass from wood boring beetles may contribute to early effects. Root length was greater away from logs during the dry season, and under logs in the wet season. Root length was correlated with microbial biomass C, soil N and soil moisture ($R = 0.36$, 0.18, and 0.27, respectively; all p values < 0.05). Microbial biomass C varied significantly among seasons but differences between positions (under vs. away) were only suggestive. Microbial C was correlated with soil N ($R = 0.35$). Surface soil on the upslope side of the logs had significantly more N and microbial biomass, likely from accumulation of leaf litter above the logs on steep slopes. We conclude that decaying wood can provide ephemeral resources that are exploited by tree roots during some seasons.

Keywords: coarse woody debris; nitrogen cycling; soil carbon; soil nitrogen; tree roots; bark beetle frass; fine root proliferation; nutrient hotspots

1. Introduction

Coarse woody (CWD) debris can directly and indirectly affect mineral nutrient availability in underlying soil through inputs of nutrients and carbon compounds in leachates, invertebrate frass and particulate decay products [1–7]. CWD as defined by Harmon et al. [3] is fallen or standing dead wood at least 10 cm diameter. Nutrient hotspots that develop in soil beneath decaying CWD may be important for maintaining forest soil fertility and provide important resources for trees [5]. As noted by Johnson et al. (2014) [8], the influence of soil nutrient hotspots and hot moments on plant growth is obscured by pooling samples, averaging values and eliminating outliers from data sets since soil nutrient availability to plants often responds non-linearly to nutrient hotspots and additions of organic matter [9]. Studies in tropical montane [10] and lowland tropical forests [11–14] show that fine roots of trees differentially proliferate in nutrient hotspots (Sayer and Banin 2016) [15]. Two studies in Australian woodlands [16,17] showed significantly higher C, N and P in soil near logs than 80 cm away, but other studies elsewhere showed decreases in soil N associated with CWD [18–21]. A companion study to ours by Zalamea et al. (2016) [22] that was conducted in different plots at Bisley and Río Chiquito located on the opposite side of the Luquillo Mountains from our site, and using somewhat different tree species found no increase in C content in their shallower soil samples (0–5 cm) under decomposing logs than the deeper strata we studied (0–10 and 10–20 cm), though Zalamea et al. (2007) [23] did find more water extractable organic matter beneath logs. Reduced soil N concentrations associated with CWD in North American temperate forests are thought to result from increased leachates with high C:N ratios which depress rates of N mineralization [18–20], while lower extractable N with large hurricane inputs of CWD in a tropical forest was attributed to increased microbial immobilization [21]. Tropical and subtropical forests have fast wood decay rates [7,24,25] and soils that are often dominated by highly weathered iron and aluminum clays that bind tightly to phosphorus and make it less available (P-fixing clays). In such tropical and subtropical sites, increases in soil C from the decay of CWD are predicted by the Century model to increase the amount of readily exchangeable phosphorus beginning several years after wood addition, thereby stimulating forest productivity [26]. CWD can also alter the physical characteristics of the underlying soil, such as pH, moisture and temperature [3,6,16,17,22].

Natural events such as storms, fires and bark beetle outbreaks as well as anthropogenic disturbances such as logging and salvage operations can deposit CWD on forest floors, potentially contributing to spatial heterogeneity of soil carbon and mineral nutrient availability [18]. Increasing sea surface temperatures due to climate change are predicted to produce more frequent high-intensity hurricanes and typhoons [27–29], while increasing winter temperatures and drought associated with climate change favor increased bark beetle outbreaks and forest fires in North America [30,31]. Given the increasing frequency of forest disturbance events, it is important to understand the effects of CWD on soil carbon, soil fertility and forest productivity.

Whether soils under CWD create hotspots of nutrient availability that are exploited by tree roots, and conversely, whether nutrient sinks under logs caused by microbial immobilization of limiting nutrients are avoided by tree roots, have received little attention. However, production of fine roots in Amazonian forest leaf litter was shown to be a response to nutrient hotspots [32], while root proliferation in response to localized fertilization has been found useful for determining which mineral nutrients are limiting to tree growth in Hawaii [13] and the Amazon [11]. These studies indicate that root length can be used as a bioindicator of where limiting resources are available in nutrient hotspots.

The objectives of this study in a tropical forest of Puerto Rico were to determine if soil N and C and tree root lengths were higher under as compared to away from decomposing logs; whether soil microbial biomass and tree root lengths in the same samples varied inversely with each other, thus suggesting competitive exclusion; whether within paired samples, tree roots were more abundant in resource hotspots; and whether there were temporal patterns in soil N, C, microbial biomass C or root length that varied with soil moisture.

2. Materials and Methods

2.1. Study Area

The El Verde Research Area of the Luquillo Experimental Forest (LEF), where this study was conducted, is on the western side of the Luquillo Mountains ($18°18'$ N; $65°49'$ W; Figure 1). The sites ranged from 350–400 m asl. Mean annual rainfall is 3500 mm with a weakly seasonal regime, and mean temperatures are 22–25 °C [32]. The area is classified as subtropical wet forest based on latitude, potential evapotranspiration and low seasonality according to the Holdridge Life Zone system [32], though the moderating effect of the surrounding ocean on temperatures make this comparable to a tropical wet climate despite the higher latitude.

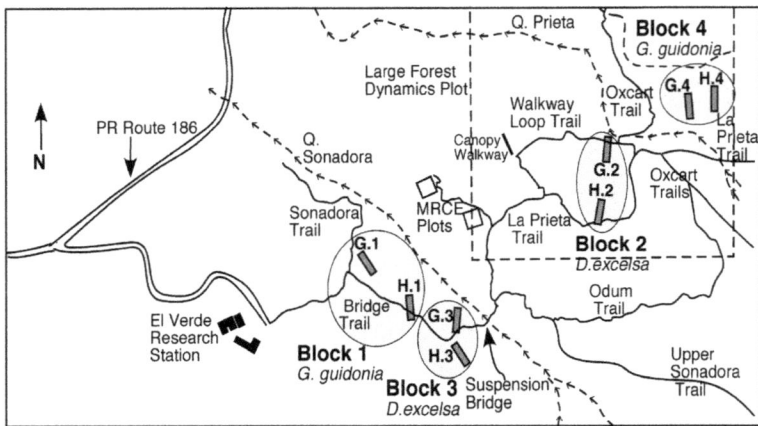

Figure 1. Section of the El Verde Research Area in the Luquillo Experimental Forest of Puerto Rico where this study was conducted, showing the four blocks, each with a pair of logs belonging to two different hurricane cohorts (Hurricane Hugo in 1989 and Hurricane Georges in 1998). The first letter of the log code refers to the hurricane (G for Georges and H for Hugo) while the number refers to the block. Logs in two of the blocks were *Dacryodes excelsa* Vahl. While the logs in the other two blocks were *Guarea guidonia* (L.) Sleumer. Direction of stream flow in the Quebrada Prieta and Quebrada Sonadora is indicated by arrows. The logs are not shown to scale.

Soils in the study area are weathered oxisol clay loams and clays (30%–55% clay, 28%–35% silt and 18%–34% sand) in the Cristal and Coloso series [33], (Table 1). These clays have a high capacity for binding to phosphorus, but the upper horizon has high soil organic matter content and moderately slow permeability. The Coloso clay series has a higher cation exchange capacity than the Cristal clay loams (43 vs. 29 CEC C^{-1}, respectively) [33]. Litter decomposes quickly and humus is usually absent between the forest floor litter and the mineral soil [34], as is typical of mull soils. Leaf litter accumulates on the upslope side of debris dams on slopes >25% [9], resulting in higher soil carbon above than below permanent debris dams [35].

Table 1. Location of study blocks and their soil types in the El Verde Research Area of the Luquillo Experimental Forest.

Location	Block	Latitude, Longitude	Soil Type
Q. Sonadora watershed	1	18°19'24'' N, 65°49'03'' W	Cristal clay-clay loam
Q. Prieta watershed in LFDP cells 02.09 and 07.10	2	18°19'30.7'' N, 65°48'56.6'' W 18°19'31'' N, 65°49'02'' W	Cristal clay-clay loam
Q. Sonadora watershed	3	18°19'24'' N, 65°49'04'' W	Cristal clay-clay loam
Q. Prieta watershed in LFDP cells 10.15 and 10.16	4	18°19'37'' N, 65°49'00'' W 18°19'27'' N, 65°49'00'' W	Coloso clay

2.2. Sampling Design

We examined the effects of decomposing logs of two tree species originating from two major hurricanes, Hugo and Georges, on soil carbon (C) nitrogen (N) content, microbial biomass C and tree root length in the underlying soil. Soil samples collected at two depths beneath the fallen logs were spatially and temporally paired with samples collected 20–50 cm away. This study on the western side of the Luquillo Mountains was conducted when the hurricane-felled logs were 0.6–1.1 and 9.6–11 years after treefall in the two hurricane cohorts, respectively, and was a precursor to a separate study on the eastern side of the Luquillo Mountains by Zalamea et al. [22,23] when their logs had decayed 6 and 15 years. This study differs in that soils were collected from 0–10 and 10–20 cm depths versus 0–5 cm in Zalamea et al. [22,23], and we used the chloroform fumigation-incubation method [36] to estimate soil microbial biomass C rather than substrate induced respiration technique [37] used in Zalamea et al. [22].

Four areas (blocks) were selected for study, each with a pair of logs of the same species that were at least 30 cm diameter, with one felled by Hurricane Hugo and one felled by Hurricane Georges (Figure 1). The two species were *Dacryodes excelsa* Vahl. (Burseraceae) and *Guarea guidonia* (L.) Sleumer (Meliaceae); there were two replicate blocks of each tree species. Fallen trees were >30–80 cm in diameter and at least half of the trunk was in contact with the ground (diameters are given in Table 2). Slopes varied from 20%–40% (Table 2). All logs were decayed by white-rot basidiomycete fungi based on the bleached appearance and spongy-fibrous texture, though rates of decay varied among logs (see Table 2 for details). We used the four-class decay system proposed by Torres (1994) [38] as the decay classification used for logs in temperate forests works poorly in the tropics where the bark can be retained on extensively decayed logs. At the time of this study, the youngest logs (0.6–2 years-old) belonging to the Georges cohort were in decay classes I: no detectable decay, and II: intact bark, sapwood partially soft, and few invading roots present. The older logs belonging to the Hugo cohort (9.5–11 years-old) were in decay II to class III: bark partially lost, sapwood soft, roots invading.

Table 2. Characteristics of logs that were used in this study. The label is comprised of the first letter of the hurricane that felled the tree, followed by the block number. Tree species were *Guarea guidonia* (Meliaceae) and *Dacryodes excelsa* (Burseraceae) but tree species were found to be similar and were not part of the model. Decay class was rated based on description of the logs were selected for study according to the classification of Torres (1994) [38]. Decay rate was classified 18 and 27 years after Hurricane Georges and Hugo, respectively: slow—retaining some bark, diameter loss less than 5 cm; moderate—no bark remaining, diameter loss more than 5 cm but wood remaining; fast—either in the form of a mound of humus invaded by roots or no visible remains). Slope (%) was measured where the logs were located.

Location	Label	Hurricane	Tree Species	Diam. (cm)	Decay Class	Decay Rate	Slope (%)
Q. Sonadora	H.1	Hugo 1989	*G. guidonia*	35–55	2	Slow	28
Q. Sonadora	G.1	Georges 1998	*G. guidonia*	38–80	1	Slow	35
Q. Prieta, LFDP cell 02.09	H.2	Hugo 1989	*D. excelsa*	35–52	3	Fast	35
Q. Prieta, LFDP cell 07.10	G.2	Georges 1998	*D. excelsa*	30–35	2	Fast	40
Q. Sonadora	H.3	Hugo 1989	*D. excelsa*	30–43	2	Moderate	20
Q. Sonadora	G.3	Georges 1998	*D. excelsa*	30–55	1	Slow	40
LFDP cell 10.16	H.4	Hugo 1989	*G. guidonia*	39–53	3	Fast	25
LFDP cell 10.15	G.4	Georges 1998	*G. guidonia*	45–70	2	Fast	25

2.3. Soil Sampling and Analysis

2.3.1. Soil Sampling

Volumetric soil cores (8 × 8 cm) were extracted to 10 cm depth using a two-part aluminum corer as shown in Figure 2. Soil cores were extracted from under logs by digging an approach trench and driving the rectangular corer in horizontally (Figure 2). Each core from under a log was spatially and temporally paired with a core extracted vertically and located 20–50 cm from the first core and away from the log. These are hence referred to as "under" and "away" positions. On each sample date, one pair of cores was extracted from each log on the uphill side of the hill slope, and a second pair was extracted on the downhill side of the same log to balance the effects of the logs acting as debris dams on steep slopes [9,35] (Figure 2). Samples were not pooled.

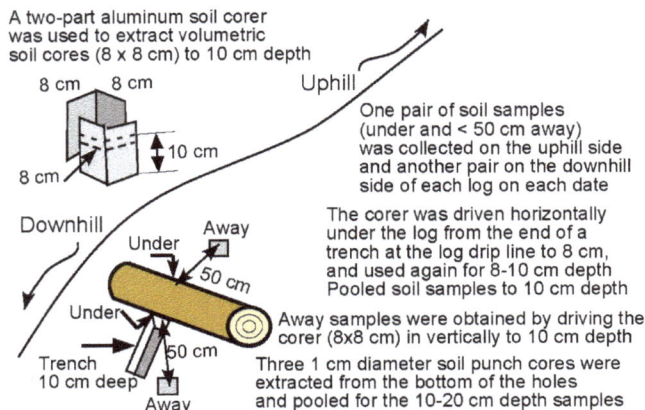

A two-part aluminum soil corer was used to extract volumetric soil cores (8 x 8 cm) to 10 cm depth

8 cm   8 cm

Uphill

10 cm

8 cm

Downhill

Away

Under

50 cm

Under

50 cm

Trench
10 cm deep

Away

One pair of soil samples (under and < 50 cm away) was collected on the uphill side and another pair on the downhill side of each log on each date

The corer was driven horizontally under the log from the end of a trench at the log drip line to 8 cm, and used again for 8-10 cm depth Pooled soil samples to 10 cm depth

Away samples were obtained by driving the corer (8x8 cm) in vertically to 10 cm depth

Three 1 cm diameter soil punch cores were extracted from the bottom of the holes and pooled for the 10-20 cm depth samples

Figure 2. A rectangular, 2-part soil corer was driven into the soil horizontally from the end of an approach trench to obtain volumetric soil samples from 0 to 10 cm depth under logs. Corer halves were used twice (0–8 and 8–10 cm) to obtain soil to 10 cm depth. Paired samples taken away from the logs were extracted vertically and located 20–50 cm away from the under-log core. Two pairs of cores were extracted on each date, one pair on the uphill side and the other on the downhill side of the log. Paired sample locations were randomly selected on each side of the log except for avoiding previously cored areas and large rocks. Soil samples from the 10–20 cm depth were extracted from the bottom of the first core holes using a punch corer.

There were three sampling times: April to early May 1999 (dry season); January 2000 (late wet season) and September 2000 (mid-wet season). These times corresponded to 0.6, 1.4 and 2 years after Hurricane Georges, and 9.5, 10.4 and 11 years after Hurricane Hugo, respectively. Samples in the same block were collected on the same day, and all samples for a particular time were collected within a 12-day period.

2.3.2. Soil Nitrogen and Carbon Content

Soil samples were weighed, hand-sorted to separate soil, roots and rocks; the roots and rocks were then weighed. A 15–20 g subsample was weighed into an aluminum pan, oven dried at 40 °C and re-weighed to determine soil % moisture. Remaining soil was air dried and stored, then oven dried at 65 °C. Total percent soil C and N were determined at the United States Forest Service (USFS) International Institute of Tropical Forestry laboratory using a LECO-2000 CNS analyzer [39] following the procedure of Vitousek and Matson [40]. Two samples of fresh bark beetle frass were also collected from different logs and analyzed for C and N using the CNS analyzer. We were unable to analyze the ammonium and nitrate we had extracted with KCl from fresh soils because of hurricane damage to facilities and infrastructure caused by Hurricane Georges.

2.3.3. Soil Biota

Root length was used as an indicator of resource hotspots. Roots were separated into coarse (>2 mm diameter) and fine (up to 2 mm diameter) from the 0–10 cm depth core samples, weighed and cut into ca. 1 cm long pieces. The decision to estimate root length was made after the first core (G.2) was processed, so those data are missing. Estimates of root lengths were made from three weighed subsamples per sample using the line-intercept method of Newmann [41]. Total root length was estimated by multiplying the mean root length per g of roots by the total weight of the roots. Analyses were made on total root length (fine plus coarse) as coarse roots were negligible.

Soil microbial biomass C was estimated using the chloroform fumigation-incubation method [36]. Soil moisture was adjusted if there was less than 55% field capacity in two 30–40 g subsamples the day after collection. One of the subsamples from each sample was fumigated with chloroform under vacuum, reinoculated with soil and incubated for 10 days in air-tight jars with NaOH for trapping the CO_2 respired; the unfumigated control was also incubated with a CO_2 trap. After 10 days, the amount of CO_2 trapped in the NaOH solution was determined by titration with HCl using phenolphthalein indicator dye. CO_2 in the NaOH stock solution was also determined using titration, and the amount subtracted from the total in the incubation traps. Respired CO_2 in the controls was subtracted from that in the fumigated samples according to the formula in Jenkins and Powlson [36].

2.4. Statistical Analyses

The design was a repeated measures nested randomized complete block. Dependent variables (soil % C, soil % N, microbial biomass C, total root length and soil moisture at 0–10 cm depth) were analyzed separately using a Repeated Measures Analysis; the 10–20 cm depth was only sampled twice for % C and % N, so those data were analyzed separately as a Repeated Measures nested randomized complete block. The independent variables (fixed effects in the models) were hurricane, position (under versus away from logs), hillslope direction (uphill or downhill), and time after Hurricane Georges (0.6, 1.4 and 2 years). Random effects of logs were nested within hurricanes. Block 2 was omitted from the root length analysis because of missing data for log G.2. All statistical analyses were performed using SAS Institute (version 9.3, [42]) software for a generalized linear mixed model using the pseudo-likelihood estimation technique via PROC GLIMMIX. In all analyses, either the normal or lognormal distribution and an identity link function best fit the data. For % N and root length at 0–10 cm, we had to add the groups = hillslope and groups = position option, respectively, to adjust for heterogeneity of variance. In all cases, the correlation between the repeated time was best fit by the unstructured covariance structure. Model selection was based on fit statistics using HQIC (see Table 3 in Results). All models used the Identity Link Function, and all possible interactions were included. No differences were found between tree species, so they were treated as replicates in all the analyses.

Pearson's Product Moment correlations were used to analyze relationships among root length, soil % N, soil % moisture and microbial biomass C. Analyses were performed in Excel. A backwards elimination multiple linear regression was performed on the April 1999 and September 2000 data from the 0–10 cm depth using SAS software [42] after determining that all the variables met assumptions of normality and homoscedasticity of variance. The dependent variable was total root length, and the independent variables were soil microbial C, soil % moisture and soil % N.

Table 3. Summary of the best fit statistical analysis models and results. Probability annotations are as follows: suggestive—sug., $p < 0.1$; * $p < 0.05$; ** $p < 0.01$; *** $p < 0.001$.

Variable (s)	Levels	Probability
0–10 cm soil horizon		
Carbon (total %) Lognormal, Identity Link function, Repeated Measures GLIMMIX, HQIC fit statistics, all possible interactions included in model (shown if significant)		
Position (Under vs. Away)	2	0.0324 *
Hurricane within Block (Hugo vs. Georges)	2	0.3264
Hillslope direction (Upslope vs. Downslope)	2	0.8188
Time (0.6, 1.4 and 2 years post-Georges)	3	0.0374 *
Block (log number, Hurricane nested within Block) true replicates	4	

Table 3. *Cont.*

Variable (s)	Levels	Probability
Nitrogen (total %) Lognormal, Identity Link function, Repeated Measures GLIMMIX, HQIC fit statistics, all possible interactions included in model (shown if significant)		
Position (Under vs. Away)	2	0.0020 **
Hurricane within Block (Hugo vs. Georges)	2	0.2177
Hillslope direction (Upslope vs. Downslope)	2	0.0420 *
Time (0.6, 1.4 and 2 years post-Georges)	3	0.6816
Block (log number, Hurricane nested within Block) true replicates	4	
Microbial biomass C (mg·kg⁻¹ soil) Gaussian, Identity Link function, Repeated Measures GLIMMIX, HQIC fit statistics, all possible interactions included (shown if significant)		
Position (Under vs. Away)	2	0.0840 sug.
Hurricane within Block (Hugo vs. Georges)	2	0.9038
Hillslope direction (Upslope vs. Downslope)	2	0.8660
Time (0.6, 1.4 and 2 years post-Georges)	3	< 0.001 ***
Block (log number, Hurricane nested within Block) true replicates	4	
Root length (total) Lognormal, Identity Link function, Repeated Measures GLIMMIX, HQIC fit statistics, all possible interactions included (shown if significant)		
Position (Under vs. Away)	2	0.9049
Hurricane within Block (Hugo vs. Georges)	2	0.0440 *
Hillslope direction (Upslope vs. Downslope)	2	0.0250 *
Time (0.6, 1.4 and 2 years post-Georges)	2	0.0182 *
Block (log number, Hurricane nested within Block) true replicates, deleted one incomplete block	3	
Position by Time interaction	1 DF	0.0368 *
10–20 cm soil horizon		
Carbon (total %) Lognormal, Identity Link function, Repeated Measures GLIMMIX, HQIC fit statistics, all possible interactions included (shown if significant)		
Position (Under vs. Away)	2	0.0273 *
Hurricane within Block (Hugo vs. Georges)	2	0.5899
Hillslope direction (Upslope vs. Downslope)	2	0.7656
Time (0.6, and 1.4 years post-Georges)	2	0.1673
Block (log number, Hurricane nested within Block) true replicates	4	
Hurricane by position interaction		0.0870 sug.
Nitrogen (total %) Lognormal, Identity Link function, Repeated Measures GLIMMIX, HQIC fit statistics, all possible interactions included (shown if significant)		
Position (Under vs. Away)	2	0.0188 *
Hurricane within Block (Hugo vs. Georges)	2	0.7820
Hillslope direction (Upslope vs. Downslope)	2	0.8711
Time (0.6, and 1.4 years post-Georges)	2	0.0579 sug.
Block (log number, Hurricane nested within Block) true replicates	4	

3. Results

3.1. Soil Nitrogen and Carbon

Soil C and N were higher under than away from logs of both hurricanes (the 1989 and 1998 cohorts; Figure 3) and hurricane did not contribute significantly to these models (Table 3). All of the C and N analysis models used a Lognormal distribution and an identity link function based on best fit statistics (Table 3). Significant differences in both C and N concentration (%) were found between under and away from log positions at the 0–10 and 10–20 cm depths (Table 3). Hillslope direction (upslope versus downslope) was not significant except for soil % N at the 0–10 cm depth; N was higher on the upslope side. In most comparisons within logs on a given date, mean C and N (mean of upslope and downslope) were higher under than away from the logs (Figure 3). Percent soil C and N differed significantly among sample dates, but only at the 0–10 cm depth (Table 3).

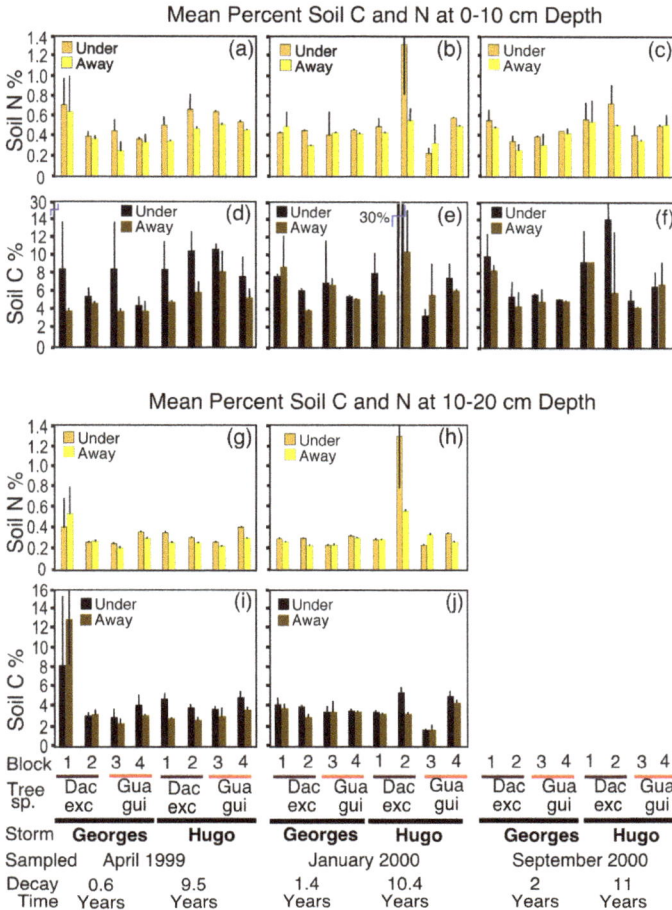

Figure 3. Mean percent soil C and N are presented for paired samples taken under decomposing logs (darker bars) versus samples taken 20–50 cm away and not under the logs (lighter bars) 0.6–2 and 9.5–11 years after Hurricanes Georges and Hugo, respectively. Tree species abbreviations are: Dac exc—*Dacryodes excelsa*, and Gua gui—*Guarea guidonia*. Data are means of two samples, one taken in the upslope direction and the other taken in the downslope direction on the same day (SD bars in black). April 1999 sampling was during the dry season, January 2000 was the end of the wet season, and September 2000 was during the wet season. (**a**–**c**) Mean soil % N at 0–10 cm depth; (**d**–**f**) Mean soil % C at 0–10 cm depth; (**g**,**h**) Mean soil % N at 10–20 cm depth; (**i**,**j**) Mean soil % C at 10–20 cm depth.

Fresh Scoletid beetle frass C and N collected from *Tabebuia heterophylla* (DC.) Britton were 45.0% and 0.4%, respectively, while frass from *Mangifera indica* L. contained 41.8% C and 1.4%. Thus, the C:N ratios were 128.4 for *Tabebuia* and 30.3 for *Mangifera*.

3.2. Soil Biota

3.2.1. Soil Microbial Biomass C

The best fit model for soil microbial biomass C was a Gaussian distribution and an identity link function. Differences in soil microbial biomass C were only suggestive between positions (Table 3). Mean microbial biomass C was generally higher under than away from logs, especially in the April 1999 dry season and January 2000 wet season samples (Figure 4) but there were no significant interactions of position with time (position time $p = 0.634$; position time hurricane $p = 0.268$). Differences in microbial C were significant among sampling dates (Table 3). Microbial biomass C was lower during the April 1999 dry season when soils were drier than in the two wet season samples (January and September 2000; Figure 4).

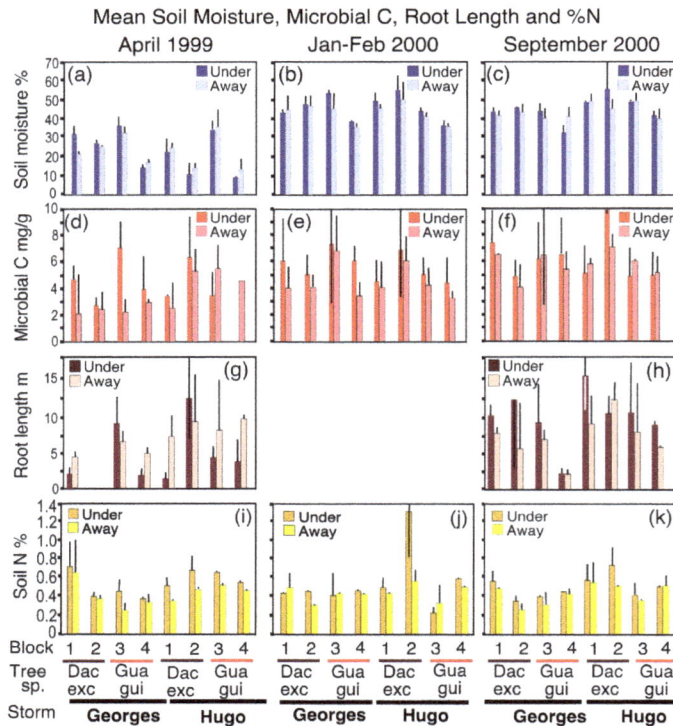

Figure 4. Mean soil % moisture, microbial biomass C, total root length and soil % N are presented for paired (8 × 8 cm) soil cores taken to 10 cm depth under decomposing logs (darker bars) versus samples taken 20–50 cm away and not under the logs (lighter bars). April 1999 sampling was during the dry season, 0.6 and 9.5 years after Hurricane Georges and Hugo, respectively. January 2000 was the end of the wet season, 1.4 and 10.4 years after Hurricanes Georges and Hugo, respectively. September 2000 was during the wet season 2 and 11 years after Hurricanes Georges and Hugo, respectively. Data are means of two samples, one taken in the upslope direction and the other taken in the downslope direction on the same day (SD bars in black). Tree species abbreviations are: Dac exc—*Dacryodes excelsa*, and Gua gui—*Guarea guidonia*. One log from each hurricane (Georges 1998) and Hugo (1989) is within each of the four blocks. (a–c): Mean soil % moisture; (d–f): Mean microbial biomass C; (g,h): Mean m root length; (i–k): Mean soil % N.

3.2.2. Roots

Most of the tree roots were magnoliod (coarse, little-branched, hairless roots) in which the finest roots ranged from 1 to 2 mm diameter. The best fit model for the data was a Lognormal distribution and an identity link function. There was a significant time by position interaction (Table 3). Root length was greater away from the logs in the April dry season, and greater under the logs in the September wet season samples (Figures 4 and 5). Hurricane was significant (Table 3) as root length was greater for older logs from the Hurricane Hugo cohort than for younger logs from the Hurricane Georges cohort (Figure 4 and comparison of Least Square Means). Although hillslope direction was also significant (Table 3), and a comparison of Least Square Means showed significantly higher root length on the upslope than the downslope side of the logs, the pattern is not readily apparent in Figures 4 and 5 because of the reversal of rooting patterns by position between the dry and rainy seasons.

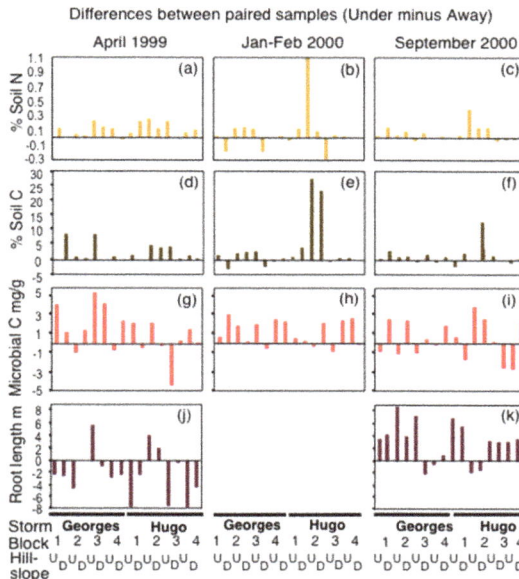

Figure 5. Percent soil N and C, microbial biomass C, and total root length are presented as differences in values between paired (8 × 8) cm soil cores taken to 10 cm depth under decomposing logs minus values for samples taken 20–50 cm away and not under the logs. Thus, bars above the zero line had greater values under than away from the log within a pair while bars below the line had lower values under than away from the log. Bars represent a single pair of samples, one pair taken in the upslope direction (Hillslope U) and the other pair taken in the downslope direction (Hillslope D) on the same day. Logs fell in either Hurricane Hugo (1989) or Hurricane Georges (1998). Two logs, one from each hurricane, were represented in each of the four blocks. April 1999 sampling was during the dry season, 0.6 and 9.5 years after Hurricane Georges and Hugo, respectively. January 2000 was the end of the wet season, 1.4 and 10.4 years after Hurricanes Georges and Hugo, respectively. September 2000 was during the wet season 2 and 11 years after Hurricanes Georges and Hugo, respectively. (**a–c**) Difference (under—away) in soil % N; (**d–f**) Difference (under—away) in soil % C; (**g–i**) Difference (under—away) in microbial biomass C; (**j,k**): Difference (under—away) in m total root length.

3.3. Soil Moisture

Soil moisture only differed significantly between seasons ($p < 0.0001$). No differences were found between under versus away positions or for the interaction between season and position. Soils were significantly drier in the April dry season than in the two wet season samples (Figure 4).

3.4. Correlations among Variables

Root length was most strongly correlated with microbial biomass ($R = 0.36$). The relationships between root length, soil % N and microbial biomass C are shown in Figure 4. Root length was significantly positively correlated with soil microbial C ($R = 0.36$), soil moisture ($R = 0.27$) and soil % N ($R = 0.18$), and all the variables were significantly positively correlated with each other ($p < 0.05$). It was therefore surprising that the best multiple linear regression model (highest adjusted R^2 of 0.12; $p = 0.016$) included all three variables, and the coefficient for microbial C changed from positive in the simpler models to negative because of the inverse relationship between root length and microbial C in the April dry season samples (Figures 4 and 5). Microbial biomass C was significantly positively correlated with soil moisture ($R = 0.47$) and % N ($R = 0.27$).

4. Discussion

4.1. Do Tree Roots Proliferate in Resource Hotspots?

Decaying CWD has been found to either augment soil nutrient content or availability through inputs of leachates, insect frass and decayed wood [1,2,4–7,17], or conversely, to diminish available nutrients though responses of soil microbial communities to inputs of decay products with high C:N ratios [18,19,21]. The question of how tree roots respond to nutrient hotspots or local nutrient sinks created by CWD has received less attention [43], though decayed wood itself has been shown to provide an important source of moisture and nutrients for tree roots [3,6,43]. Trees respond readily to patches that contain limiting resources by proliferating fine roots disproportionately in resource hotspots [8,10–14]. Nutrient hotspots associated with CWD may thus play a critical role in maintaining forest productivity [12]. In this study, we used closely paired soil samples taken under and away from decaying logs belonging to two hurricane cohorts to determine if soil nitrogen, carbon, moisture, microbial biomass and tree root lengths were higher or lower under the CWD. Further, we examined whether the patterns of soil N, C and moisture changed through time and season, and whether roots proliferated more in patches with high nitrogen or moisture, and whether roots avoided soil patches with high microbial biomass since microbes can compete with roots for limiting nutrients [21].

4.2. Effects on Soil Nitrogen and Carbon

Unlike the studies by Zalamea et al. [22] and Bantle et al. [1,2], we did not find differences in soil % C or % N between tree species. One of our tree species was shared with the Zalamea et al. study (*D. excelsa*) while our *G. guidonia* is in the same family and is nearly indistinguishable in quality from *S. macrophylla* that was used in Zalamea et al. [22,23]. Fresh wood of *D. excelsa* and *S. macrophylla* (and presumably, *G. guidonia*) has similar N concentrations [22]. Sites and species were confounded in the Zalamea et al. study [22], and while they included soil characteristics as covariates in their ANOVAs to remove effects due to differences between their two sites; it is possible that not all of the site soil differences were accounted for and that differences in clay content may have contributed to what was attributed to differences in tree species. In our study, three of the four blocks were located on the same soil type and all were at the same field site. We found significantly higher % C under than away from logs at both depths (0–10 and 10–20 cm) and at all sample times. However, the subsequent study by Zalamea et al. [22,23] 7 and 15 years after Hurricanes Hugo and Georges did not find any differences between under and away positions in soil % C at 0–5 cm depth in the same forest type located at a different site on the eastern side of the Luquillo Mountains. Clay content is generally well correlated with soil organic C (SOC) [44] and the clay contents of soils in the Zalamea et al. [22] sites (Bisley and Río Chiquito) were lower than at our El Verde site (20%–22% versus 30%–55%). In addition, the soils in the Zalamea et al. study [22,23] were collected only to 5 cm depth; surface soils contain less clay and are also more influenced by inputs of organic matter from leaf litter than are deeper soil horizons. We therefore think that differences in sampling depth and clay content can explain the disparity between our results and those of Zalamea et al. [22,23]. Clay mineralogy is also an important determinant of SOC [44], but all of the highly weathered soils in both the Zalamea et al. [22,23] and this study were high in aluminum and iron [33]. Zinn et al. [45] noted that aluminum and iron cations are the most important interlayer mineral binders of dissolved organic matter (DOM), and Kaiser and Guggenberger [46] showed sorption to the mineral matrix strongly preserved DOM. In contrast, Kayahara et al. [47] did not find greater amounts of organically complexed Fe and Al in soils below forest floors with woody debris versus those without wood in the boreal forest of British Colombia despite finding more total C and greater C in the humic, fulvic and polyphenol fractions of the overlying forest floors containing wood.

While we do not know how stable the C originating from our logs is and thus how much the logs contribute to C sequestration in soils, Kalbitz et al. [48] showed that sorption of recalcitrant compounds was four times greater than labile compounds. Interestingly, one of our logs (H.2) showed high % C (14%–33% at 0–10 cm depth on two of the three sample dates and 4.7% at 10–20 cm depth in January 2000, and % C was also high away from the log in one of the two January 2000 samples. Log H.2 was *D. excelsa*, a species that had historically been used in the El Verde Research Area for making charcoal, so we suspected this log fell onto an old charcoal pit. Soils around the marked log were examined for charcoal, which is known to cause elevated soil C, but none was found. We therefore postulate instead that the paired cores collected in January and September of 2000 from log H.2 coincided with the C imprint of a previously decomposed log whereas the April 1999 core samples did not. The highest % C values in the upper 10 cm of soil both under and away from log H.2 in January 2000 corresponded to our lowest soil bulk density measurements, which suggests an abundance of light fraction soil C. Log G.1 also had remarkably high soil % C and % N, but primarily at the lower, 10–20 cm depth. This suggests that the legacy of the former decaying log near G.1 dated earlier than the legacy of the decayed log associated with H.2, and that light fraction C in the surface soil near G.1 was lost through decomposition and leaching. Our data indicate that 20% of randomly placed soil cores at our site may fall on C and N-rich hotspots that are the legacy of decomposed prior CWD. Consistent with our results, Gutiérrez del Arroyo and Silver [49] found that ten years after a simulated hurricane debris addition experiment at our site, soil C and N were elevated relative to control plots in both surface and deep soils, and that both light fraction organic matter and organic molecules complexed with minerals in the heavy soil fractions were elevated. Both Harmon et al. [3] and Spears et al. [20] noted difficulty in detecting effects of decaying logs on the underlying soil in Oregon in northwestern USA, likely because of overlapping legacies from previously decomposed CWD. The fast rates of CWD decomposition at our wet tropical site (Table 2) and wet tropical forests in general [24] are probably correlated with fast rates of carbon loss belowground, which makes the imprint of prior treefalls on soil C and N in our forest clearer than in the cool temperate forests of Oregon with slower decomposition [3,20].

Our finding of more C under logs only 0.6 years after Hurricane Georges under logs from Georges was surprising as little decomposition was observed. Furthermore, we did not find significant differences in % C in soils beneath logs from two different hurricane cohorts spaced 9 years apart. However, Bantle et al. [1] found concentrations of dissolved organic C (DOC) were 5–10 times higher in runoff from logs than in throughfall even during early decomposition in a temperate beech forest (3 year-old logs). The products of fungal white rot in wood are primarily CO_2 and DOC [34]. Zalamea et al. [23] found significantly more water-soluble organic matter under than away from their decomposing logs, consistent with DOC of Bantle et al. [1], despite finding no differences in total soil C.

We found a significant effect of season on soil % C at 0–10 cm depth, with increases in the dry season (see graphical abstract). Turner et al. [50] found a similar increase in soil C during the dry season in a more strongly seasonal forest on Barro Colorado Island in Panama. However, we did not observe a significant increase in soil N whereas the study in Panama did [50].

Much of the early C and N inputs to soil from fallen logs in Puerto Rico may come from bark beetle frass [21,38,51], and large deposits of frass were noted under our Georges logs beginning 5 months after they had fallen. We also found especially high microbial biomass C in addition to higher C and N under versus away from Georges logs in that sampling (Figure 5). The few studies on the ecological roles of bark beetles, and effects of xylophagous beetle frass on nutrient cycling [4,5,7,51] indicate rapid decay of the inner bark, which contains the highest concentration of nitrogen. Showalter [4] showed rapid loss of inner bark in hardwood logs was related to invertebrate activity and was independent of climate across a broad geographic gradient in North America. Similar to our study, in Australia, Lindsay and Cunningham [16] and Goldin and Hutchinson [17] found significantly higher C, N and P in soil near logs than 80 cm away. Also consistent with our observations, Cobb et al. [51] found in boreal forest of Canada that xylophagous beetles in a conifer forest that had burned followed by salvage logging had significantly increased soil C and N at the bases of dead trees via frass and chewing debris, and a positive and significant relationship between larval abundance and total soil N, presumably due to inputs of frass. They also found that addition of these beetle frass and chewings to soil increase microbial respiration by more than three-fold; furthermore, frass additions decreased germination of two of their three test plant species (as most of the nitrogen returned via frass is immobilized by microbial biomass and is thus not immediately available to plants). There are two mechanisms by which saproxylic organisms can affect N release as decomposition proceeds. They can accelerate release of N immobilized in microbial biomass through indirect interactions with cord-forming fungi [52,53]. They can also promote N_2 fixation by providing conditions conducive to prokaryote activity in their guts as well as in comminuted wood [54]. Results from Asmus [55] suggest that frass maybe a preferred substrate for N_2 fixing bacteria. Thus, our findings contribute to a greater understanding of the potential effects of CWD on soil carbon and nutrient cycling following both natural and anthropogenic forest disturbances.

Further contributing to early impacts of logs on soil nutrient cycling, wood decomposes more quickly in tropical forests [24,25] such as the one at El Verde in Puerto Rico than in temperate forest studied by Bantle et al. [1]. Three of our four logs in blocks 2 and 4 could only be identified by the pink-tipped poles marking their drip lines at both ends of the logs and the fourth log was reduced to a 2 cm deep humus deposit with roots after 18 and 27 years of decomposition for Georges and Hugo cohorts, respectively. These CWD turynover times are consistent with those from elsewhere in the neotropics [24,25]. Lang and Knight [56] proposed a faster wood decomposition in the tropics than in temperate zones because a warm and humid forest floor in the tropics would favor invertebrate activity through the year. In contrast, Torres [38,57] argued that high moisture content, the cool temperatures (caused by cloud cover, rainfall and winds) and a low animal diversity on the logs in the wet forest in Puerto Rico seem to retard wood decay in this habitat. Here, we argue that hurricane conditions can significantly alter both abiotic (changes in light and moisture level conditions) and biotic conditions that can affect the rate of recovery and nutrient cycling in the forest via decomposition processes [58]. Our observations in this study would support the contention that dry exposed wood generated by hurricane disturbance favors saproxylic species [57], which in turn can initiate the decay process. It has been suggested that wood comminution and bark fragmentation caused by these organisms might have increased nutrient loss through leaching [59,60].

Our finding of significantly higher % N in soils at two depths under than away from logs is concordant with the pattern of % C, which is not surprising as soil C and N are usually correlated and C and N mineralization are coupled [61,62]. Bantle et al. [2] found a net release of N from decaying beech logs during early decomposition in temperate forest in Germany, with the largest component being dissolved organic N (DON). Hillslope direction was also significant for the 0–10 cm depth, with % N (but not % C) being higher on the upslope side of the logs. This result is consistent with logs acting as terrestrial debris dams that trap leaf litter on their upslope side [9,35], especially in April 1999 and January 2000 (Figure 5a,b). The dates with high surface soil % N coincide with peaks in leaf litterfall in our site [63]. Tropical forest leaf litter at El Verde in Puerto Rico has high concentrations of N (1% in González et al. and Silver et al. [64,65]; 1.5% in Lodge et al. [66]) relative to temperate coniferous forest litter, as is typical of wet lowland and low elevation montane tropical forests, and net mineralization usually begins immediately and proceeds steadily rather than being preceded by immobilization phase in the litter during decomposition [67,68]. It is thus not surprising that soil % N was higher in the upper soil horizon on the upslope side of our logs.

4.3. Effects on Biota

4.3.1. Root Responses

Root length was greater away from the logs during the April dry season, but the pattern reversed in the September wet season leading to a significant time by position interaction. Although the shifting of roots away from soil under logs in the dry season corresponds to when soil microbial biomass C was highest under logs, this might be interpreted as competitive exclusion of roots for nutrients by soil microbial biomass, the correlation between roots and soil microbial biomass C was modest ($R = 0.36$). The correlation between root length and soil moisture was higher ($R = 0.47$) while the correlation with soil % N was smaller (0.18); the correlation of microbial C with soil % moisture ($R = 0.47$) was higher than with soil % N ($R = 0.27$). The change in the microbial C coefficient from positive to negative in the best-fit multiple linear regression model that included all independent variables was informative and consistent with the inverse relationship between microbial C and root length in the 0.6 year dry season samples from Georges logs. These patterns are consistent with root proliferation where nutrients were more available. On the other hand, mean root length was higher under than away from logs whereas mean soil microbial biomass C did not differ between positions in September (wet season), which is not consistent with the competitive exclusion hypothesis. However, inputs of bark beetle frass were greatest during the first sampling in the dry season, so the inverse correlation between roots and microbial C in the dry season may be related to effects of early frass deposition. Root lengths were also significantly higher on the upslope sides of logs, concordant with significantly higher % N on the upslope side in the upper 10 cm of soil, suggesting that roots may proliferate where N is more available. Availability of other nutrients, such as P, may have contributed to the rooting patterns since % N is correlated with % C, and C is known to increase P-availability in weathered tropical clay soils [26]. Soil moisture is another resource that can increase root proliferation [3,43]. Although soil moisture was slightly higher under than away from Georges cohort logs in the dry season and the January 2000 wet season sampling (Figures 4 and 5), the differences were slight. One of the lowest soil moistures was recorded for a Hugo log in Block 2 (Figure 4a), which corresponded to some of the highest root lengths; this pattern is not consistent with moisture being a limiting resource that determines root proliferation. Soil microbial C and soil % N were also high in those samples, so nutrient availability also may not have been limiting in the wet season under H.2. Competitive displacement can only occur where critical resources are limited.

4.3.2. Soil Microbial Biomass Responses

We found the strongest differences in soil microbial C between seasons, with values being significantly lower during the April dry than in the two wet season samplings. The soil was so dry at the beginning of the April sampling that one soil sample did not lose weight on oven drying, and some of the others probably lost water of hydration rather than free water. Fungal biomass [68] and the microbial community [69,70] in wet tropical forest of Puerto Rico were shown to be highly sensitive to drying, though community effects were most evident in the very driest soils.

Soil microbial biomass was often (suggestively) higher under than away from logs at the 0–10 cm depth, especially for Georges logs 0.6 years after the hurricane in the April dry season samples. This coincided with lower root lengths under than away from the Georges logs in the April dry season samples. Zalamea et al. [22] did not find differences in microbial biomass between positions, but they did not sample during the early stages of decomposition. Two possible mechanisms seem plausible for explaining this pattern: competition between microbial biomass and plant roots for limiting nutrients and allelophathic effects of beetle frass on roots. As noted above in the discussion of soil % C, the input of Scoletid beetle frass may have stimulated the soil microbial biomass immobilization of N that was observed by Zimmerman et al. [21] beginning 4–6 months after Hurricane Hugo. Cobb et al. [35] similarly found higher microbial activity with input of xylophagous beetle frass and chewings in temperate conifer forest. The C:N ratio of the beetle frass and chewings in the Cobb et al. [55] study was 84–55 while the Scoletid beetle frass we analyzed had C:N ratios of 30–128; all higher than ratios >12–24 that are associated with N immobilization [62,71]. Frass extracts of bark beetle larvae have been attributed to potent allelochemicals that might inhibit germination and seedling growth [72]. Thus, the results of our study would suggest a conservative mechanism of recovery and nutrient retention following a hurricane in this tropical forest that is similar to that observed in the boreal forest of Canada after post fire disturbance. The observation of greater root length away than underneath the decaying logs could potentially be explained by allochemical inhibition by secondary compounds contained in bark beetle frass on roots.

In addition to DOC contributions to soil C from bark and sapwood leachates, frass from bark beetles in the Scoletideae may have contributed to early increases in soil C below our logs. The fourth author recorded copious amounts of frass from Scoletid bark beetles underneath our recently fallen logs five months after Hurricane Georges. Our analyses of two representative samples of Scoletid beetle frass had 42%–45% C. Because Scoletid beetle frass has a fine texture resembling sawdust due to comminution by the beetles and processing in their guts and a high C:N ratio, it is likely to be decomposed quickly by fungi. Our observations of higher soil microbial biomass under than away from Georges logs 7 months after falling (Figure 5) are consistent with the N immobilization by soil microbial biomass and the slowing of forest canopy cover beginning 4–6 months after Hurricane Hugo found by Zimmerman et al. [21]. In a canopy trimming experiment conducted at our field site in El Verde, Silver (personal communication, 11 March 2014) found increases in soil respiration for several years after forest plots were subjected to a simulated hurricane as compared to control plots. Further, Xianbin Liu (personal communication, 12 January 2016) found more labile C in soils of the hurricane simulation plots for at least 1.5 years after the second application of trimming. Thus, it is likely that at least some of the C originating from coarse woody debris from disturbances such as tropical cyclones is labile and readily decomposed once it reaches the soil and is not sequestered. This labile carbon, however, can contribute to microbial immobilization of nutrients, leading to temporarily lower nutrient availability for tree roots.

5. Conclusions

Tropical cyclonic storms, bark beetle outbreaks, fires and anthropogenic disturbances deposit CWD on forest floors, which significantly alters the soil C and N dynamics, and could potentially alter soil fertility and forest productivity. Decomposing logs from two hurricanes spaced 9 years apart left a significant signature on the underlying soil, even as early as 0.6 years after the trees fell. Percent soil C and N were significantly higher under than near logs at both 0–10 and 10–20 cm depths at all sample times. Legacies of previously decomposed logs in the soil C and N were also detected. These signatures of decomposing logs on soil C and nutrients contribute to high spatial heterogeneity in soil characteristics, and lead to nutrient hotspots as well as sinks. The close spatial pairing (<50 cm apart) of soil samples taken under and away from decomposing logs in this study was effective in reducing background variation sufficiently to detect the effects of the logs on soil C, N, microbial biomass C and roots, and was also comparable to root 'choice experiments'. Early increases in soil microbial C under logs felled by Hurricane Georges were detected 0.6 years after the hurricane, and may have been stimulated by inputs of insect frass with high C:N ratios from Scoletid bark beetles in addition to leachates from bark and sapwood. The soils beneath the logs in these early, dry season samples may have been a nutrient sink caused by microbial nutrient immobilization. Correspondingly, roots were less abundant under the logs than away from the logs at that time. Soils on the upslope side of the logs differed only from the downslope side for soil % N and total root length in the upper 0–10 cm, likely because of N fluxes from leaf litter accumulations on the upslope sides of logs on steep slopes and corresponding proliferation of fine roots where N was more available. This also indicates that root proliferation was tracking nutrient availability in transient hotspots. The longevity of the C in soils originating from the decomposing logs was not determined, but spatial anomalies (hotspots) in some our samples suggest that higher soil C may persist after any remains of a decomposed log have disappeared. Evidence from other studies in the same area also suggest, however, that part of the soil C originating from coarse woody debris is labile and is respired quickly by soil microbial biomass as CO_2. The proliferation of roots under decaying logs in the wet season but not the dry season highlights the importance of maintaining spatial heterogeneity in management of coarse woody debris following natural disturbances and logging plus salvage operations in order to allow tree roots to track and exploit nutrient and resource hotspots as they change through time.

Acknowledgments: This research was performed under grants DEB-0218039 and 1239764 from the National Science Foundation to the Institute of Tropical Ecosystem Studies, University of Puerto Rico, and the United States Department of Agriculture, Forest Service, International Institute of Tropical Forestry as part of the Long-Term Ecological Research Program in the Luquillo Experimental Forest. Additional support for G. González was provided by the Luquillo LCZO grant (EAR-1331841). We are thankful to Maria M. Rivera, Mary Jeane Sánchez, Maysaá Ittayem, Edwin Lopéz, Carmen Marrero, Marybelís Santiago and other staff at the International Institute of Tropical Forestry—Soil Laboratory who helped process soil samples; and Xiaoming Zou for advice on measuring soil microbial biomass. We thank the statistician for the United States Department of Agriculture, Forest Service, Northern Research Station, John Stanovick, for statistical analyses and review, and Walter Shortle and Tana Wood for pre-reviews that improved the manuscript. Additional support was provided by the University of Puerto Rico, Rio Piedras.

Author Contributions: D. Jean Lodge and Naomi Clum conceived designed the experiment, and Naomi Clum located the logs for this study and conducted the April 1999 sampling with help from D. Jean Lodge; D. Jean Lodge conducted the January 2000 sampling and Dirk Winter conducted the September 2000 sampling with assistance from D. Jean Lodge. Dirk Winter corrected the data sets and coordinated soil analyses. D. Jean Lodge, Dirk Winter and Grizelle González wrote the paper.

Conflicts of Interest: The first and third authors are United States Government Employees, and since this manuscript was prepared as part of their official duties, it is in the public domain and is not subject to copyright.

References

1. Bantle, A.; Borken, W.; Ellerbrock, R.H.; Schulze, E.D.; Weisser, W.W.; Matzner, E. Quantity and quality of dissolved organic carbon released from coarse woody debris of different tree species in the early phase of decomposition. *For. Ecol. Manag.* **2014**, *329*, 287–294. [CrossRef]
2. Bantle, A.; Borken, W.; Matzner, E. Dissolved nitrogen release from coarse woody debris of different tree species in the early phase of decomposition. *For. Ecol. Manag.* **2014**, *334*, 277–283. [CrossRef]
3. Harmon, M.E.; Franklin, J.F.; Swanson, F.J.; Sollins, P.; Gregory, S.V.; Lattin, J.D.; Anderson, N.H.; Cline, S.P.; Aumen, N.G.; Sedell, J.R.; et al. Ecology of coarse woody debris in temperate ecosystems. *Adv. Ecol. Res.* **1986**, *15*, 133–302.
4. Edmonds, R.L.; Eglitis, A. The role of Douglas-fir beetle and wood borers in the decomposition of and nutrient release from Douglas-fir logs. *Can. J. For. Res.* **1989**, *19*, 853–859. [CrossRef]
5. Schowalter, T.D. Heterogeneity and nutrient dynamics of oak (*Quercus*) logs during the first 2 years of decomposition. *Can. J. For. Res.* **1992**, *22*, 161–166. [CrossRef]
6. Stevens, V. *The Ecological Role of Coarse Woody Debris: An Overview of the Ecological Importance CWD in BC Forests*; Work Paper 30; British Columbia Ministry of Forests: Victoria, BC, Canada, 1997.
7. Ricker, M.C.; Graeme Lockaby, B.; Blosser, G.D.; Conner, W.H. Rapid wood decay and nutrient mineraliztion in an old-growth bottomland hardwood forest. *Biogeochemistry* **2016**, *127*, 323–338. [CrossRef]
8. Johnson, D.W.; Woodward, C.; Meadows, M.W. A three-dimensional view of nutrient hotspots in a Sierra Nevada forest soil. *Soil Sci. Soc. Am. J.* **2014**, *78*, S225–S236. [CrossRef]
9. Lodge, D.J.; McDowell, W.H.; Macy, J.; Ward, S.K.; Leisso, R.; Claudio Campos, K.; Kuhnert, K. Distribution and role of mat-forming saprobic basidiomycetes in a tropical forest. In *Ecology of Saprobic Basidiomycetes*; Boddy, L., Frankland, J.C., Eds.; Academic Press, Elsevier Ltd.: Amsterdam, The Netherlands, 2008; pp. 195–208.
10. Stewart, C.G. A test of nutrient limitation in two tropical montane forests using root ingrowth cores. *Biotropica* **2000**, *32*, 369–373. [CrossRef]

11. Cuevas, E.; Medina, E. Nutrient dynamics within Amazonian forests. II. Fine root growth, nutrient availability, and leaf litter decomposition. *Oecologia* **1988**, *76*, 222–235. [CrossRef]
12. Haines, B. Impact of leaf-cutting ants on vegetation development at Barro Colorado Island. In *Tropical Ecological Systems: Trends in Terrestrial and Aquatic Research*; Golley, F.G., Medina, E., Eds.; Springer: New York, NY, USA, 1975; pp. 99–101.
13. Raich, J.W.; Riley, R.H.; Vitousek, P.M. Use of root-ingrowth cores to assess nutrient limitations in forest ecosystems. *Can. J. For. Res.* **1994**, *24*, 2135–2138. [CrossRef]
14. St. John, T.V.; Coleman, D.C.; Reid, C.P. Growth and spatial distribution of nutrient-absorbing organs: Selective exploitation of soil heterogeneity. *Plant Soil* **1983**, *71*, 487–493. [CrossRef]
15. Sayer, E.J.; Banin, L.F. Tree nutrient cycling in tropical forest—Lessons from fertilization experiments. In *Tropical Tree Physiology: Adaptation and Responses in a Changing Environment*; Goldstein, G., Santiago, L.S., Eds.; Tree Physiology Volume 6; Springer: New York, NY, USA, 2016; pp. 275–297.
16. Goldin, S.R.; Hutchinson, M.F. Coarse woody debris modifies surface soils of degraded temperate eucalypt woodlands. *Plant Soil* **2013**, *370*, 461–469. [CrossRef]
17. Lindsay, A.E.; Cunningham, S.A. Native grass establishment in grassy woodlands with nutrient-enriched soil and exotic grass invasion. *Restor. Ecol.* **2011**, *19*, 131–140. [CrossRef]
18. Hafner, S.D.; Groffman, P.M. Soil nitrogen cycling under litter and coarse woody debris in a mixed forest in New York State. *Soil Biol. Biochem.* **2005**, *37*, 2159–2162. [CrossRef]
19. Kwak, J.H.; Chang, S.X.; Naeth, M.A.; Schaaf, W. Coarse woody debris extract decreases nitrogen availability in two reclaimed oil sands soils in Canada. *Ecol. Eng.* **2015**, *84*, 13–21. [CrossRef]
20. Spears, J.D.H.; Holub, S.M.; Harmon, M.E.; Lajtha, K. The influence of decomposing logs on soil biology and nutrient cycling in an old-growth mixed coniferous forest in Oregon, USA. *Can. J. For. Res.* **2003**, *33*, 2193–2201. [CrossRef]
21. Zimmerman, J.K.; Pulliam, W.M.; Lodge, D.J.; Quiñones-Orfila, V.; Fetcher, N.; Guzman-Grajáles, S.; Parrotta, J.A.; Asbury, C.E.; Walker, L.R.; Waide, R.B. Nitrogen immobilization by decomposing woody debris and the recovery of tropical wet forest from hurricane damage. *Oikos* **1995**, *72*, 314–322. [CrossRef]
22. Zalamea, M.; González, G.; Lodge, D.J. Physical, chemical and biological properties of soil under decaying wood in a tropical wet forest in Puerto Rico. *Forests* **2016**, *7*, 168. [CrossRef]
23. Zalamea, M.; González, G.; Ping, C.L.; Michaelson, G. Soil organic matter dynamics under decaying Wood in a subtropical wet Forest: Effect of tree species and decay stage. *Plant Soil* **2007**, *296*, 173–185. [CrossRef]
24. Clark, D.B.; Clark, D.A.; Brown, S.; Oberbauer, S.F.; Veldkamp, E. Stocks and flows of coarse woody debris across a tropical rain forest nutrient and topography gradient. *For. Ecol. Manag.* **2002**, *164*, 237–248. [CrossRef]
25. Delaney, M.; Brown, S.; Lugo, A.E.; Torres-Lezama, A.; Quintero, N.B. The quantity and turnover of dead wood in permanent forest plots in six life zones of Venezuela. *Biotropica* **1998**, *30*, 2–11. [CrossRef]
26. Sanford, R.L., Jr.; Parton, W.J.; Ojima, D.S.; Lodge, D.J. Hurricane effects on soil organic matter dynamics and forest production in the Luquillo Experimental Forest, Puerto Rico: Results of simulation modeling. *Biotropica* **1991**, *23*, 364–372. [CrossRef]
27. Emanuel, K.A. Increasing destructiveness of tropical cyclones over the past 30 years. *Nature* **2005**, *436*, 686–688. [CrossRef] [PubMed]
28. Knutson, T.R.; McBride, J.L.; Chan, J.; Emanuel, K.; Holland, G.; Landsea, C.; Held, I.; Kossin, J.P.; Srivastava, A.K.; Sugi, M. Tropical cyclones and climate change. *Nat. Geosci.* **2010**, *3*, 157–163. [CrossRef]
29. Mei, W.; Xie, S.P. Intensification of landfalling typhoons over the northwest Pacific since the late 1970s. *Nat. Geosci.* **2016**, *9*, 753–757. [CrossRef]
30. Bentz, B. Climate Change and bark beetles of the western United States and Canada: Direct and indirect effects. *Bioscience* **2010**, *60*, 602–613. [CrossRef]
31. Dale, V.H.; Joyce, L.A.; McNulty, S.; Neilson, R.P.; Ayres, M.P.; Flannigan, M.D.; Hanson, P.J.; Irland, L.C.; Lugo, A.E.; Peterson, C.J.; et al. Climate change and forest disturbance. *Bioscience* **2001**, *51*, 723–734. [CrossRef]
32. Brown, S.; Lugo, A.E.; Silander, S.; Liegel, L. *Research History and Opportunities in the Luquillo Experimental Forest*; General Technical Report, SO-44; United States Forest Service: Washington, DC, USA, 1983.
33. Soil Survey Staff. *Order 1 Soil Survey of the Luquillo Long-Term Ecological Research Grid, Puerto Rico*; United States Department of Agriculture, National Resources Conservation Service: Lincoln, NE, USA, 1995.

34. Lodge, D.J.; Cantrell, S.A.; González, G. Effects of canopy opening and debris deposition on fungal connectivity, phosphorus movement between litter cohorts and mass loss. *For. Ecol. Manag.* **2014**, *332*, 11–21. [CrossRef]

35. Shanley, J.B.; Lodge, D.J.; Krabbenhoft, D.P.; Olson, M.L.; McDowell, W.H. New and old mercury fluxes from mercury amendments to Puerto Rico soil columns. In Presented at the American Geophysical Union Fall 2008 Meetings, San Francisco, CA, USA, 15–19 December 2008; Available online: http://adsabs.harvard.edu/abs/2008AGUFM.B13C0457S (accessed on 17 August 2009).

36. Jenkinson, D.S.; Powlson, D.S. The effects of biocidal treatments on metabolism in soil—V: A method for measuring soil biomass. *Soil Biol. Biochem.* **1976**, *8*, 209–213. [CrossRef]

37. Lin, Q.; Brookes, P.C. An evaluation of the substrate-induced respiration method. *Soil Biol. Biochem.* **1999**, *31*, 1969–1983. [CrossRef]

38. Torres, J.A. Wood decomposition of *Cyrilla racemiflora* in a tropical montane forest. *Biotropica* **1994**, *26*, 124–140. [CrossRef]

39. Tabatabai, M.A.; Bremner, J.M. Automated instruments for determination of total carbon, nitrogen, and sulfur in soils by combustion techniques. In *Soil Analysis, Modern Instruments Techniques*; Marcel Dekker, Inc.: New York, NY, USA, 1991; pp. 261–286.

40. Vitousek, P.M.; Matson, P.A. Mechanisms of nitrogen retention in forest ecosystems: A field experiment. *Science* **1984**, *225*, 51–52. [CrossRef] [PubMed]

41. Newmann, E.I. A method of estimating the total length of root in a sample. *J. App. Ecol.* **1966**, *3*, 139–145. [CrossRef]

42. SAS Institute Inc. *SAS/STAT® 9.3 User's Guide*; SAS Institute Inc.: Cary, NC, USA, 2011.

43. Harvey, A.E.; Jurgensen, M.F.; Larssen, M.J. *Effects of Soil Organic Matter on Regeneration in Northern Rocky Mountain Forests*; GTR-PNW-163; USDA Forest Service, Pacific Northwest Research Station: Portland, OR, USA, 1983; pp. 239–242.

44. Price, S.P.; Bradford, M.A.; Ashton, M.S. Characterizing organic carbon stocks and flows in forest soils. In *Managing Forest Carbon in a Changing Environment*; Ashton, M.S., Tyrrell, M.L., Spalding, D., Gentry, B., Eds.; Springer Science and Business Media: New York, NY, USA, 2012; pp. 7–30.

45. Zinn, Y.L.; Lal, R.; Bigham, J.M.; Resck, D.V.S. Edaphic controls on soil organic carbon retention in the Brazilian Cerrado: Texture and mineralogy. *Soil Sci. Soc. Am. J.* **2007**, *71*, 1204–1214. [CrossRef]

46. Kaiser, K.; Guggenberger, G. The role of DOM sorption to mineral surfaces in the preservation of organic matter in soils. *Org. Geochem.* **2000**, *31*, 711–725. [CrossRef]

47. Kayahara, G.J.; Klinka, K.; Lavkulich, L.M. Effects of decaying wood on eluviation, podzolization, acidification and nutrition in soils with different moisture regimes. *Environ. Monit. Assess.* **1991**, *39*, 485–492. [CrossRef] [PubMed]

48. Kalbitz, K.; Schwesig, D.; Rethemeyer, J.; Matzner, E. Stabilization of dissolved organic matter by sorption to the mineral soil. *Soil Bio. Biochem.* **2005**, *37*, 1319–1331. [CrossRef]

49. Gutiérrez del Arroyo, O.; Silver, W.L. How deep does disturbance go? The legacy of hurricanes on tropical forest soil biogeochemistry. In Proceedings of the American Geophysical Union Meeting, San Francisco, CA, USA, 12–16 December 2016; Available online: https://agu.confex.com/agu/fm16/meetingapp.cgi/Paper/168890 (accessed on 21 October 2016).

50. Turner, B.L.; Yavitt, J.B.; Harms, K.E.; Garcia, M.N.; Wright, S.J. Seasonal changes in soil organic matter after a decade of nutrient addition in a lowland tropical forest. *Biogeochemistry* **2015**, *123*, 221–235. [CrossRef]

51. Cobb, T.P.; Hannam, K.D.; Kishchuk, B.E.; Langor, D.W.; Quideau, S.A.; Spence, J.R. Wood-feeding beetles and soil nutrient cycling in burned forests: Implications of post-fire salvage logging. *Agric. For. Entomol.* **2010**, *12*, 9–17. [CrossRef]

52. Boddy, L.; Watkinson, S.C. Wood decomposition, higher fungi, and their role in nutrient redistribution. *Can. J. Bot.* **1995**, *73*, S1377–S1383. [CrossRef]

53. Boddy, L. Saprotrophic cord-forming fungi: Meeting the challenge of heterogeneous environments. *Mycologia* **1999**, *91*, 13–32. [CrossRef]

54. Ulyshen, M.D. Insect-mediated nitrogen dynamics in decomposing wood. *Ecol. Entomol.* **2015**, *40* (Suppl. 1), 97–112. [CrossRef]

55. Ausmus, B.S. Regulation of wood decomposition rates by arthropod and annelid populations. *Ecol. Bull.* **1977**, *25*, 180–192.

56. Lang, G.E.; Knight, D.H. Decay rates for boles of tropical trees in Panama. *Biotropica* **1979**, *11*, 316–317. [CrossRef]

57. Torres, J.A.; González, G. Wood Decomposition of *Cyrilla racemiflora* (Cyrillaceae) in Puerto Rican Dry and Wet Forests: A 13-year Case Study. *Biotropica* **2005**, *37*, 452–456. [CrossRef]

58. Shiels, A.; González, G.; Lodge, D.J.; Willig, M.R.; Zimmerman, J.K. Cascading effects of canopy opening and debris deposition from a large-scale hurricane experiment in a tropical rainforest. *Bioscience* **2015**, *65*, 871–881. [CrossRef]

59. Bouget, C.; Duelli, P. The effects of windthrow on forest insect communities: A literature review. *Biol. Conserv.* **2004**, *118*, 281–299. [CrossRef]

60. Swift, M.J.; Boddy, L. Animal-microbial interactions in wood decomposition. In *Invertebrate-Microbial Interactions*; Anderson, J.M., Rayner, A.D.M., Walton, D.W.H., Eds.; Cambridge University Press: Cambridge, UK, 1984; pp. 89–131.

61. Winsor, G.W.; Pollard, A.G. Carbon-nitrogen relationships in soil. IV. Mineralization of carbon and nitrogen. *J. Sci. Food Agric.* **1956**, *7*, 618–624. [CrossRef]

62. Cleveland, C.C.; Liptzin, D. C:N:P stoichiometry in soil: Is there a 'Redfield' ratio for the microbial biomass? *Biogeochemistry* **2007**, *85*, 235–252. [CrossRef]

63. Zalamea, M.; González, G. Leaf fall phenology in a subtropical wet forest in Puerto Rico: From species to community patterns. *Biotropica* **2008**, *40*, 295–304. [CrossRef]

64. González, G.; Lodge, D.J.; Richardson, B.A.; Richardson, M.J. A canopy trimming experiment in Puerto Rico: The response of litter decomposition and nutrient release to canopy opening and debris deposition in a subtropical wet forest. *For. Ecol. Manag.* **2014**, *332*, 32–46. [CrossRef]

65. Silver, W.L.; Hall, S.J.; González, G. Differential effects of canopy trimming and litter deposition on litterfall and nutrient dynamics in a wet subtropical forest. *For. Ecol. Manag.* **2014**, *332*, 47–55. [CrossRef]

66. Lodge, D.J.; Scatena, F.N.; Asbury, C.E.; Sánchez, M.J. Fine litterfall and related nutrient inputs resulting from Hurricane Hugo in subtropical wet and lower montane tropical forest of Puerto Rico. *Biotropica* **1991**, *23*, 336–342. [CrossRef]

67. Zou, X.; Zucca, C.P.; Waide, R.B.; McDowell, W.H. Long-term influence of deforestation on tree species composition and litter dynamics of a tropical rain forest in Puerto Rico. *For. Ecol. Manag.* **1995**, *78*, 147–157. [CrossRef]

68. Lodge, D.J. Nutrient cycling by fungi in wet tropical forests. In *Aspects of Tropical Mycology*; BMS Symposium Series 19; Isaac, S., Frankland, J.C., Watling, R., Whalley, A.J.S., Eds.; Cambridge University Press: Cambridge, UK, 1993; pp. 37–57.

69. Bouskill, N.J.; Lim, H.C.; Borglin, S.; Salve, R.; Wood, T.E.; Silver, W.L. Pre-exposure to drought increases the resistance of tropical forest soil bacterial communities to extended drought. *ISME J.* **2013**, *7*, 384–394. [CrossRef] [PubMed]

70. Bouskill, N.J.; Wood, T.E.; Baran, R.; Ye, Z.; Bowen, B.P.; Lim, H.C.; Zhou, J.; Van Norstrand, J.D.; Nico, P.; Northern, T.R.; et al. Belowground response to drought in a tropical forest soil. I. Changes in microbial functional potential and metabolism. *Front. Microbiol.* **2016**, *7*, 525. [CrossRef] [PubMed]

71. Paul, E.A.; Clark, E.E. *Soil Microbiology and Biochemistry*; Academic Press: San Diego, CA, USA, 1996.

72. Khan, D.; Sahito, Z.A.; Dawar, S.; Zaki, M.J. Frass of saproxylic-cerambycid larvae from dead twigs of *Acacia stenophylla* A. Cunn. EX. Benth. and its effects on germination and seedling growth of *Lactuca sativa* L. var. grand rapids. *Int. J. Biol. Biotechnol.* **2016**, *13*, 461–470.

forests

MDPI

Article

Growth Characteristics of Ectomycorrhizal Seedlings of *Quercus glauca, Quercus salicina, Quercus myrsinaefolia,* and *Castanopsis cuspidata* Planted in Calcareous Soil

Masazumi Kayama [1,*,†] **and Takashi Yamanaka** [2]

1   Kyushu Research Center, Forestry and Forest Products Research Institute, Kumamoto 860-0862, Japan
2   Forestry and Forest Products Research Institute, Tsukuba 305-8687, Japan; yamanaka@affrc.go.jp
*   Correspondence: kayama@affrc.go.jp; Tel.: +81-29-838-6313; Fax: +81-29-838-6654
†   Present address: Japan International Research Center for Agricultural Sciences, Ibaraki 305-8686, Japan.

Academic Editors: Scott X. Chang and Xiangyang Sun
Received: 29 August 2016; Accepted: 31 October 2016; Published: 5 November 2016

Abstract: To verify the acclimation capacity of evergreen Fagaceae species on calcareous soil, we compared ecophysiological traits between *Quercus glauca* Thunb., *Q. salicina* Blume, *Q. myrsinaefolia* Blume, and *Castanopsis cuspidata* (Thunb.) Schottky as typical woody species from southwestern Japan. We also examined the inoculation effects of the ectomycorrhizal (ECM) fungi *Astraeus hygrometricus* and *Scleroderma citrinum,* and planted seedlings in calcareous soil collected from a limestone quarry. We measured growth, ectomycorrhizal colonization, photosynthetic rate, and concentrations of nutrients in plant organs for *A. hygrometricus*-inoculated, *S. citrinum*-inoculated, and non-ECM seedlings. Six months after planting on calcareous soil, seedlings of the three *Quercus* species inoculated with *A. hygrometricus* were larger than non-ECM seedlings, especially *Q. salicina,* which showed the greatest increase in dry mass. The dry mass of *C. cuspidata* seedlings was inferior to that of the three *Quercus* species. In the nutrient-uptake analysis, phosphorus, manganese, and iron uptakes were suppressed in calcareous soil for each Fagaceae species. However, seedlings of Fagaceae species that showed better growth had increased concentrations of phosphorus in roots. We concluded that seedlings of *Q. salicina* and *Q. glauca* inoculated with *A. hygrometricus* were best suited to calcareous soil and were considered as useful species for the reforestation in limestone quarries.

Keywords: oak; ectomycorrhizal fungi; limestone; photosynthetic capacity; nutrient physiology

1. Introduction

Limestone occurs throughout East Asia [1]. Calcareous soil, which is formed by weathering of limestone, shows the pH range 7.5–8.5 owing to the buffering effects of the large amounts of calcium carbonate [2,3]. Under such conditions, the uptake of micronutrients (e.g., iron (Fe), manganese (Mn), copper (Cu) and zinc (Zn)), which exhibit low availability at high pH, is suppressed [3,4]. Additionally, carbonate in calcareous soil causes lime-induced chlorosis as a consequence of Fe deficiency [2,3]. Moreover, phosphorus (P) is bound by calcium (Ca), and the uptake of P is also suppressed [3,5]. Accordingly, some plant species exhibit poor growth on calcareous soils [5–8]. Limestone has been quarried at many sites worldwide [9], but once quarrying has finished, natural revegetation of such sites is slow [10,11]. A limestone quarry site in Japan is currently being rehabilitated to a forest stand [12]. The calcareous soil at the quarry has a high pH and large amounts of Ca [12]. Seed from a range of plant species from an adjacent natural vegetation site was dispersed across the quarry site; however, growth of germinated plants was slow [13]. Thus, limestone quarry sites require artificial reforestation for environmental conservation of such sites.

Some plant species have high adaptability to calcareous soil because these species can obtain micronutrients and P from such soil [7]. Evergreen broad-leaved forests are distributed across warm-temperature zones in southern parts of China and southwestern Japan [14–16], and several *Quercus* species are distributed widely across limestone regions [17–19]. In particular, *Q. glauca* Thunb. is abundant on calcareous soils in Japan and China [17–23], but, in contrast, the genus *Castanopsis* is not distributed generally on such soils [18,20,24]. However, the distributions of *Quercus* and *Castanopsis* in East Asia have not been widely examined for their ability to grow on calcareous soil.

Fagaceae species can form a symbiotic interaction with ectomycorrhizal (ECM) fungi [25,26]. On calcareous soil, exudation of organic acids from ECM fungi and the dissolution of insoluble P is important for the uptake of P [27,28]. In the case of Fe, exudation of organic acids, such as citric acid, results in the chelation of Fe, leading to more available for plant uptake [29]. The communities of ECM fungi in calcareous soil can differ from those in non-calcareous soil [30,31], although cosmopolitan ECM fungi, such as genera *Astraeus* and *Scleroderma*, are distributed on calcareous soil [32,33].

The removal of limestone at a quarry site disturbs the topsoil. The abundance of the species of ECM fungi was poor at a disturbed site compared with an undisturbed site [34]. It is predicted that when non-ECM seedlings of Fagaceae species are planted into limestone, colonization of ECM fungi may be poor. When woody species inoculated with ECM fungi are planted on calcareous soil, several species exhibit tolerance to these nutrient-limiting soil conditions [33]. The beneficial effects of inoculation of ECM fungi for plant growth on sites with a poor diversity of ECM fungi have been shown previously [35]. However, few studies have examined the inoculation of seedlings with ECM fungi prior to growing in calcareous soil. We expected that evergreen Fagaceae species inoculated with ECM fungi may be effective for growth on limestone quarry sites. The aim of our research was to verify the acclimation capacity of *Quercus* and *Castanopsis* on calcareous soil. Moreover, we examined the inoculation effects of ECM fungi and compared these between *Quercus* and *Castanopsis* species. We examined the ecophysiological traits of seedlings of *Quercus* and *Castanopsis*, specifically (1) the growth characteristics of the seedlings; (2) symbiosis of ECM fungi on short roots; and (3) concentrations of elements in plant organs. These parameters were compared between ECM-inoculated and non-inoculated seedlings. Our findings may be applied in the artificial reforestation of limestone quarries using Fagaceae seedlings inoculated with ECM fungi.

2. Materials and Methods

2.1. Plants and ECM Fungi

We used *Quercus glauca* Thunb., *Quercus salicina* Blume, *Quercus myrsinaefolia* Blume, and *Castanopsis cuspidata* (Thunb.) Schottky. The three *Quercus* species are distributed across limestone regions in southwestern Japan, whereas *C. cuspidata* is not present in these regions [17–19,21]. Acorns of the four Fagaceae species were collected from an evergreen broad-leaved forest located in the central part of Kyushu, Southwestern Japan in December 2007 for *Q. glauca* and *Q. salicina*, and in December 2008 for *Q. myrsinaefolia* and *C. cuspidata*. To eliminate differences in genetic characteristics caused by edaphic factors, the forest-collected acorns of the four Fagaceae species were from an area with non-calcareous soil. The acorns were stored in a refrigerator at 4 °C prior to sowing.

Information about species of ECM fungi distributed in calcareous soil in Japan has rarely been reported [36]. However, reports from other countries indicate that *Astraeus hygrometricus* (Pers.) Morgan and *Scleroderma citrinum* Pers. can form symbiotic interactions in calcareous soil [31,32]. In addition, *A. hygrometricus* and *S. citrinum* are typical ECM fungi from Fagaceae forests in Japan [37,38], and we found previously that these species could form symbiotic relationships with Fagaceae species [39,40]. Thus, we chose *A. hygrometricus* and *S. citrinum* for the inoculation of Fagaceae seedlings.

In July 2008, fruiting bodies of *A. hygrometricus* and *S. citrinum* were collected from a slope with exposed mineral soil on the edge of a secondary evergreen, broad-leaved forest. This site was the same as that for the collection of acorns. Several fragments (3–7 mm) were cut off from the fruiting bodies and placed on agar medium. The medium consisted of 10 g·L^{-1} glucose, 10 g·L^{-1} malt extract, and 1 g·L^{-1} bactopeptone [41]. Fragments of *A. hygrometricus* and *S. citrinum* were incubated at 25 °C for 2 weeks in darkness. We confirmed the developed mycelial colony of *A. hygrometricus* and *S. citrinum* based on descriptions from previous literatures [38,41]. We used isolated mycelia for inoculation.

2.2. Preparation of Pot Experiments

Acorns of Fagaceae species were germinated in June 2008 for *Q. glauca* and *Q. salicina*, and in June 2009 for *Q. myrsinaefolia* and *C. cuspidata*. The acorns were surface sterilized with 30% H$_2$O$_2$ for 20 min, and rinsed 4 times with sterilized distilled water. Acorns were placed on a sterilized perlite and akadama (red ball earth, originated from subsoil in Kanto region, Japan) soil (1:1) in a free-draining container (width × length × depth = 31 cm × 21 cm × 8 cm). After germination, inoculation of ECM was performed in August of the same year.

Isolated mycelia of *A. hygrometricus* and *S. citrinum* were cultured in 40 mL of Ohta liquid medium [42] in 100 mL flasks at 24 °C for 2 weeks in darkness. The incubated mycelia of *A. hygrometricus* and *S. citrinum* were aseptically fragmented using a stirrer (DX, As One Co., Osaka, Japan) at 7000 rpm for 20 s. The suspension of hyphal fragments was poured into a 50 mL plastic tube and centrifuged at 2320 g for 10 min to collect the hyphal fragments, which were then mixed with sterilized distilled water and centrifuged at 2320 g for 10 min. After the supernatant was removed, the procedure was repeated. The hyphal fragments were resuspended in sterilized distilled water, and 48 seedlings each of *Q. glauca*, *Q. salicina*, *Q. myrsinaefolia*, and *C. cuspidata* were inoculated with *A. hygrometricus* or *S. citrinum*. Prior to the inoculation, seedlings were removed from their containers, and the soil was washed from their roots with sterilized distilled water. On the exposed roots of seedlings, 1 mL of hyphal suspension, which contained 1 mg of hyphal fragments, was added. In addition, 24 seedlings each of *Q. glauca*, *Q. salicina*, *Q. myrsinaefolia*, and *C. cuspidata* were prepared similarly but without inoculation of ECM fungi (non-ECM controls).

All seedlings were planted into Ray Leach Containers (164 mL, SC10, Stuewe and Sons, Tangent, OR, USA) filled with stabilized perlite: akadama soil (1:1). The pH of the perlite: akadama soil was 6.1. The seedlings were grown in an enclosed, air-conditioned (25.0–35.0 °C) greenhouse at the Forestry and Forest Products Research Institute (FFPRI), Tsukuba until April 2009 for *Q. glauca* and *Q. salicina.*, and until April 2010 for *Q. myrsinaefolia* and *C. cuspidata*. The greenhouse was sealed to prevent the seedlings from becoming contaminated by airborne spores or hyphal fragments of mycorrhizal fungi. No fertilizers were applied to the seedlings, and each pot was irrigated with 20 mL of distilled water per day. We also confirmed the contamination of ECM root visually, and the control seedlings contaminated with ECM on their roots were eliminated from the trial.

2.3. Soil Collection and Analysis

In January 2009, we collected calcareous soil from the B horizon of a limestone quarry (Todaka Mining Co., Tsukumi, Oita Pref., Japan, 33°03′ N, 131°48′ E, 580 m a.s.l.). The color of the soil was dark red (5YR 3/6), and it showed high viscosity owing to a high content of clay [43]. Collected soil was sieved (2 mm) at the same site and collected in 32 sandbags. After transportation, soil was kept in a storehouse in Kyushu Research Center, FFPRI.

The pot experiments were carried out by dividing them into two groups, in May 2009 for *Q. glauca* and *Q. salicina* and in May 2010 for *Q. myrsinaefolia* and *C. cuspidata*. Before planting for each year, the collected soil was placed in autoclave bags and sterilized using a wet cycle of an autoclave at 120 °C for 20 min. (KTS-2346B, Alp Co., Tokyo, Japan) to remove all other mycorrhizal species. After sterilization, five soil samples were collected from five bags for each year.

We measured the soil properties including pH and the concentration of carbon (C), nitrogen (N), exchanged P, base cations (Ca, magnesium (Mg), potassium (K) and sodium (Na)), Fe, and Mn. Soil pH was measured using a pH meter (SG2, Mettler Toledo, Zürich, Switzerland); 10 g of soil was mixed with 25 mL of distilled water and shaken for 1 h before the reading was taken [44].

After measuring the soil pH, samples were dried at 105 °C for 24 h for further analysis. Dried soil samples were used to determine the concentration of C and N using a NC analyzer (Sumigraph NC-220F, Sumika Chemical Analysis Service, Tokyo, Japan). Exchanged P was extracted by the Olsen method [45] by shaking for 1 h. P concentration in the soil solution was determined by the molybdenum blue method [46] using a spectrophotometer (UV-2500PC, Shimadzu, Kyoto, Japan).

The concentrations of exchanged base cations and Mn were determined; 2.5 g of dry soil was mixed with 50 mL of 1 M ammonium acetate solution and shaken for 1 h prior to analysis [45]. Solutions for analysis of exchanged Fe were obtained by adding 50 mL of 1 M KCl to 2.5 g of dry soil and shaking for 1 h [45]. The base cations, Fe, and Mn concentrations in the soil solution were analyzed using an inductivity coupled plasma analyser (ICPE-9000, Shimadzu, Kyoto, Japan).

2.4. Establishment of the Pot Experiment

A total of 48 mycorrhizal seedlings and 24 non-mycorrhizal seedlings of the four Fagaceae species were transported to Kyushu Research Center, FFPRI in May. A total of 24 mycorrhizal seedlings and 12 non-mycorrhizal seedlings each of *Q. glauca*, *Q. salicina*, *Q. myrsinaefolia*, and *C. cuspidata* were transplanted into free-draining pots (depth: 30 cm, diameter: 15 cm, volume: 3.7 L) filled with calcareous soil. The remaining seedlings, i.e., 24 mycorrhizal seedlings and 12 non-mycorrhizal seedlings of the four Fafaceae species, were sampled at this time. Fagaceae seedlings were raised for 6 months in a naturally illuminated phytotron (Koito Co., Tokyo, Japan) at Kyushu Research Center, FFPRI. The phytotron had three rooms, and seedlings of *Q. glauca*, *Q. salicina*, *Q. myrsinaefolia*, and *C. cuspidata* were inoculated with *A. hygrometricus* (abbreviated to AH), *S. citrinum* (SC), and non-ECM (NE), the seedlings were then put into the three rooms. Four seedlings of each species inoculated with AH, SC and NE were placed in each room.

After establishment, each pot was irrigated with 60 mL of distilled water per day. The average daily photosynthetically active radiation (PAR) during the experimental period was 39.6 mol·m^{-2}·day^{-1} in 2009, and 37.5 mol·m^{-2}·day^{-1} in 2010, calculated using data on solar radiation from the Japan Meteorological Agency [47] and the ratio of solar radiation to PAR [48]. Air temperature in the phytotron rooms was controlled at 30 °C (day) and 25 °C (night) from July to September, and 25 °C (day) and 20 °C (night) from May to June and in October. These temperatures were based on air temperature in Kumamoto city, Japan [47]. Relative humidity was controlled at 75%. During the experiment, some contamination by airborne spores was noted. The position of seedlings was changed at regular intervals (every month), and the phytotron rooms were disinfected using ethanol at this time.

2.5. Measurement of Seedling Growth

To determine the growth characteristics of seedlings of *Q. glauca*, *Q. salicina*, *Q. myrsinaefolia*, and *C. cuspidata*, we measured the dry mass of leaves, stems and branches, and roots. Twelve AH, SC, and NE seedlings each for the four tree species were harvested in May (first sampling) and November (second sampling). The roots of harvested seedlings were washed twice with tap water to remove soil, and then washed with distilled water. The washed seedlings were divided into leaves, stems and branches, and root components. Each component was put into its own envelope and oven-dried at 60 °C for 4 days. The dry mass of each component was determined. From the values of dry mass of each component at the first and second samplings, we calculated relative dry mass as follows:

$$\text{Dry mass at second sampling/Dry mass at first sampling} \times 100 \tag{1}$$

For the relative dry mass, 100 was set as the value at the first sampling. We also measured specific leaf area (SLA [7]) for each species. We used the SLA value to convert the concentration of nutrients into area based values.

2.6. Measurement of the Rate of Colonization of ECM Fungi

To measure the proportion of roots with ECM fungal colonization, we selected five large clusters of roots from each of seedling, and over 500 short roots (<5 mm in length) that diverged from them. This proportion was determined before drying. For assessment of ECM fungi, roots of seedlings were harvested at the first and second samplings and carefully washed free of soil under gently flowing water. Cluster roots were soaked in distilled water and observed at 10–40× magnification using a stereomicroscope (STZ-40TBIT, Shimadzu, Kyoto, Japan). ECM short roots of *A. hygrometricus* and *S. citrinum* were confirmed by the color, form, and size of the ECM fungi, the presence of rhizomorphs and the absence of root hairs [38,40,49]. The numbers of colonized and uncolonized short roots were counted, and we calculated the percentage of ECM fungal colonization from the proportion of colonized short roots to the total number of short roots [50]. We also examined the relationship between growth (total dry mass) and ECM fungal colonization.

2.7. Photosynthetic Capacity

The photosynthetic rate at light saturation (P_{sat}) was measured in all seedlings in October. Third leaves counted from the top of the seedling were used for the measurement of P_{sat} and stomatal conductance (g_s). Measurements were made using a portable gas analyzer (LI-6400, LI-COR Biosciences, Lincoln, NE, USA) under steady-state conditions at an ambient CO_2 concentration of 37.0 Pa. The LED light source was adjusted to a saturation light level of 1700 $\mu mol \cdot m^{-2} \cdot s^{-1} \cdot PPF$.

2.8. Analysis of Element Concentrations in Plants

We measured the concentrations of N, P, K, Ca, Mg, Fe, and Mn in leaves and roots at the first and second samplings. Dried samples were ground to a fine powder using a sample mill (WB-1; Osaka Chemical Co., Osaka, Japan). The concentration of N was determined using a CN CORDER analyzer (MT-600, Yanako New Science Inc., Kyoto, Japan). The remaining samples were digested by the HNO_3-HCl-H_2O_2 method [51]. The concentration of P was determined by the molybdenum blue method using a spectrophotometer. Concentrations of K, Ca, Mg, Fe, and Mn were analyzed using an atomic absorption spectrophotometer (Z-2310, Hitachi High-Technologies Co., Tokyo, Japan).

The concentrations of each nutrient at the first and second samplings were calculated as relative nutrient concentration, as follows:

$$\text{Concentration at second sampling/Concentration at first sampling} \times 100 \qquad (2)$$

For the relative concentrations, 100 was set as the value from the first sampling. Based on relative nutrient concentration and relative dry mass, we examined vector analysis in accordance with Valentine and Allen [52] and Scagel [53]. Changes in dry mass and the concentration of nutrients were plotted for AH, SC, and NE seedlings of the four Fagaceae species. The interpretation of vector shifts, as described in Valentine and Allen [52], are shown in Figure 1 as a reference.

We also examined the effects of nutrients on photosynthetic rate. In general, photosynthetic rate is closely related to N [7,54] and P [55,56]. Correlation analysis was performed between P_{sat} and the concentrations of N and P in leaves at the second sampling. For the concentrations of N and P, we used the SLA area-based values.

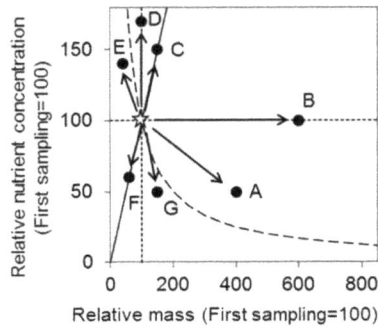

Vector shift	Concentration	Content	Mass	Interpretation
A	Decrease	Increase	Increase	Nutrient non-limiting **Dilution** due to growth
B	Unchanged	Increase	Increase	Nutrient non-limiting **Sufficiency** on growth
C	Increase	Increase	Increase	**Enhancement** of nutrient uptake, related to change in mass
D	Increase	Increase	Unchanged	**Luxury consumption** Little effect on growth
E	Increase	Decrease	Decrease	**Toxicity** of elevated nutrient Decrease of growth
F	Decrease	Decrease	Decrease	**Nutrient deficiency** due to low availability
G	Decrease	Decrease	Increase	**Efficient use** of very little nutrient

Figure 1. Graphical representation of interpretation comparing changes in dry mass and concentrations of nutrients grown in different media types. The content isograms represent combinations of dry mass and concentration required for constant content per unit dry mass. The arrows indicate the direction in which each interpretation holds (adapted from Valentine and Allen [48] and Scagel [49]). The curve indicated by the broken line shows the border to estimate increase or decrease of content. A plot located above the curve shows increase of content.

2.9. Statistical Analysis

All parameters were analysed for normality and homoscedasticity using the Shapiro–Wilk test with Sigma Plot 12.3 (Systat Software Inc., San Jose, CA, USA,). Some parameters did not show normality and homoscedasticity; therefore, we used nonparametric tests by Kyplot 3.0 (Kyens Lab. Inc., Tokyo, Japan). The data of soil chemical properties were compared between 2009 and 2010 using Mann-Whitney U test. Dry mass of the whole plant and each organ, ECM fungal colonization, P_{sat} and g_s, and concentrations of elements in leaves and roots were examined using a Steel–Dwass test. The mean values of total dry mass and ECM fungal colonization for *Q. glauca*, *Q. salicina*, *Q. myrsinaefolia*, and *C. cuspidata* were compared among the first and second samplings of AH, SC, and NE seedlings. The values of dry mass of each organ, P_{sat}, g_s and concentrations of elements in leaves and roots of the four Fagaceae species were compared among AH, SC and NE seedlings. Different letters are used to indicate statistically significant differences at $p < 0.05$. The mean values followed by the same letter showed no significant differences.

To examine the relationship between total dry mass and ECM fungal colonization, we conducted a Spearman's rank correlation. Additionally, we examined the relationship between P_{sat} and nutrients in leaves by the same method. For the data from the second sampling, we also compared each parameter of the seedlings from each of the three phytotron rooms by a Kruskal-Wallis test. There were no significant differences among the three rooms.

3. Results

3.1. Soil Chemical Properties

Table 1 lists the chemical properties of the calcareous soil. The pH of the calcareous soil was almost neutral (7.01). The Ca concentration was high. Comparing the two experimental years for calcareous soil, there was no significant difference.

Table 1. Chemical properties of calcareous soil used in this experiment (mean $\pm$ SD, $n = 5$).

Year	pH	C (g·kg^{-1})	N (g·kg^{-1})	P [1] (mg·kg^{-1})
2009	7.00 ± 0.03	26.5 ± 2.6	0.962 ± 0.082	59.2 ± 11.3
2010	7.02 ± 0.02	28.6 ± 1.1	0.935 ± 0.089	52.1 ± 10.6
	Ca^2 (g·kg^{-1})	Mg^2 (mg·kg^{-1})	K^2 (mg·kg^{-1})	Na^2 (mg·kg^{-1})
2009	2.64 ± 0.05	62.9 ± 1.3	26.0 ± 1.8	117 ± 6
2010	2.56 ± 0.05	62.1 ± 1.6	25.8 ± 0.8	115 ± 1
	Fe^3 (mg·kg^{-1})	Mn^2 (mg·kg^{-1})		
2009	8.34 ± 0.58	4.93 ± 0.59		
2010	9.52 ± 5.14	4.81 ± 0.68		

Mean values of each parameter were analyzed by Mann-Whitney U test (no significance; $p > 0.05$). [1] Extracted by the Olsen method; [2] Extracted in 1 M ammonium acetate; [3] Extracted in 1 M potassium chloride.

3.2. Growth Characteristics

At the first sampling stage before the experiment, the total dry mass of three *Quercus* species showed little difference among AH, SC, and NE seedlings (Figure 2). For *C. cuspidata*; the total dry mass of AH and SC seedlings showed significant increases compared with NE seedlings.

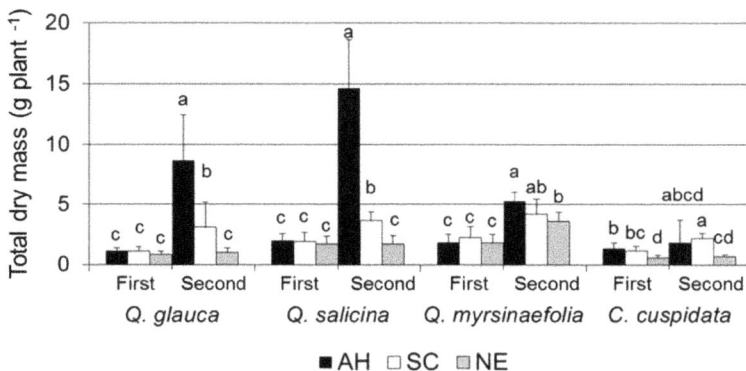

Figure 2. Total dry mass for seedlings of four Fagaceae species grown in a calcareous soil (g·plant^{-1}, mean $\pm$ SD, $n = 12$). Different letters indicate significant effects according to a Steel-Dwass test ($p < 0.05$). Mean values followed by the same letter showed no significant difference ($p > 0.05$). AH, seedlings inoculated with *Astraeus hygrometricus*; SC, seedlings inoculated with *Scleroderma citrinum*; NE, non-ectomycorrhizal seedlings.

At the second sampling, the total dry mass was significantly increased in AH and SC seedlings of *Q. glauca*, *Q. salicina*, and *Q. myrsinaefolia*. For NE seedlings of *Q. myrsinaefolia*, the total dry mass increased significantly. In contrast, other NE seedlings did not show a significant increase in the total dry mass in *Q. glauca*, *Q. salicina*, and *C. cuspidata*. AH seedlings of *C. cuspidata* did not show significant increases in the total dry mass.

Comparing AH, SC, and NE seedlings at the second sampling, the total dry mass and dry mass of each organ were significantly higher in AH seedlings of *Q. glauca*, *Q. salicina*, and *Q. myrsinaefolia* than those in NE seedlings (Figures 2 and 3). Additionally, SC seedlings had significantly higher dry mass for each organ compared with NE seedlings of *Q. glauca*, *Q. salicina*, and *C. cuspidata*. For the SC seedlings of *Q. myrsinaefolia* and AH seedlings of *C. cuspidata*, the dry mass of each organ was not higher than that of NE seedlings. Comparing the AH seedlings of the four Fagaceae species, dry mass was highest for *Q. salicina* and lowest for *C. cuspidata*.

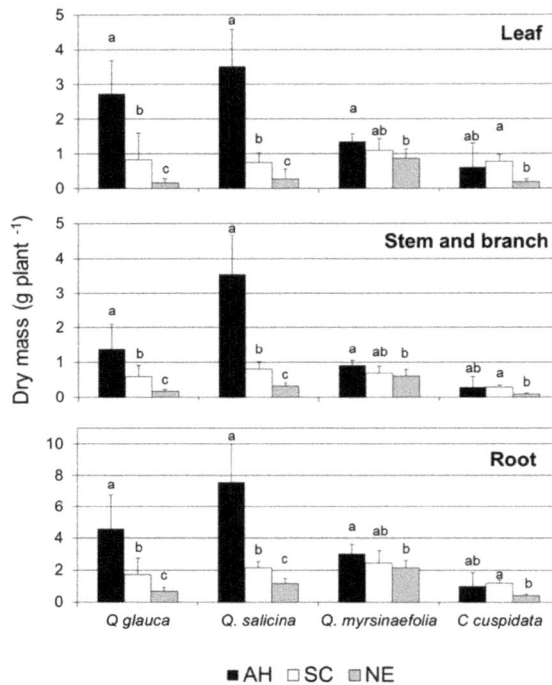

Figure 3. Dry mass of each organ (leaf, stem and branch, and root) for seedlings of four Fagaceae species grown in a calcareous soil at the second sampling (g·plant^{-1}, mean ± SD, $n = 12$). Different letters indicate significant effects according to a Steel–Dwass test ($p < 0.05$). Mean values followed by the same letter showed no significant difference ($p > 0.05$).

SLA of the four Fagaceae species was 85.2 cm^2·g^{-1} for *Q. glauca*, 95.6 cm^2·g^{-1} for *Q. salicina*, 87.7 cm^2· g^{-1} for *Q. myrsinaefolia*, and 162.2 cm^2·g^{-1} for *C. cuspidata*. There were no significant difference among AH, SC, and NE seedlings for each species.

3.3. The Percentage of ECM Colonization

At the first sampling, the percentages of ECM colonization in roots of AH seedlings were 69% for *Q. glauca* and *C. cuspidata*, 83% for *Q. salicina*, and 55% for *Q. myrsinaefolia* (Figure 4). In SC seedlings, these values were significantly lower than those of AH seedlings except for *Q. myrsinaefolia*. The percentage of ECM colonization of SC seedlings was around 40% for *Q. glauca*, *Q. myrsinaefolia*, and *C. cuspidate*, whereas this value was 54% for *Q. salicina*.

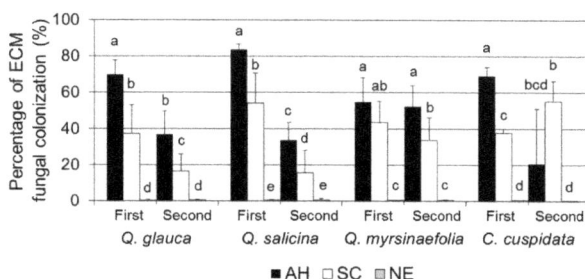

Figure 4. Percentage of ectomycorrhizal fungal colonization for seedlings of four Fagaceae species grown in a calcareous soil (%, mean $\pm$ SD, $n = 12$). Different letters indicate significant effects according to a Steel–Dwass test ($p < 0.05$). Mean values followed by the same letter showed no significant difference ($p > 0.05$).

After the second sampling, the percentages of ECM colonization of AH seedlings decreased significantly for *Q. glauca*, *Q. salicina*, and *C. cuspidata*. In addition, the percentages of ECM colonization decreased significantly for SC seedlings of *Q. glauca* and *Q. salicina*. The percentages of ECM colonization of the three *Quercus* species were significantly higher for AH seedlings than those of SC seedlings. In contrast, the percentage of ECM colonization of SC seedlings increased significantly to 55% for *C. cuspidata*. In half of the AH seedlings of *C. cuspidata*, the percentage of ECM colonization decreased to <1%. For *Q. myrsinaefolia*, the percentage of ECM colonization of AH and SC seedlings did not change after the experiment.

In NE seedlings of the four Fagaceae species, the percentage of ECM colonization was <1% at the two sampling points. In addition, these values of NE seedlings were significantly lower than those of AH and SC seedlings.

In terms of the relationship between total dry mass and the percentage of ECM colonization, we found a significant positive correlation for each Fagaceae species at the second sampling (Table 2, $p < 0.001$). Moreover, *Q. glauca* and *C. cuspidata* showed a significant positive difference at the first sampling ($p < 0.05$).

Table 2. Spearman's correlation coefficients (R) for seedlings of four Fagaceae species grown in a calcareous soil ($n = 12$).

	Q. glauca		*Q. salicina*		*Q. myrsinaefolia*		*C. cuspidata*	
	R	p	R	p	R	p	R	p
	0.826	<0.001 *	0.876	<0.001 *	0.638	<0.001 *	0.776	<0.001 *
Elements	*Q. glauca*		*Q. salicina*		*Q. myrsinaefolia*		*C. cuspidata*	
	R	p	R	p	R	p	R	p
N	0.217	0.199	−0.350	0.038 *	−0.394	0.020 *	0.138	0.415
P	0.030	0.859	0.435	0.010 *	0.605	<0.001 *	0.777	<0.001 *

The upper row shows the relationships between total dry mass and ectomycorrhizal (ECM) fungal colonization. The lower row shows the relationship between photosynthetic rate at light saturation (P_{sat}) and area-based concentrations of N and P in leaves at the second sampling. * shows a significant correlation at $p < 0.05$.

3.4. Photosynthetic Capacity

The P_{sat} of *Q. glauca* and *Q. salicina* was significantly higher in AH seedlings than in SC and NE seedlings (Figure 5). In *Q. myrsinaefolia*, the P_{sat} in AH and SC seedlings was significantly higher than that in NE seedlings. The P_{sat} of *C. cuspidata* was significantly higher in SC seedlings than in NE seedlings. Comparing the four Fagaceae species, P_{sat} was highest in AH seedlings of *Q. glauca* and lowest in NE seedlings of *C. cuspidata*.

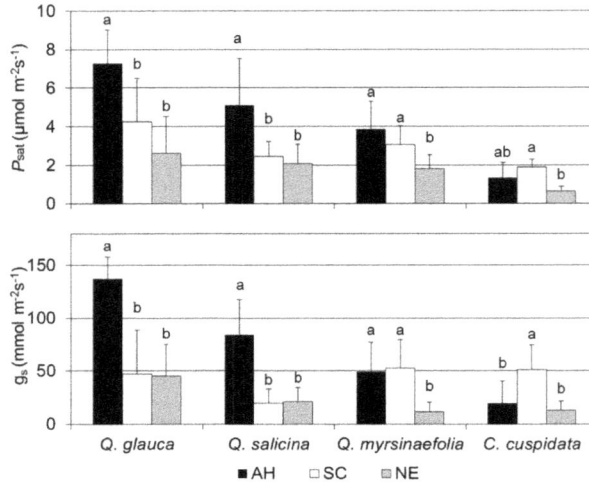

Figure 5. Photosynthetic rate at light saturation (P_{sat}, $\mu mol \cdot m^{-2} \cdot s^{-1}$) and stomatal conductance (g_s, $mol \cdot m^{-2} \cdot s^{-1}$) for seedlings of four Fagaceae species grown in a calcareous soil (mean $\pm$ SD, $n = 12$). Different letters indicate significant effects according to a Steel–Dwass test ($p < 0.05$). Mean values followed by the same letter showed no significant difference ($p > 0.05$). P_{sat} and g_s for *Q. glauca* and *Q. salicina* were measured in October 2009, and for *Q. myrsinaefolia* and *C. cuspidata* in October 2010.

A similar trend was observed for g_s as for P_{sat}, and g_s of *Q. glauca* and *Q. salicina* was significantly higher in AH seedlings than in SC and NE seedlings. g_s in AH and SC seedlings of *Q. myrsinaefolia* was significantly higher than that in NE seedlings. In *C. cuspidata*, the g_s in SC seedlings was higher than that in AH and NE seedlings. In particular, the g_s in AH seedlings of *Q. glauca* was higher compared with other seedlings.

3.5. Element Concentrations in Leaves

The concentrations of elements in leaves at the second sampling are shown in Table 3. The concentration of P was significantly higher in AH seedlings of *Q. salicina* and *Q. myrsinaefolia* than in SC seedlings. In particular, AH seedlings of *Q. salicina* showed the highest P concentration between the four Fagaceae species. In *C. cuspidata*, the concentration of P was highest for SC seedlings. In contrast, other elements did not show high concentrations in AH seedlings despite their greater growth. The concentration of N in AH seedlings was significantly lower for *Q. salicina* and *C. cuspidata* than in SC and NE seedlings. Similarly, the concentration of K was significantly lower in AH seedlings of *Q. glauca* and *Q. myrsinaefolia* compared with in SC and NE seedlings. Moreover, NE seedlings had the highest Mg concentration in *Q. salicina* and *C. cuspidata*.

For the relationship between the concentrations of N or P and P_{sat}, we found a significant positive correlation between the concentrations of P and P_{sat} of *Q. salicina*, *Q. myrsinaefolia*, and *C. cuspidata* (Table 2, $p < 0.05$). In contrast, there were significant negative correlations between concentration of N and P_{sat} of *Q. salicina* and *Q. myrsinaefolia* ($p < 0.05$).

Table 3. Concentrations of elements (N, P, K, Ca, Mg, Fe, and Mn) in leaves of seedlings of four Fagaceae species grown in a calcareous soil sampled at the second sampling (mean ± SD, $n = 12$).

Element, Inoculation		*Q. glauca*	*Q. salicina*	*Q. myrsinaefolia*	*C. cuspidata*
N (mg·g^{-1})	AH	10.7 ± 3.4 a	9.4 ± 2.2 b	9.2 ± 2.8 a	9.7 ± 1.8 b
	SC	11.1 ± 1.5 a	13.2 ± 2.3 a	11.2 ± 3.8 a	15.4 ± 3.4 a
	NE	10.5 ± 2.8 a	15.3 ± 2.0 a	11.0 ± 2.41 a	12.2 ± 1.10 a
P (mg·g^{-1})	AH	1.43 ± 0.33 a	1.98 ± 0.30 a	1.81 ± 0.27 a	0.84 ± 0.33 a
	SC	1.40 ± 0.28 a	1.34 ± 0.35 b	1.41 ± 0.12 b	1.11 ± 0.19 a
	NE	1.35 ± 0.52 a	1.41 ± 0.62 ab	0.69 ± 0.29 c	0.30 ± 0.12 b
K (mg·g^{-1})	AH	5.54 ± 0.80 b	4.52 ± 1.12 a	3.91 ± 0.50 b	5.20 ± 2.12 ab
	SC	8.58 ± 1.37 a	5.39 ± 0.83 a	4.78 ± 1.48 b	3.96 ± 0.65 b
	NE	8.65 ± 1.15 a	6.17 ± 2.22 a	7.15 ± 1.18 a	5.81 ± 1.18 a
Ca (mg·g^{-1})	AH	25.3 ± 7.1 a	19.6 ± 6.5 a	26.6 ± 3.3 a	15.9 ± 2.8 a
	SC	19.2 ± 5.8 a	14.9 ± 3.2 a	26.8 ± 5.1 a	16.2 ± 2.7 a
	NE	20.6 ± 4.8 a	14.6 ± 4.2 a	23.8 ± 3.30 a	17.3 ± 5.11 a
Mg (mg·g^{-1})	AH	1.40 ± 0.43 a	1.12 ± 0.35 b	2.30 ± 0.31 a	1.86 ± 0.65 ab
	SC	1.43 ± 0.40 a	1.23 ± 0.20 b	2.47 ± 0.40 a	1.36 ± 0.25 b
	NE	1.88 ± 0.60 a	3.50 ± 2.66 a	2.68 ± 0.41 a	2.18 ± 0.52 a
Fe (μg·g^{-1})	AH	102 ± 20 a	77 ± 12 b	85 ± 19 a	155 ± 139 a
	SC	110 ± 44 a	159 ± 51 a	79 ± 17 a	105 ± 26 a
	NE	182 ± 137 a	215 ± 139 a	94 ± 21 a	180 ± 115 a
Mn (μg·g^{-1})	AH	204 ± 31 b	202 ± 76 ab	273 ± 65 a	454 ± 239 a
	SC	140 ± 67 b	163 ± 73 b	200 ± 115 a	167 ± 58 b
	NE	434 ± 189 a	363 ± 151 a	206 ± 67 a	507 ± 222 a

Different letters indicate significant effects according to a Steel-Dwass test ($p < 0.05$). Mean values followed by the same letter showed no significant difference ($p > 0.05$). AH, seedlings inoculated with *Astraeus hygrometricus*; SC, seedlings inoculated with *Scleroderma citrinum*; NE, non-ectomycorrhizal seedlings.

3.6. Element Concentrations and Contents in Roots

The concentrations of elements in roots at the second sampling are shown in Table 4. The concentrations of P in *Q. glauca*, *Q. salicina*, and *Q. myrsinaefolia* were significantly higher in AH seedlings than in NE seedlings. AH seedlings also exhibited high concentrations of Fe and Mn in *Q. myrsinaefolia* than in SC seedlings. In *C. cuspidata*, the concentrations of N and P were highest in SC seedlings. The concentration of Ca was significantly lower in SC seedlings of *Q. myrsinaefolia* and *C. cuspidata*. In addition, the concentration of Fe in SC seedlings of *Q. salicina* and *Q. myrsinaefolia* was significantly lower. In NE seedlings of *Q. myrsinaefolia*, the concentration of Mg was significantly higher than in AH and SC seedlings. Across the four Fagaceae species, the concentration of P was the highest for AH seedlings of *Q. salicina*.

Table 4. Concentrations of elements (N, P, K, Ca, Mg, Fe, and Mn) in roots of seedlings of four Fagaceae species grown in a calcareous soil sampled at the second sampling (mean $\pm$ SD, $n = 12$).

Element, Inoculation		Q. glauca		Q. salicina		Q. myrsinaefolia		C. cuspidata	
N ($mg \cdot g^{-1}$)	AH	8.3 ± 1.8	a	9.8 ± 1.7	a	7.5 ± 1.9	a	7.5 ± 1.9	b
	SC	7.3 ± 1.3	a	7.6 ± 0.8	b	6.3 ± 1.9	a	12.2 ± 1.4	a
	NE	6.8 ± 1.6	a	7.7 ± 1.6	ab	5.5 ± 1.8	a	5.8 ± 1.1	b
P ($mg \cdot g^{-1}$)	AH	2.34 ± 0.49	a	3.76 ± 0.70	a	2.63 ± 0.74	a	1.20 ± 0.92	ab
	SC	1.84 ± 0.76	ab	1.21 ± 0.30	b	1.95 ± 1.02	ab	2.45 ± 0.75	a
	NE	1.13 ± 0.29	b	1.17 ± 0.29	b	0.97 ± 0.17	b	0.39 ± 0.17	b
K ($mg \cdot g^{-1}$)	AH	2.74 ± 0.42	a	2.65 ± 0.42	a	2.75 ± 0.36	a	2.17 ± 0.50	a
	SC	2.70 ± 1.40	a	2.28 ± 0.45	a	2.21 ± 0.42	a	2.44 ± 0.34	a
	NE	2.07 ± 0.54	a	2.49 ± 0.60	a	2.52 ± 0.42	a	1.96 ± 0.37	a
Ca ($mg \cdot g^{-1}$)	AH	14.8 ± 2.3	a	15.0 ± 2.2	a	13.4 ± 1.0	a	15.3 ± 2.3	a
	SC	13.8 ± 2.2	a	13.8 ± 1.1	a	11.7 ± 1.0	b	12.5 ± 1.2	b
	NE	17.0 ± 2.2	a	15.7 ± 1.6	a	13.5 ± 1.7	ab	15.1 ± 1.81	a
Mg ($mg \cdot g^{-1}$)	AH	1.85 ± 0.21	ab	1.64 ± 0.23	a	1.54 ± 0.18	b	1.46 ± 0.18	a
	SC	1.70 ± 0.59	b	1.62 ± 0.27	a	1.51 ± 0.28	b	1.25 ± 0.12	bc
	NE	2.10 ± 0.32	a	1.71 ± 0.26	a	1.94 ± 0.19	a	1.36 ± 0.16	ab
Fe ($\mu g \cdot g^{-1}$)	AH	0.99 ± 0.49	a	1.45 ± 0.53	a	1.38 ± 0.42	a	0.76 ± 0.87	ab
	SC	0.82 ± 0.33	a	0.76 ± 0.19	b	0.64 ± 0.20	b	0.69 ± 0.26	a
	NE	1.49 ± 0.77	a	1.32 ± 0.33	a	0.98 ± 0.42	ab	0.38 ± 0.18	b
Mn ($\mu g \cdot g^{-1}$)	AH	63 ± 17	b	70 ± 20	a	39 ± 9	a	31 ± 13	a
	SC	73 ± 11	b	67 ± 18	a	28 ± 6	b	33 ± 8	a
	NE	138 ± 78	a	80 ± 20	a	34 ± 10	ab	24 ± 5	a

Different letters indicate significant effects according to a Steel-Dwass test ($p < 0.05$). Mean values followed by the same letter showed no significant difference ($p > 0.05$).

3.7. Vector Analysis between Relative Dry Mass and Nutrient Concentration

The relationships between relative leaf dry mass and relative concentrations in leaves are shown in Figure 6. For the relationship of Ca, uptake of each Fagaceae species showed enhanced or luxury consumption. In contrast, Mn showed dilution or efficient use by each Fagaceae species. For N, most plots showed dilution or sufficiency, whereas those for SC seedlings of *Q. salicina* and SC and NE seedlings of *C. cuspidate* showed enhanced uptake. In the case of NE seedlings of *Q. salicina*, the plot showed toxicity.

For P and Fe in the leaves, most plots of each Fagaceae species showed dilution or efficient use for leaves and roots. For P in leaves of NE seedlings of *Q. glauca* and *Q. salicina*, and Fe in leaves of NE seedlings of *Q. salicina*, the plots showed nutrient deficiency. The trend of K in leaves showed various patterns. Plots of AH seedling of *Q. glauca* and *Q. salicina* showed sufficiency, but the plots of *C. cuspidata* and AH and SC seedlings of *Q. myrsinaefolia* showed dilution. Other plots showed enhanced or luxury consumption. For the relationship of Mg, the plots showed sufficiency or dilution for each Fagaceae species. In the case of NE seedlings of *Q. salicina*, the plot showed toxicity.

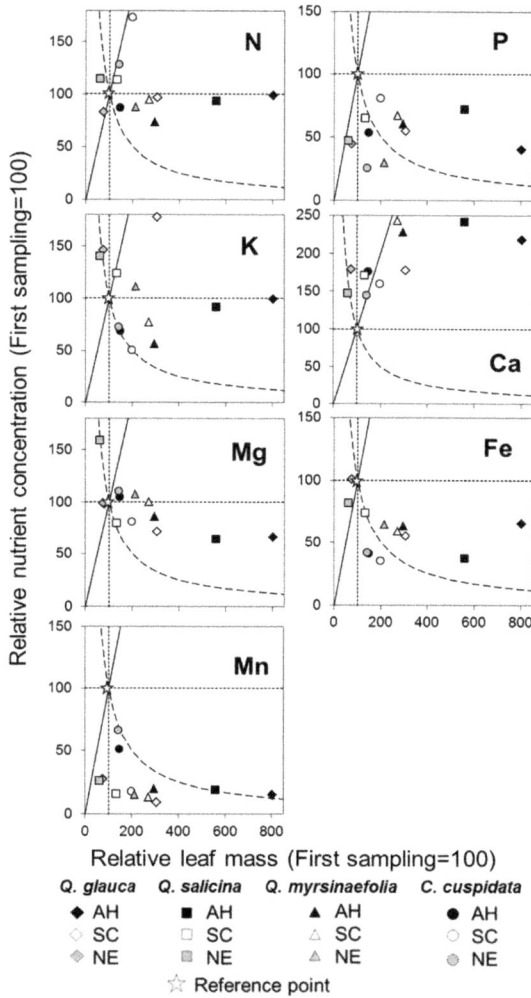

Figure 6. Comparison of relative nutrient concentration in leaves and leaf dry masses for seedlings of four Fagaceae species grown in a calcareous soil. The content isograms represent the combination of dry mass and concentration required for these values at the second sampling relative to the first sampling. A value of 100 represents the dry mass and concentration at the first sampling.

4. Discussion

4.1. Growth and Photosynthesis

After 6 months of cultivation in calcareous soil, we confirmed that symbiosis with ECM fungi was essential for growth acceleration in calcareous soil (Figure 2). Symbiosis with *A. hygrometricus* contributed to growth acceleration in the three *Quercus* species compared with that of *S. citrinum*. In *C. cuspidata*, only *S. citrinum* inoculation showed high values of dry mass of each organ (Figure 3). In general, compatibility with ECM fungi and effects on growth acceleration are different among woody species [57,58]. Unlike *Quercus* species, *C. cuspidata* may have better compatibility with *S. citrinum*. We have already confirmed that *C. cuspidata* inoculated with *S. citrinum* showed accelerated growth [40].

Moreover, AH seedlings of *Q. glauca* and *Q. salicina* showed the highest values of P_{sat} and g_s (Figure 5). Woody species inoculated with ECM fungi exhibit accelerated water uptake from external hyphae [59]. AH seedlings probably had accelerated water uptake, and stomatal closure did not occur. Accordingly, the P_{sat} and g_s were high in AH seedlings. Similar trends were shown for AH and SC seedlings of *Q. myrsinaefolia*, which had high values for the percentage of ECM colonization (Figures 4 and 5).

Comparing AH seedlings, *Q. salicina* showed remarkable growth acceleration (Figure 2). *Q. salicina* have high acclimation capacity in calcareous soil, and symbiosis with *A. hygrometricus* can take place in calcareous soil. *Q. salicina* grows at high altitudes in the karst region in southwestern Japan [17,18,21]. Our results reflected the distribution of *Q. salicina*.

Compared with *Q. salicina* and *Q. glauca*, growth acceleration by inoculation with ECM fungi was not obvious for *Q. myrsinaefolia* (Figure 2). Acclimation capacity in calcareous soil was lower for *Q. myrsinaefolia* than for *Q. salicina* and *Q. glauca*. In contrast, the total dry mass for *C. cuspidata* was lowest even in SC seedlings (Figure 2). The leaves of *C. cuspidata* yellowed after transplantation, indicating the occurrence of lime-induced chlorosis [3]. In addition, lime-induced chlorosis can readily appear in calcifuge plants [7]. Based on growth and lime-induced chlorosis, *C. cuspidata* was considered as a calcifuge plant. Conversely, *C. cuspidata* inoculated with ECM fungi accelerated its growth in acidic soil [40], indicating that this species has a preference for an acidic environment.

4.2. Nutrient Relationships

For the nutrients in calcareous soil, the concentration of exchangeable Ca was 2.6 g·kg^{-1} (Table 1). The values of soil pH and Ca concentration differ among calcareous soils in Japan [12,13,60–63], with ranges of 5.8 to 8.3 for pH and 0.8 to 12.1 g·kg^{-1} for Ca. Compared with other regions, the soil used for our experiment did not have high soil pH, and did not exhibit an extreme excess of Ca. In contrast, the concentration of N in calcareous soil used in our experiment was lower than those in other regions [60–63]. However, the concentration of N in leaves of the four Fagaceae species was not in the deficient range compared with other woody species grown in calcareous soil [13,63]. For other nutrients, the concentrations of K and Mg were higher than typical forest soil in this region [64]. There was no previous information available on P in calcareous soil in Japan from analyses by the Olsen method. Compared with the data from other countries, the concentration of P was higher in calcareous soil used in our experiment [65,66]. Thus, the concentrations of N, P, K, and Mg in calcareous soil in our experiment were not insufficient.

AH and SC seedlings of the four Fagaceae species did not show high N concentrations in leaves (Table 3). Thus, uptake of N in leaves is not related to ECM colonization. Comparing traits among *Quercus* species, the concentration of N in roots was higher for *Q. salicina* and *Q. glauca* than that for *Q. myrsinaefolia* (Table 4). *Q. salicina* and *Q. glauca* had higher adaptability to various poor habitats [19,67]. *Q. glauca* and *Q. salicina* have high capacity to acquire N from calcareous soil. In contrast, the habitats of natural forests of *Q. myrsinaefolia* include deep, well-drained soil [68]. In the karst region, *Q. myrsinaefolia* has been found on a site with deep calcareous soil [19]. Differences in acclimation capacity for *Q. myrsinaefolia* may originate from the inherent nutrient requirements. There was a similar case for *Picea abies*, which has low adaptability to calcareous soil [69], and suppressed uptake of N in roots increased in calcareous soil [70].

On the uptake of P in roots, NE seedlings showed low P concentrations compared with AH seedlings of *Q. glauca*, *Q. salicina* and *Q. myrsinaefolia* (Table 4). Thus, uptake of P is linked with symbiosis with ECM fungi. On calcareous soil, symbiosis with ECM fungi is important for the uptake of P [27,28]. Symbiosis between *Quercus* species and *A. hygrometricus* produced a high capacity for the acquisition of P.

Meanwhile, we confirmed that the uptake of P, Fe and Mn in leaves was suppressed in each Fagaceae species (Table 3, Figure 6). Uptake of micronutrients (e.g., Fe, Mn, Cu, and Zn) and P is suppressed by low availability at high pH in calcareous soil [3–5,70]. Our results suggested that uptake of Mn is suppressed for the Fagaceae species in calcareous soil. For P and Fe in leaves, AH seedlings of three *Quercus* species exhibited increased contents; however, concentration was decreased and showed evidence of dilution (Figure 6). This trend suggests that the uptake of P and Fe in leaves is suppressed even after inoculation with *A. hygrometricus*. In particular, NE seedlings of *Q. glauca* and *Q. salicina* showed a P deficiency in leaves (Figure 6). We considered that NE seedlings of *Q. glauca* and *Q. salicina* suffered from serious nutrient suppression due to lower availability. *C. cuspidata* showed an obvious decrease of the concentration of Fe in leaves, and showed efficient use of little Fe (Figure 6). When calcifuge woody plants are inoculated with ECM fungi, the effects of the fungi are probably quite limited in calcareous soil.

In contrast, Ca in leaves and roots showed a high value (Tables 3 and 4) compared with seedlings of Fagaceae species grown on acidic colluvial soil [40]. Moreover, many plots showed enhancement of uptake or luxury consumption (Figure 6). However, the actual concentration of Ca in plant organs was almost the same among AH, SC, and NE seedlings for each Fagaceae species (Tables 3 and 4). Thus, we considered that the differences in Ca were not the main factor for growth differences among AH, SC, and NE seedlings.

4.3. Relationships among Variables

In examining the factors affecting differences in growth between AH, SC and NE seedlings, ECM colonization was considered one of the important factors. The percentages of ECM colonization were higher in AH seedlings of the three *Quercus* species and in SC seedlings of *C. cuspidata* (Figure 4). Additionally, there were positive relationships between total dry mass and ECM colonization (Table 2). Thus, different percentages of ECM colonization were probably related to differences in growth.

With respect to the relationships between growth and nutrient uptake, N and P in roots showed enhancement of uptake in AH seedlings of *Q. salicina* (Table 4). The storage of N and P in belowground organs has a significant role in plant growth, and plants with high N and P in roots show high growth rates [71–73]. Based on these results, it appeared that the amount of stored N and P resulted in the high growth of AH seedlings of *Q. salicina*.

The photosynthetic rate showed positive correlations between P_{sat} and the concentrations of P in leaves of *Q. salicina*, *Q. myrsinaefolia*, and *C. cuspidata* (Table 2). In general, the photosynthetic rate is closely related to P concentration [55,56]. Thus, the concentration of P in leaves is considered as an important factor that affects the photosynthetic rate. In contrast, *Q. glauca* did not show a positive correlation between P_{sat} and the concentration of P (Table 2). Other factors that affect differences in P_{sat} include high value of g_s (Figure 5). AH seedlings of *Q. glauca* could absorb water through developed hyphae (Figure 4); as a result, the stomata stay open and absorption of CO_2 through the stomata is also increased.

The concentration of N in leaves showed a negative correlation with P_{sat} for *Q. salicina* and *Q. myrsinaefolia* (Table 2), and our results contradicted the general trend [7,54]. The cause of this negative trend was probably related to the high concentration of N in leaves and the low values of P_{sat} in NE seedlings. In the case of NE seedlings, they had no hyphae of ECM fungi and a lower capacity for water uptake. The stomata of NE seedlings may have closed in order to avoid water stress; as a result, NE seedlings showed lower P_{sat} and g_s and decreased absorption of CO_2 through stomata. Similar trends were shown for NE seedlings of Fagaceae species planted on acidic soil [40].

With respect to the relationship among nutrients for *C. cuspidata* and AH seedlings of *Q. myrsinaefolia*, the concentrations of K were decreased in leaves (Table 3) and were diluted (Figure 6). High Ca content in soil causes ion-antagonism with K [70]. As a result, the suppression of was confirmed for several woody species (*Picea abies* and several Lauraceae species), in which growth was suppressed on calcareous soil [13,70]. Compared with AH seedlings of *Q. glauca* and *Q. salicina*, growth levels of *C. cuspidata* and AH seedlings of *Q. myrsinaefolia* were inferior. Woody species that show less adaptability to calcareous soil can probably suppress uptake of K.

5. Conclusions

When seedlings of the four Fagaceae species were planted on calcareous soil, the uptake of various nutrients, such as P, Fe, and Mn in leaves was suppressed. However, Fagaceae seedlings inoculated with ECM fungi showed accelerated growth. In particular, Fagaceae seedlings inoculated with ECM fungi had increased concentration of P in roots, and this trait was important for growth acceleration. Additionally, the concentration of P in leaves was an important factor and affected the differences in photosynthetic rate. The most suitable ECM fungi differed among the four Fagaceae species, and higher growth levels were shown from inoculation with *A. hygrometricus* for the three *Quercus* species, and with *S. citrinum* for *C. cuspidata*. Growth levels of the seedlings inoculated with ECM fungi were different among the four Fagaceae species, and *C. cuspidata*, which showed the lowest growth, was considered as a calcifuge plant. From their growth characteristics, we concluded that *Q. salicina* and *Q. glauca* inoculated with *A. hygrometricus* showed the highest growth levels, and these species are most suitable for growth on calcareous soil.

These findings are expected to contribute to new methods for reforestation of limestone quarries in warm-temperature zones in East Asia. In particular, *Q. salicina* and *Q. glauca* inoculated with *A. hygrometricus* may be useful species for reforestation. The effects on growth acceleration without fertilization are advantageous for reforestation in limestone environments. It should be noted that our experiment used sterilized calcareous soil; therefore, further field cultivation experiments with ECM-fungi inoculated plants will be needed.

Acknowledgments: We thank researchers of the Kyushu Research Center, FFPRI, for their encouragement. We also thank the staff of Todaka Mining Co. for the collection of calcareous soil. We thank Yosuke Matsuda for his cooperation in the collection of ECM fungi. The phytotron experiment was carried out with the cooperation of Nahoko Aoki, to whom we express our thanks. Moreover, we thank Yoshimi Narimatsu and Takeo Kokuryo for analysis of Fagaceae seedlings. The soil analyses were carried out at JIRCAS. Thanks are also due to the staff of JIRCAS and Medical English Service, Kyoto, and Forestry Division, for English proofreading. This research was supported by a grant from the Ministry of Agriculture, Forestry and Fishery (Research and Development Projects for Application in Promoting New Policy of Agriculture, Forestry and Fishery, No. 1904: Development of Regeneration Technology for Leading Artificial Coniferous Forest to Broadleaf Forests). In addition, this study was supported by a grant-in-aid for Scientific Research from the Japan Society of Promotion of Science (No. 20780124: Physiological Characteristics of Trees Grown on a Limestone Region and Application to Revegetation Technology of Limestone Quarry).

Author Contributions: Masazumi Kayama conceived the experiments. Takashi Yamanaka raised seedlings of all Fagaceae species and inoculated them with ectomycorrhizal fungi. Masazumi Kayama performed the experiments, measured photosynthetic rates, and analyzed various nutrients. Masazumi Kayama and Takashi Yamanaka discussed the results, and co-wrote the paper.

Conflicts of Interest: The authors declare no conflict of interest.

References

1. Ford, D.C.; Williams, P.W. *Karst Geomorphology and Hydrology*; Chapman & Hall: London, UK, 1989; p. 620.
2. Haleem, A.A.; Loeppert, R.H.; Anderson, W.B. Role of soil carbonate and iron oxide in iron nutrition of soybean in calcareous soils of Egypt and the United States. In *Iron Nutrition in Soils and Plants*; Abadía, J., Ed.; Kluwer Academic Publishers: Dordrecht, The Netherland, 1995; pp. 307–314.
3. Marschner, H. *Mineral Nutrition of Higher Plants*; Academic Press: New York, NY, USA, 1995; p. 889.
4. Chen, Y.; Barak, P. Iron nutrition of plants in calcareous soils. *Adv. Agron.* **1982**, *35*, 217–240.
5. Tyler, G. Mineral nutrient limitation of calcifuge plants in phosphate sufficient limestone soil. *Ann. Bot.* **1996**, *77*, 649–656. [CrossRef]
6. Jefferies, R.L.; Willis, A.J. Studies on the calcicole-calcifuge habit: II. The influence of calcium on the growth and establishment of four species in soil and sand cultures. *J. Ecol.* **1964**, *52*, 691–707. [CrossRef]
7. Larcher, W. *Physiological Plant Ecology*, 4th ed.; Springer: Berlin, Germany, 2003; p. 513.

8. Zohlen, A.; Tyler, G. Soluble inorganic tissue phosphorus and calcicole-calcifuge behaviour of plants. *Ann. Bot.* **2004**, *94*, 427–432. [CrossRef]
9. Oates, T. Lime and limestone. In *Kirk-Othmer Encyclopedia of Chemical Technology*; Kirk-Othmer, Ed.; John Wiley and Sons: Hoboken, NJ, USA, 2010; pp. 1–53.
10. Ursic, K.A.; Kenkel, N.C.; Larson, D.W. Revegetation dynamics of cliff faces in abandoned limestone quarries. *J. Appl. Ecol.* **1997**, *34*, 289–303. [CrossRef]
11. Duan, W.J.; Ren, H.; Fu, S.L.; Wang, J.; Yang, L.; Zhang, J. Natural recovery of different areas of a deserted quarry in south China. *J. Environ. Sci.* **2008**, *20*, 476–481. [CrossRef]
12. Kayama, M. Growth characteristics of Lauraceae tree species planted on limestone quarry. *Ann. Rep. Kyushu Res. Cen. FFPRI* **2011**, *23*, 16–17. (In Japanese)
13. Kayama, M. Growth characteristics of four Lauraceae tree species planted on calcareous and brown forest soils. *Kyushu J. For. Res.* **2013**, *66*, 63–66. (In Japanese)
14. Kira, T. Forest ecosystems of east and Southeast Asia in a global perspective. *Ecol. Res.* **1991**, *6*, 185–200. [CrossRef]
15. Fang, J.Y.; Song, Y.C.; Liu, H.Y.; Piao, S.L. Vegetation-climate relationship and its application in the division of vegetation zone in China. *Acta Bot. Sin.* **2002**, *44*, 1105–1122.
16. Menitsky, Y.L. *Oaks of Asia*; Science Publishers: Enfield, CT, USA, 2005; p. 549.
17. Yamanaka, T. The limestone vegetation in middle Kyushu. *Res. Rep. Kochi Univ. Nat. Sci.* **1966**, *15*, 1–9. (In Japanese with English summary)
18. Yamanaka, T. The forest and scrub vegetation in limestone areas of Shikoku, Japan. *Vegetatio* **1969**, *19*, 286–307. [CrossRef]
19. Miyawaki, A. *Vegetation of Japan, Chugoku*; Shibundo: Tokyo, Japan, 1983; p. 540. (In Japanese with English summary)
20. Xu, Z.R. A study of the vegetation and floristic affinity of the limestone forest in southern and southwestern China. *Ann. Mo. Bot. Gard.* **1995**, *82*, 570–580.
21. Miyawaki, A. *Vegetation of Japan, Kyushu*; Shibundo: Tokyo, Japan, 1981; p. 484. (In Japanese with English summary)
22. Liu, J.M. The reproductive and regenerative counter measures of the main woody species in Maolan karst forest. *Sci. Silvae Sin.* **2000**, *36*, 115–122. (In Chinese with English summary)
23. Yao, Y.Q.; Zhang, Z.H.; Liang, S.C.; Bi, X.L.; Li, G.R.; Hu, G. Structure of *Cyclobalanopsis glauca* population on karst hills of Guilin. *J. Zhejiang For. Sci. Technol.* **2008**, *28*, 8–11. (In Chinese with English summary)
24. Feroz, S.M.; Hagihara, A. Comparative studies on community ecology of two types of subtropical forest grown in silicate and limestone habitats in the northern part of Okinawa Island, Japan. *Taiwania* **2008**, *53*, 139–149.
25. Newman, E.I.; Reddell, P. The distribution of mycorrhizas among families of vascular plants. *New Phytol.* **1987**, *106*, 745–751. [CrossRef]
26. Wang, B.; Qiu, Y.-L. Phylogenic distribution and evolution of mycorrhizas in land plants. *Mycorrhiza* **2006**, *16*, 299–363. [CrossRef]
27. Wallander, H. Uptake of P from apatite by *Pinus sylvestris* seedlings colonised by different ectomycorrhizal fungi. *Plant Soil* **2000**, *218*, 249–256. [CrossRef]
28. Ström, L.; Owen, A.G.; Godbold, D.L.; Jones, D.L. Organic acid behaviour in a calcareous soil: Sorption reactions and biodegradation rates. *Soil Biol. Biochem.* **2001**, *33*, 2125–2133. [CrossRef]
29. Tyler, G. Some ecophysiological and historical approaches to species richness and calcicole/calcifuge behavior—Contribution to a debate. *Folia Geobot.* **2003**, *38*, 419–428. [CrossRef]
30. Van der Heijden, E.W.; Vosatka, M. Mycorrhizal association of *Salix repens* L. communities in succession of dune ecosystems. II. Mycorrhizal dynamics and interactions of ectomycorrhizal and arbuscular mycorrhizal fungi. *Can. J. Bot.* **1999**, *77*, 1833–1841. [CrossRef]
31. Wiemken, V.; Laczko, E.; Ineichen, K.; Boller, T. Effects of elevated carbon dioxide and nitrogen fertilization on mycorrhizal fine roots and the soil microbial community in beech-spruce ecosystems on siliceous and calcareous soil. *Microb. Ecol.* **2001**, *42*, 126–135.
32. Kennedy, K.H.; Maxwell, J.F.; Lumyong, S. Fire and the production of *Astraeus odoratus* (Basidiomycetes) sporocarps in deciduous dipterocarp-oak forests in northern Thailand. *Maejo Int. J. Sci. Technol.* **2012**, *6*, 483–504.

33. Lapeyrie, F. The role of ectomycorrhizal fungi in calcareous soil tolerance by "symbiocalcicole" woody plants. *Ann. Sci. For.* **1990**, *21*, 579–589. [CrossRef]

34. Taste, F.P.; Schmidt, M.G.; Berch, S.M.; Bulmer, C.; Egger, K.N. Effects of ectomycorrhizal inoculants on survival and growth of interior Douglas-fir seedlings on reforestation sites and partially rehabilitated landings. *Can. J. For. Res.* **2004**, *34*, 2074–2088. [CrossRef]

35. Marx, D.H.; Maul, S.B.; Cordell, C.E. Application of specific ectomycorrhizal fungi in world forestry. In *Frontiers in Industrial Mycology*; Leatham, G.F., Ed.; Chapman and Hall: New York, NY, USA, 1992; pp. 78–98.

36. Endo, M. Higher fungi of the Hirao-karst topography. *Res. Bull. Cult. Asset Kitakyushu* **1973**, *13*, 53–56. (In Japanese)

37. Yoshimi, S. Taxonomic study of the Japanese taxa of *Scleroderma* Pers. *Jpn. J. Mycol.* **2002**, *43*, 3–18. (In Japanese with English summary)

38. Fangfuk, W.; Okada, K.; Petchang, R.; To-anun, C.; Fukuda, M.; Yamada, A. In vitro mycorrhization of edible *Astraeus* mushrooms and their morphological characterization. *Mycoscience* **2010**, *51*, 234–241. [CrossRef]

39. Makita, N.; Hirano, Y.; Yamanaka, T.; Yoshimura, K.; Kosugi, Y. Ectomycorrhizal-fungal colonization induces physio-morphological changes in *Quercus serrata* leaves and roots. *J. Plant Nutr. Soil Sci.* **2012**, *17*, 900–906. [CrossRef]

40. Kayama, M.; Yamanaka, T. Growth characteristics of ectomycorrhizal seedlings of *Quercus glauca*, *Quercus salicina*, and *Castanopsis cuspidata* planted on acidic soil. *Trees* **2014**, *28*, 569–583. [CrossRef]

41. Akama, K.; Okabe, H.; Yamanaka, T. Growth of ectomycorrhizal fungi on various culture media. *Bull. FFPRI* **2008**, *7*, 165–181. (In Japanese with English summary)

42. Ohta, A. A new medium for mycelial growth of mycorrhizal fungi. *Trans. Mycol. Soc. Jpn.* **1990**, *31*, 323–334.

43. Oita Prefecture. *Fundamental land classification survey, Usuki and Hotojima*; Fuji Micro-service Center: Kumamoto, Japan, 1980; p. 51. (In Japanese)

44. Van Reeuwijk, L.P. *Procedures for Soil Analysis*, 6th ed.; International Soil Reference and Information Centre: Wagningen, The Netherlands, 2002; p. 100.

45. Sparks, D.L.; Page, A.L.; Helmke, P.A.; Loeppert, R.H.; Soltanpour, P.N.; Tabatabai, M.A.; Johnson, C.T.; Sumner, M.E. *Methods of Soil Analysis, Part 3. Chemical Methods*; Soil Science Society of America Inc.: Madison, WI, USA, 1996; p. 1390.

46. American Public Health Association; American Water Works Association; Water Environment Federation. *Standard Methods for the Examination of Water and Wastewater*, 20th ed.; American Public Health Association: Washington, DC, USA, 1998; p. 1220.

47. Japan Meteorological Agency. Climate Statistics. Available online: http://www.data.jma.go.jp/obd/stats/etrn/index.php (access on 31 October 2010).

48. Iwasaki, T.; Okuyama, T.; Sakuratani, T. Climatic estimation of photosynthetically active radiation (PAR) in Kanto and Tokai districts. *Bull. Natl. Inst. Agro Environ. Sci.* **1986**, *1*, 1–35. (In Japanese with English summary)

49. Mohan, V.; Natarajan, K.; Ingleby, K. Anatomical studies of ectomycorrhizas. III. The ectomycorrhizas produced by *Rhizopogon luteolus* and *Scleroderma citrinum* on *Pinus patula*. *Mycorrhiza* **1993**, *3*, 51–56. [CrossRef]

50. Quoreshi, A.M.; Timmer, V.R. Exponential fertilization increases nutrient uptake and ectomycorrhizal development of black spruce seedlings. *Can. J. For. Res.* **1998**, *28*, 674–682. [CrossRef]

51. Goto, S. Digestion method. In *Manual of Plant Nutrition*; Editorial Committee of Methods for Experiments in Plant Nutrition. Hakuyusha: Tokyo, Japan, 1990; pp. 125–128. (In Japanese)

52. Valentine, D.W.; Allen, H.L. Foliar responses to fertilization identify nutrient limitation in loblolly pine. *Can. J. For. Res.* **1990**, *20*, 144–151. [CrossRef]

53. Scagel, C.F. Growth and nutrient use of ericaceous plants grown in media amended with sphagnum moss peat or coir dust. *HortScience* **2003**, *38*, 46–54.

54. Evans, J.R. Photosynthesis and nitrogen relationships in leaves of C_3 plants. *Oecologia* **1989**, *78*, 9–19. [CrossRef]

55. Raaimakers, D.; Boot, R.G.A.; Dijkstra, P.; Pot, S.; Pons, T. Photosynthetic rates in relation to leaf phosphorus content in pioneer versus climax tropical rainforest trees. *Oecologia* **1995**, *102*, 120–125. [CrossRef]

56. Bown, H.E.; Watt, M.S.; Clinton, P.W.; Mason, E.G.; Richardson, B. Partitioning concurrent influences of nitrogen and phosphorus supply on photosynthetic model parameters of *Pinus radiata*. *Tree Physiol.* **2007**, *27*, 335–344. [CrossRef]

57. Parladé, J.; Peta, J.; Alvarez, I.F. Inoculation of containerized *Pseudotsuga menziesii* and *Pinus pinaster* seedlings with spores of five species of ectomycorrhizal fungi. *Mycorrhiza* **1996**, *6*, 237–245. [CrossRef]

58. Chen, Y.L.; Dell, B.; Malajczuk, N. Effects of *Scleroderma* spore density and age on mycorrhiza formation and growth of containerized *Eucalyptus globulus* and *E. urophylla* seedlings. *New For.* **2006**, *31*, 453–467. [CrossRef]

59. Lehto, T.; Zwiazek, J.J. Ectomycorrhizas and water relations of trees: A review. *Mycorrhiza* **2011**, *21*, 71–90. [CrossRef]

60. Morita, S.; Nagata, S. The causes and prevention of the leaf dying of the melon in the calcareous soil. *Jpn. J. Soil Sci. Plant Nutr.* **1995**, *66*, 217–222. (In Japanese with English summary)

61. Terai, A. The effects of organic fertilizer application on limestone-derived soils in Miyakojima Islands, Okinawa. *Ryukoku J. Econ. Stud.* **2008**, *47*, 125–138. (In Japanese)

62. Kayama, M. Concentrations of nutrients for trees grown on limestone region. *Ann. Rep. Kyushu Res. Cen. FFPRI* **2009**, *21*, 9–10. (In Japanese)

63. Kayama, M.; Aoki, N.; Hirayama, D.; Tateno, R.; Kawaji, M.; Yoneda, T. Growth characteristics of *Quercus miyagii* and *Q. glauca* var. *amamiana* seedlings planted on two types of soil. *Kyushu J. For. Res.* **2012**, *65*, 64–67. (In Japanese)

64. Maesako, S.; Yonemaru, S.; Sakai, M. Comparative studies of the physical and chemical properties of surface soils in a sugi plantation and it's two adjacent broad-leaved forest stands. *Bull. Kagoshima Prefect. For. Exp. Stn.* **2002**, *7*, 14–22. (In Japanese)

65. Sardans, J.; Rodà, F.; Peñuelas, J. Phosphorus limitation and competitive capacities of *Pinus halepensis* and *Quercus ilex* subsp. *rotundifolia* on different soils. *Plant Ecol.* **2004**, *174*, 305–317.

66. Pascual, S.; Olarieta, J.R.; Rodoríguez-Ochoa, R. Development of *Queucus ilex* is related to soil phosphorus availability on shallow calcareous soils. *New For.* **2012**, *43*, 805–814. [CrossRef]

67. Ito, S.; Ohtsuka, K.; Yamashita, T. Ecological distribution of seven evergreen *Quercus* species. *Veg. Sci.* **2007**, *24*, 53–63.

68. Miyawaki, A. *Vegetation of Japan, Kanto*; Shibundo: Tokyo, Japan, 1986; p. 641. (In Japanese with English summary)

69. Ewald, J. Ecological background of crown condition, growth and nutritional status of *Picea abies* (L.) Karst. In the Bavarian Alps. *Eur. J. For. Res.* **2005**, *124*, 9–18. [CrossRef]

70. Baier, R.; Ettl, R.; Hahn, C.; Göttlein, A. Early development and nutrition of Norway spruce (*Picea abies* (L.) Karst.) seedlings on different seedbeds in the Bavarian limestone alps—A bioassay. *Ann. For. Sci.* **2006**, *63*, 339–348. [CrossRef]

71. Volenec, J.J.; Ourry, A.; Joern, B.C. A role for nitrogen reserves in forage regrowth and stress tolerance. *Physiol. Plant.* **1996**, *97*, 185–193. [CrossRef]

72. Gloser, V. The consequences of lower nitrogen availability in autumn for internal nitrogen reserves and spring growth of *Calamagrostis epigejos*. *Plant Ecol.* **2005**, *179*, 119–126. [CrossRef]

73. Peri, P.L.; Gargaglione, V.; Pastur, G.M. Above- and belowground nutrients storage and biomass accumulation in marginal *Nothofagus antarctica* forests in southern Patagonia. *For. Ecol. Manag.* **2008**, *255*, 2502–2511. [CrossRef]

forests

Article

Spatial Distribution of Soil Nitrogen, Phosphorus and Potassium Stocks in Moso Bamboo Forests in Subtropical China

Xiaolu Tang, Mingpeng Xia, Fengying Guan * and Shaohui Fan *

Key laboratory of Bamboo and Rattan, International Centre for Bamboo and Rattan, Beijing 100102, China; lxtt2010@163.cm (X.T.); mp_xia@yahoo.com (M.X.)

* Correspondence: guanfy@icbr.ac.cn (F.G.); fansh@icbr.ac.cn (S.F.); Tel.: +86-10-8478-9810 (F.G.); +86-10-8478-9820 (S.F.); Fax: +86-10-8478-9726 (F.G. & S.F.)

Academic Editors: Scott X. Chang and Xiangyang Sun
Received: 15 August 2016; Accepted: 31 October 2016; Published: 7 November 2016

Abstract: Moso bamboo is famous for fast growth and biomass accumulation, as well as high annual output for timber and bamboo shoots. These high outputs require high nutrient inputs to maintain and improve stand productivity. Soil nitrogen (N), phosphorus (P), and potassium (K) are important macronutrients for plant growth and productivity. Due to high variability of soils, analysing spatial patterns of soil N, P, and K stocks is necessary for scientific nutrient management of Moso bamboo forests. In this study, soils were sampled from 138 locations across Yong'an City and ordinary kriging was applied for spatial interpolation of soil N, P, and K stocks within 0–60 cm. The nugget-to-sill ratio suggested a strong spatial dependence for soil N stock and a moderate spatial dependence for soil P and K stocks, indicating that soil N stock was mainly controlled by intrinsic factors while soil P and K stocks were controlled by both intrinsic and extrinsic factors. Different spatial patterns were observed for soil N, P, and K stocks across the study area, indicating that fertilizations with different ratios of N:P:K should be applied for different sites to maintain and improve stand productivity. The total soil N, P, and K stocks within 0–60 cm were 0.624, 0.020, and 0.583 Tg, respectively, indicating soils were important pools for N, P, and K.

Keywords: ordinary kriging; geostatistical analysis; spatial variability; Moso bamboo

1. Introduction

Soil nitrogen (N), phosphorus (P), and potassium (K) are important macronutrients which can limit or co-limit plant growth [1,2]. Human activities, such as fertilization, reclamation, and weeding, have greatly affected the biogeochemical cycling of N, P, and K, thereby altering the pattern, magnitude, and extent of nutrient limitation on land [3]. Effective, efficient, and site-specific management and estimation of soil N, P, and K have attracted great interests for scientists looking to improve nutrient input efficiency, thus increasing stand productivity and reducing environmental risks [4]. Due to various climatic conditions [5], parent materials [6], topography [7], vegetation types [8], soil texture [9], and land use [10], soils are characterized by a highly spatial and temporal variability. This has made the accurate estimation of spatial nutrient content difficult. However, rational soil management requires a deep understanding of the spatiotemporal variability of soil N, P, K, and their maps. For example, areas of particular concern, such as nutrient deficiency, can be identified. Additionally, soil N and P levels are closely related to soil organic carbon cycling [11], which may lead to dynamic effects on greenhouse gas emissions that potentially result in feedback to global climate change. Therefore, it is necessary to improve our understanding of the spatial distribution of soil N, P, and K stocks when evaluating current and potential soil productivity, identifying potential environmental protections, and assessing climate change [12].

Fast developments of geostatistical techniques, such as kriging [11], have significantly advanced the estimation of spatial variability of soil N, P, and K stocks. The most common interpolation techniques calculate the unmeasured property at a given place using measured neighbours data with a weighted mean [13]. The core of geostatistics is that experimental semivariograms are calculated to analyse the spatial autocorrelation of the edaphic variables and determine the range of spatial dependence [13]. The spatial data structure, choice of variogram models, search radius, and number of closest neighbouring points can determine the performance of Kriging approaches. Compared to other interpolation approaches (e.g., Inverse Distance Weighting and splines methods [14,15]), ordinary kriging (OK) can provide more accurate estimates of the spatial distribution of soil nutrients because: (1) OK can provide the best linear unbiased estimates and information on the distribution of the estimation error [15]; (2) OK has relatively strong statistical advantages [16]. Additionally, OK is easy to conduct with high accuracy compared to other kriging approaches, such as cokriging, universal kriging, and factorial kriging. For example, cokriging requires an additional correlated covariant, which leads to the substantial increase of field and lab work and does not increase the statistical power [15]. Thus, OK has been widely used in spatial interpolation for different soil properties. For example, Liu et al. [11] predicted the spatial patterns of soil N and P stocks for the whole Loess Plateau using OK, and total N and P stocks amounted to 0.217 and 0.205 Pg (1 Pg = 10^{15} g), respectively. Martín et al. [17] predicted the spatial distribution of soil carbon stocks within 0–30 cm with OK across Spain.

Bamboo is an important forest type in Southern China, representing an area of 6.16 million ha, more than 70% of which are Moso bamboo (*Phyllostachys heterocycla* (Carr.) Mitford *cv. Pubescens*) forests. Moso bamboo forests are famous for their rapid growth and fast biomass accumulation. However, annual timber harvest and bamboo shoots remove a large amount of nutrients, which means that Moso bamboo forests require more nutrient input compared to other forest types in order to maintain high stand productivity. Consequently, intensive management with fertilization, reclamation, and regular understory removal is becoming increasingly popular in Southern China, especially in the main bamboo producing provinces, such as Zhejiang and Fujian Provinces [18–20]. These activities have significantly changed the nutrient levels by affecting the microbial processes, soil structures, and chemical compositions [21,22]. Although many studies focus on the spatial patterns of soil N, P, and K stocks in different ecosystems [23,24], site specific maps of soil N, P, and K stocks for scientific management of Moso bamboo forests are still lacking in our study area. In contrast, a better understanding of the spatial distribution of soil N, P, and K stocks is strongly required for land management and for maintaining and improving stand productivity of Moso bamboo forests. Therefore, the objectives of this study were to: (1) predict the spatial variability of soil N, P, and K stocks of Moso bamboo forests in Yong'an City, China and (2) estimate total soil N, P, and K stocks within 0–60 cm.

2. Materials and Methods

2.1. Study Area

The study was conducted in the Moso bamboo forests (Figure 1, polygons) across Yong'an City, Fujian Province, Southern China (117°56′–117°47′ E, 25°33′–26°12′ N). It has a subtropical southeast monsoon climate with a mean annual temperature of 19.3 °C, while the lowest temperature is −11°C and the highest temperature is 40 °C. The average annual precipitation is about 1600 mm [25,26]. The elevation of the study area ranged from 580 m to 1605 m above sea level [25]. The accumulated temperature of $\geq$10 °C is 4520–5800 °C, lasting for 225–250 days, and the relative humidity is about 80% [25]. The main soil type is Oxisol. Yong'an City has a forest cover of 82%, including an area of 5.85 × 10^4 ha of Moso bamboo forests [26]. Moso bamboo forests are mainly distributed below 800 m, most of which are pure stands and sometimes mixed with *Keteleeria cyclolepis* Flous., *Cunninghamia lanceolata* (Lamb.) Hook., *Myrica rubra* (Lour.) S. et Zucc., *Choerospondias axillaris* (Roxb.)

Burtt et Hill., *Liriodendron chinense* (Roxb.) Burtt et Hill., and *Schima Superba* Gardn. et Champ., etc. To improve the stand productivity and increase the income, fertilization treatment has been widely applied in most of the Moso bamboo forests. The fertilization treatment was conducted across the whole study area using the same protocol established by the local Forest Bureau. Similar fertilization types—organic fertilizers with N, P, and K addition, were offered every year by the Forest Bureau because the residents can get an economic subsidy if they follow the protocol and use the recommended fertilizers.

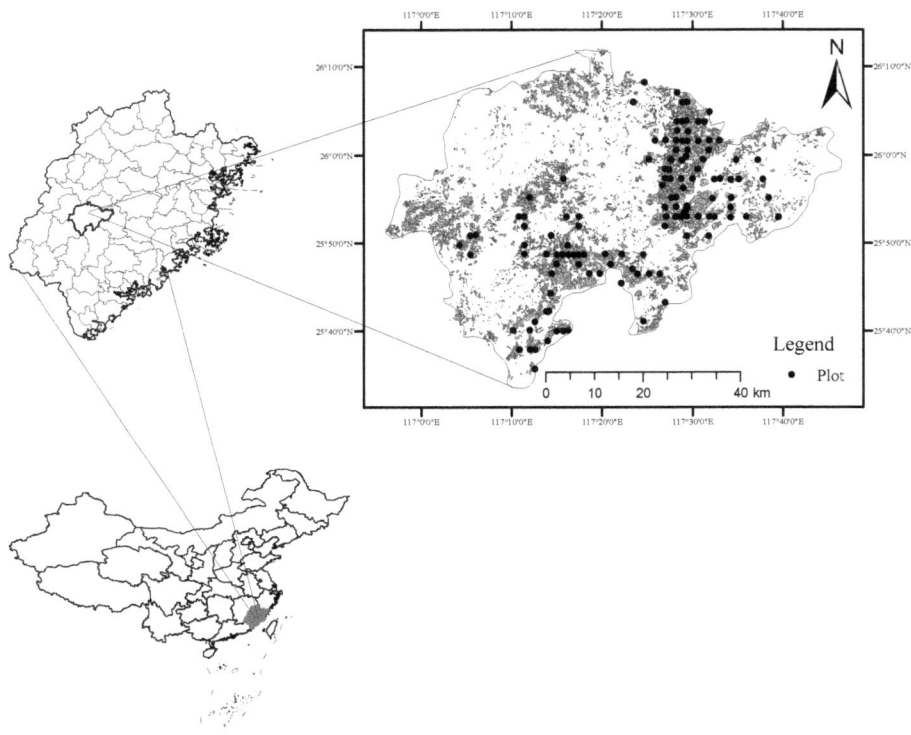

Figure 1. A map of the study area showing the distribution of bamboo forests (polygons) and sampling locations (solid circle).

2.2. Soil Sampling

The representative soils were sampled in the sub-compartments of forest resource management sites in Fujian province, China. These sample sites are established by the local Forest Bureau for soil mapping according to the protocol of forest management inventory [27] (Figure 1). In the targeted sub-compartments, a total of 138 sites were selected and a cluster of three circular plots with a size of 33.3 m^2 were established for each site. Due to the unfavourable access of sampling plots, most of the samples were collected in the East and Southern portions of the study area, which covered 78% of the total Moso bamboo forests. Thus, it is acceptable with these plots to study the spatial distribution of soil N, P, and K stock across the whole study area using the unbiased interpolation approach of OK (see below).

In each plot centre, soils were sampled down to 60 cm by the consideration of three layers: 0–20 cm, 20–40 cm, and 40–60 cm. In the field, soil samples in the sample layer were mixed for one cluster of the three circular plots. In the laboratory, soil samples were air-dried at room temperature and prepared for sieving through 2-mm and 0.15-mm for N, P, and K content analysis. Identifiable plant residues and root materials were removed during sieving. Because the majority of bamboo roots were distributed within 0–40 cm [28], soil samples down to 60 cm were able to meet the research purpose to study spatial patterns of soil N, P, and K stocks. Soil bulk density was determined by a cutting ring approach for each soil layer [29]. During the implementation of fertilization, stones and rocks were removed in Moso bamboo forests. As a result, few stones and rocks were observed in core samples, thus, we did not correct for gravel content. Additionally, site information about elevation, coordinate, soil depth, soil type, soil organic matter content, and bamboo diameters were recorded and determined.

Soil N content (g kg^{-1}) was analysed using the Kjeldahl digestion procedure (5 mL concentrated H$_2$SO$_4$, 1 g of soil passed through a 0.15-mm sieve, heated for 90 min, titrated with 0.2 mol L^{-1} standardized Na$_2$B$_4$O$_7$). Soil total phosphorus concentration (g kg^{-1}) was determined by alkaline digestion (NaOH 0.2 g, <0.25 mm soil 0.25 g, 300 °C for 15 min and then 750 °C for 15 min) followed by molybdate colorimetric measurement. Soil K content (g kg^{-1}) was determined by sodium hydroxide flame photometer (NaOH 0.2 g, <0.25 mm soil 0.25 g, 300 °C for 15 min and then 750 °C for 15 min, 80 °C water 10 mL). All the analyses followed the national standard protocol of China [29].

Thus, soil N, P, and K stocks were calculated as [30–32]:

$$SON \text{ (or P or K) stock} = SON \text{ (P,K)} \times BD \times D/10 \tag{1}$$

where *SON* (or P or K) are the soil N, P, and K contents (g kg^{-1}); *BD* is bulk density (g cm^{-3}); *D* is the depth of soil layer (cm). Total soil N, P, and K stocks within 0–60 cm were summed for all soil layers.

2.3. Statistical and Geostatistical Analyses

A standard statistical analysis (mean, standard deviation, first quartile, third quartile, etc.) was conducted in *R* to illustrate trends of the soil N, P, and K stocks [33]. The coefficient of variation (CV) was used to describe the degree of general variation. To meet the assumption of normality for geostatistical analysis, the raw data was log-transformed and transformed back by weighted mean in GS + 10.0 [34]. OK was performed in GS + 10.0 and the distribution maps of soil N, P, and K stocks were produced in ArcGIS 10.2 [35].

2.4. OK

In the last several decades, many spatial interpolation approaches have been developed, such as kriging; OK normally performed better and is easier to apply compared to other approaches, thereby it has been widely used in the spatial interpolation of soil properties [11,17,23]. In the OK process, semivariograms are used to describe the spatial autocorrelation and provide parameters for optimal spatial interpolation based on the theory of regionalized variables [36]. Well-known theoretical models (e.g., exponential, Gaussian, and spherical models) are commonly used to fit the experimental semivariograms [37]. The experimental semivariograms are expressed as a function of the distance between the sampled points and calculate the integrity of spatial continuity in one or multiple directions using the following expression [38]:

$$\gamma(h) = \frac{1}{2N(h)} \sum_{i=1}^{N(h)} [z(x_i) - z(x_{i+h})]^2 \tag{2}$$

where $z(x_i)$, and $z(x_{i+h})$ are values of z at locations x_i and x_{i+h}, respectively; h is the lag, and $N(h)$ is the number of pairs of sample points separated by h. In this study, spherical, exponential, linear and Gaussian models were used to attempt to describe the semivariograms of soil N, P, and K stocks within 0–60 cm, then the best models determined by the smallest residuals and the largest determination coefficients were selected for spatial interpolation.

In the semevariograms, three major parameters were derived—the nugget (C_0), the sill ($C + C_0$), and the range (A_0)—in order to identify the spatial structure of soil N, P, and K stocks at a given scale. The sill ($C + C_0$) represents the total variation and the ratio of nugget and sill is considered as a criterion to classify the spatial dependence [17]. The range (A_0) represents the separation distance when semivariogram is stabilized, beyond which the measured data are spatially independent [39]. More details about semivariograms and the kriging approach can be found in Goovaerts [37]. In our study, OK was applied for the spatial patterns of soil N, P, and K stocks. The most likely value which could be expected in a particular grid cell when using m neighbouring observations was defined as [17]:

$$R\left(x\right) = \sum_{j=1}^{m} \delta_j z\left(x_i\right) \tag{3}$$

where δ_j means the optimal weight under the condition of $\sum \delta_j = 1$, $m = 16$ in this study.

2.5. Model Validation

To evaluate the prediction accuracy of soil N, P, and K stocks, leave-one-out cross-validation was used [11,17,40]. In the validation process, one datum was omitted and this value was estimated by the remaining data. Thereafter, the estimated value was compared with the real value of the omitted point [17]. This process was repeated for all the observations. Four commonly used indices (i.e., absolute mean error (AME), mean error (ME), root mean square error (RMSE), and determination coefficient (R^2)) were used to compare the interpolation accuracy. These indices were calculated as follows [11,15]:

$$AME = \frac{1}{n} \sum_{i=1}^{n} |(P_i - M_i)| \tag{4}$$

$$ME = \frac{1}{n} \sum_{i=1}^{n} (P_i - M_i) \tag{5}$$

$$RMSE = \sqrt{\frac{1}{n} \sum_{i=1}^{n} (P_i - M_i)^2} \tag{6}$$

$$R^2 = 1 - \frac{\sum_{i=1}^{n} (P_i - M_i)^2}{\sum_{i=1}^{n} (M_i - \overline{M})^2} \tag{7}$$

where P_i, M_i, and $\overline{M}$ are predicted values, measured values, and the mean values of the measured data, respectively.

3. Results and Discussion

3.1. Descriptive Statistics

The summary of statistics of soil N, P, and K stocks within 0–60 cm is shown in Table 1. N stock in Moso bamboo forest increased from a minimum of 5.30 Mg ha^{-1} to a maximum of 20.20 Mg ha^{-1} with an average of 11.57 Mg ha^{-1}. These values fell within the range of N stocks in different forest types across China (1.0–49.9 Mg ha^{-1}) [41]. However, the mean N stock was even higher than the average N stock of China's forests within one meter (8.4 Mg ha^{-1}) [41] and the Moso bamboo in Jiangxi Province (2.82 Mg ha^{-1} within 60 cm) [42]. These differences were mainly attributed to the fertilization application in Moso bamboo forests in the study area, whereas no stand management occurs in most of China's forests.

Soil P stock ranged from 0.18 to 0.68 Mg ha^{-1} and the mean value was 0.38 Mg ha^{-1}. Although this range lied within the reported values of main Chinese soil types, the mean value was much lower than the average across China (6 Mg ha^{-1} within 60 cm) [43]. Similarly, average soil K was 10.80 Mg ha^{-1} and the lowest soil K stock was 5.73 Mg ha^{-1} and highest soil K stock was 18.72 Mg ha^{-1}, which were lower than soil K stocks in subtropical forests in China [44]. These low values of soil P and K stocks may be related to the high output of P and K due to the high output of timber and bamboo shoots every year. In Moso bamboo forests, P is the most limited nutrient for stand productivity [4]. Although fertilization was applied in Moso bamboo forests, P and K input could not compensate for the P and K output from the stand. This result further indicated that more P and K fertilization should be applied in Moso bamboo forest in order to maintain and improve stand productivity.

Table 1. Summary of statistics of soil N, P, and K stocks within 0–60 cm (Mg ha^{-1}).

Nutrient	Mean	Minimum	Maximum	Median	SD	CV (%)	1st Qu	3rd Qu	Skewness	Kurtosis	p of S-W
N Stock	11.57	5.30	20.20	10.98	3.23	27.92	8.81	13.66	0.42	−0.33	0.053
P Stock	0.38	0.18	0.68	0.36	0.11	29.10	0.30	0.45	0.60	−0.14	0.010
K Stock	10.80	5.73	18.72	10.81	3.15	29.17	8.11	13.03	0.32	−0.72	0.014

SD = standard deviation; CV = coefficient of variance; 1st Qu = 25% quartile; 3rd = 75% quartile; S-W test = Shapiro-Wilk test.

A CV of 10% indicates a low variability and 10%–90% indicates a moderate variability, and CV >90% indicates higher variability [45]. In this study, the CVs of soil N, P, and K stocks were 28%–29%, indicating a moderate variability of soil N, P, and K stocks within 0–60 cm. p values of Shapiro-Wilk test were 0.053 for N stock, 0.010 for P stock, and 0.014 for K stock, indicating a normal distribution of N stock and a non-normal distribution of P and K stocks at a significance level of 0.05. However, in order to make the comparison of spatial interpolation of soil N, P, and K stocks consistent, a natural log-transformation was conducted for the three stocks in order to meet the assumption of normal distribution.

3.2. Geostatistical Analysis of Soil N, P, and K Stocks

The semevariograms are presented in Figure 2 and their parameters are shown in Table 2. The exponential model performed best for soil N and P stocks while the spherical model was best for soil K stock based on the smallest residuals and the highest determination coefficient. These models produced high determination coefficients, ranging from 0.64 to 0.74.

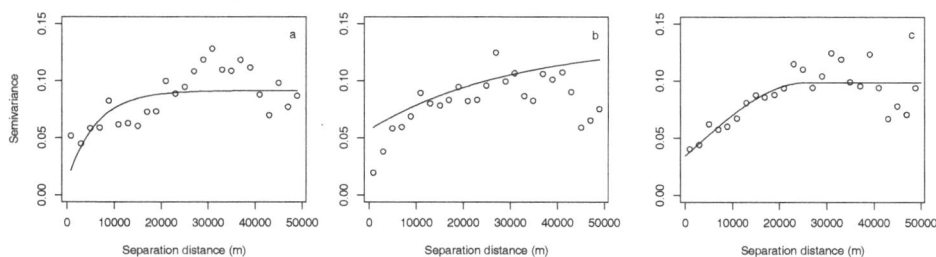

Figure 2. Experimental semivariograms of soil N (a), P (b), and K (c) stocks.

Table 2. Models and their parameters fitted for the semivariograms of soil N, P, and K stocks within 0–60 cm.

Nutrient	Models	Nugget (C_0)	Sill ($C_0 + C$)	Nugget/Sill	Range (A_0, m)	Determination Coefficient	Residuals
N stock	Exponential	0.0111	0.0912	12.17	6100	0.642	0.00458
P stock	Exponential	0.0570	0.1340	42.54	30,570	0.741	0.00110
K stock	Spherical	0.0343	0.0986	34.79	25,700	0.643	0.00469

Nugget values present undetectable experimental errors, field variation within the minimum sampling space, and inherent variability [23]. In this study, nugget values were lowest for soil N stock (0.0111), and highest for soil P stock (0.0570). These positive values suggested a positive nugget effect, a sampling error, or random and inherent variability of soil N, P, and K stocks [23]. Sill values represent total spatial variation [11]. The sill values ranged from 0.0912 for N stock to 0.1340 for P stock.

The ratio of nugget-to-sill represents a spatial dependence. If the ratio is lower than 25%, it indicates a strong spatial dependence. If the ratio is higher than 75%, it indicates a weak spatial dependence, while a ratio between 25% and 75% indicates a moderate spatial dependence [46]. A strong spatial dependence is attributed to soil intrinsic properties, such as soil parent material, soil texture, topography, and vegetation [23,47]. A weak spatial dependence indicates that the spatial variability is mainly regulated by extrinsic variations, such as soil fertilization and cultivation practices [23,46]. Therefore, a moderate spatial dependence is controlled by both intrinsic and extrinsic factors. In this study, the nugget-to-sill ratio was 12% for soil N stock, indicating that soil N stock in the study area was mainly controlled by intrinsic factors and that external factors exerted little effects on soil N stock. This may be related to N fertilization application in Moso bamboo forests in that the N input from fertilization can meet the requirement of N output from timber and bamboo shoot harvest. This conclusion was also consistent with the above result (Table 1) that N stock in this study area was higher than other forest types. The nugget-to-sill ratios were 43% for soil P stock and 35% for soil K stock, indicating a moderate spatial dependence that was controlled by both intrinsic and extrinsic factors. This was evidenced by local stand management and complex topography. Intensive managements, such as fertilization and weeding, were widely conducted in Moso bamboo forests in order to improve timber and bamboo shoot output in the study area [19,25]. Meanwhile, the study area was characterized by a high variability of elevations varying from 580 to 1605 m above sea level [25], which could directly affect the soil processes and nutrient contents [11]. In the study area, intensive fertilization treatment was conducted every year, which was expected to be the dominant factor that led to the change of the spatial variability of soil N, P, and K stocks. However, our results (nugget-to-sill ratio) indicated that intrinsic factors played a more important role in controlling spatial patterns of soil N, P, and K stocks.

The ranges (A_0) indicate different influence zones of environmental factors at different scales [23]. Within the range, soil properties are not spatially independent, while beyond the range, soil properties were spatially independent [39]. In this study, the smallest range was found for soil N stock (6100 m), and increased to 25,700 m for soil K stock and 30,570 m for soil P stock. The results indicated that soil N stock was more heterogeneous compared to soil P and K stocks, which was associated with soil processes because soil N stock was mainly controlled by intrinsic factors (see above). Generally, the spatial range was larger than the sampling intervals (Figure 1), which suggested that the sampling strategy in this study was appropriate for studying spatial patterns of soil N, P, and K stocks. However, only a small number of sampling plots were located in the northwest of the study area due to a low cover of bamboo forests. In contrast, the majority of the sampling sites were located in the southern and eastern part of the study area, where most of the bamboo forests were distributed, thus the sampling strategy could meet our research purpose and could predict the spatial distributions of soil N, P, and K stocks accurately.

3.3. Cross-Validation of OK

The predicted values of OK were plotted against the measured values to evaluate the interpolation performance (Figure 3). The linear model and 1:1 dashed line intersected for soil N, P, and K stocks. Before the intersection, the linear model overestimated soil N, P, and K stocks, and vice versa. This conclusion supported the findings of many previous studies because of the nature of the algorithms of OK, which aimed to achieve unbiased predictions of mean values [11,15]. AME, ME, and RMSE were calculated and presented in Table 3. The closer that the AME, ME, and RMSE values are to zero, the better the model performed. ME of soil N, P, and K stocks ranged from −0.0022 to

−0.2115, which were close to zero, indicating that OK produced relatively unbiased values for spatial interpolation. The ME values were negative, illustrating that OK generally underestimated soil N, P, and K stocks.

Table 3. Cross-validation indices for ordinary kriging (OK).

Nutrient	AME	ME	RMSE
N stock	2.1506	−0.2115	2.5557
P stock	0.0626	−0.0022	0.0797
K stock	2.0250	−0.1634	2.3942

AME = absolute mean error; ME = mean error; RMSE = root mean square error.

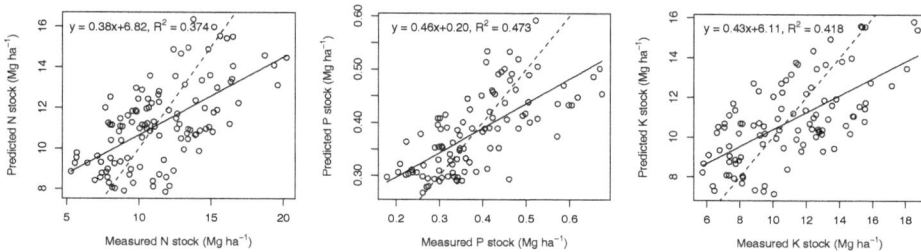

Figure 3. Cross-validation of OK interpolation for soil N, P, and K stocks.

Determination coefficients ranged from 0.37 to 0.47 (Figure 3), which may appear relatively low, but they were similar to many previous studies regarding the spatial interpolation of soil properties [47,48]. This problem is associated with the dataset, which was not sampled with a probabilistic design [48]. For example, the majority of sampling plots were allocated in the southern and eastern parts of Yong'an City in the current study (Figure 1). Therefore, a dataset with a probabilistic design is recommended for spatial interpolation of soil properties. Another possible explanation might be the strong local variation of soil N, P, and K stocks caused by the variability in environmental conditions and fertilization practices. On the other hand, the low correlation between the predicted and measured data indicates that a better methodology, such as one using Artificial Neural Network and Random Forest, should be developed to improve the accuracy of spatial interpolation of soil N, P, and K stocks, including consideration of both intrinsic and extrinsic factors.

3.4. Spatial Prediction of Soil N, ,P and K Stocks

The spatial patterns of soil N, P, and K stocks predicted by OK are shown in Figure 4. Predicted soil N, P, and K stocks ranged from 7.64 Mg ha^{-1} to 16.02 Mg ha^{-1}, from 0.19 Mg ha^{-1} to 0.63 Mg ha^{-1}, and from 6.85 Mg ha^{-1} to 17.13 Mg ha^{-1}, respectively. The lowest soil N, P, and K stocks were found in the middle or southern part of study area, where the city centre was located with a high population, indicating that human activities led to a significant decrease of soil N, P, and K stocks. The highest soil N stock was observed in the east of the study area, the highest soil P stock was found in the southwest and north, and the highest soil K stock was in the north. However, plant growth was constrained by the most limited nutrient. These different spatial distribution patterns suggested that fertilizers with different N:P:K ratios should be applied to maintain and improve the stand productivity. For instance, a relatively lower ratio of P but higher K ratio fertilizer should be applied in the south of the study area (Figure 4b,c), while a relatively lower ratio of N fertilization should be implemented in the east (Figure 4a). Further studies are strongly encouraged to study the N:P:K ratio for the whole study area in order to manage Moso bamboo forests scientifically. The total soil N, P, and K stocks of the whole study area were 0.624, 0.020, and 0.583 Tg (1 Tg = 1 × 10¹² g), respectively, indicating soils were important pools of soil N, P, and K.

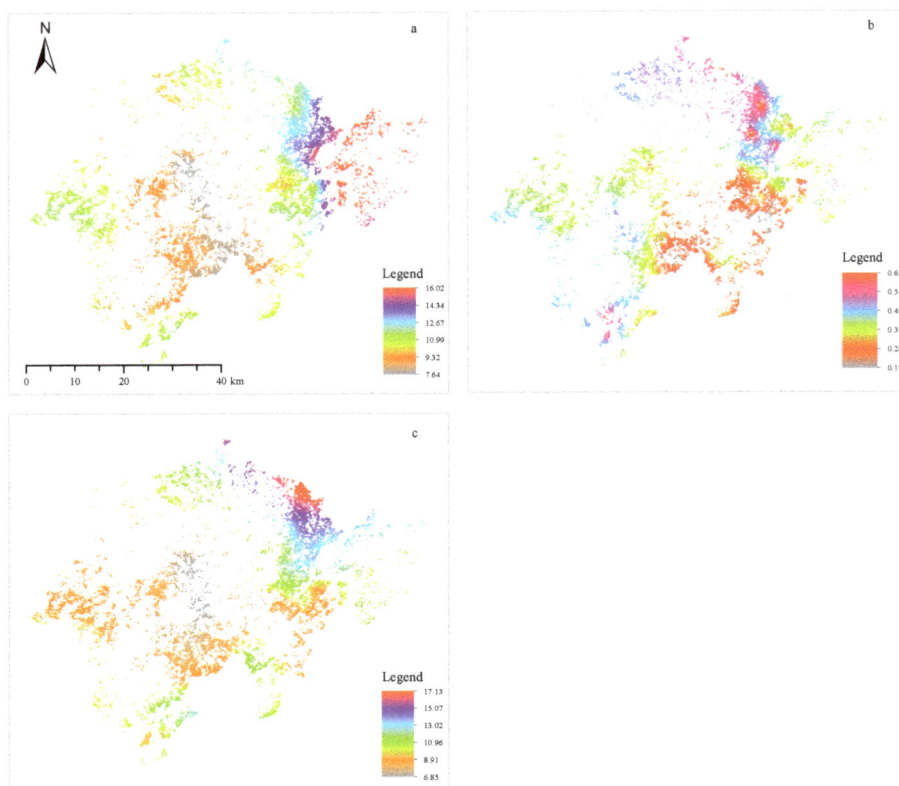

Figure 4. Spatial distribution of soil N (**a**), P (**b**), and K (**c**) stocks.

4. Conclusions

In this study, the current status and spatial interpolation of soil N, P, and K stocks within 0–60 cm were analysed using the measured data from 138 locations in Moso bamboo forests in Yong'an City, China. OK was applied for spatial interpolation of soil N, P, and K stocks across the whole study area. Exponential and spherical models performed well in describing the spatial distribution of soil N, P, and K stocks with determination coefficient from 0.64 to 0.74, and cross-validation demonstrated that OK performed well for the spatial interpolation of soil N, P, and K stocks. Soil N stocks showed a strong spatial dependence, indicating soil N stocks were mainly controlled by intrinsic factors. Soil P and K stocks showed a moderate spatial dependence, suggesting that soil P and K stocks were controlled by both intrinsic (e.g., soil parent material, soil texture, topography) and extrinsic (e.g., soil fertilization and cultivation practices) factors. Soil N, P, and K stocks showed different spatial patterns across the whole study area, indicating that fertilizers with different ratios of N:P:K should be applied for different sites.

Acknowledgments: This study was supported by special research fund of International Centre for Bamboo and Rattan (1632013010) and "Twelfth Five-year" National Technology Support Program (2012BAD23B04). We thank Yanfang Gao for support with fieldwork and Ping He for the technical support.

Author Contributions: Xiaolu Tang, Fengying Guan, and Shaohui Fan conceived and designed the experiments; Xiaolu Tang, Mingpeng Xia, and Shaohui Fan analysed the data; Xiaolu Tang, Mingpeng Xia, Fengying Guan, and Shaohui Fan wrote the paper. All authors contributed to the review of the manuscript.

Conflicts of Interest: The authors declare no conflict of interest.

Abbreviations

The following abbreviations are used in this manuscript:

AME	Absolute Mean Error
BD	Bulk Density
D	Depth
K	Potassium
ME	Mean Error
N	Nitrogen
OK	Ordinary Kriging
P	Phosphorus
RMSE	Root Mean Square Error

References

1. Li, Y.; Niu, S.L.; Yu, G.R. Aggravated phosphorus limitation on biomass production under increasing nitrogen loading: A meta-analysis. *Glob. Chang. Biol.* **2016**, *22*, 934–943. [CrossRef]
2. Tripler, C.E.; Kaushal, S.S.; Likens, G.E.; Walter, M.T. Patterns in potassium dynamics in forest ecosystems. *Ecol. Lett.* **2006**, *9*, 451–466. [CrossRef]
3. Marklein, A.R.; Houlton, B.Z. Nitrogen inputs accelerate phosphorus cycling rates across a wide variety of terrestrial ecosystems. *New Phytol.* **2012**, *193*, 696–704. [CrossRef]
4. Du, M.Y.; Fan, S.H.; Liu, G.L.; Feng, H.Y.; Guo, B.H.; Tang, X.L. Stoichiometric characteristics of carbon, nitrogen and phosphorus in *Phyllostachys pubescens* forests of China. *Chin. J. Plant Ecol.* **2016**, *40*, 760–774. (In Chinese with English Abstract)
5. Patil, R.H.; Laegdsmand, M.; Olesen, J.E.; Porter, J.R. Effect of soil warming and rainfall patterns on soil N cycling in Northern Europe. *Agric. Ecosyst. Environ.* **2010**, *139*, 195–205. [CrossRef]
6. Lin, J.S.; Shi, X.Z.; Lu, X.X.; Yu, D.S.; Wang, H.J.; Zhao, Y.C.; Sun, W.X. Storage and spatial variation of phosphorus in paddy soils of China. *Pedosphere* **2009**, *19*, 790–798. [CrossRef]
7. Rezaei, S.A.; Gilkes, R.J. The effects of landscape attributes and plant community on soil chemical properties in rangelands. *Geoderma* **2005**, *125*, 167–176. [CrossRef]
8. Rodríguez, A.; Durán, J.; Fernández-Palacios, J.M.; Gallardo, A. Spatial pattern and scale of soil N and P fractions under the influence of a leguminous shrub in a *Pinus canariensis* forest. *Geoderma* **2009**, *151*, 303–310. [CrossRef]
9. Gami, S.K.; Lauren, J.G.; Duxbury, J.M. Influence of soil texture and cultivation on carbon and nitrogen levels in soils of the eastern Indo-Gangetic Plains. *Geoderma* **2009**, *153*, 304–311. [CrossRef]
10. Ross, D.J.; Tate, K.R.; Scott, N.A.; Feltham, C.W. Land-use change: Effects on soil carbon, nitrogen and phosphorus pools and fluxes in three adjacent ecosystems. *Soil Biol. Biochem.* **1999**, *31*, 803–813. [CrossRef]
11. Liu, Z.P.; Shao, M.A.; Wang, Y.Q. Spatial patterns of soil total nitrogen and soil total phosphorus across the entire Loess Plateau region of China. *Geoderma* **2013**, *197–198*, 67–78. [CrossRef]
12. Jennings, E.; Allott, N.; Pierson, D.C.; Schneiderman, E.M.; Lenihan, D.; Samuelsson, P.; Taylor, D. Impact of climate change on phosphorus loading from a grassland catchment: Implication for future management. *Water Res.* **2009**, *43*, 4316–4326. [CrossRef]
13. Yasrebi, J.; Saffari, M.; Fathi, H.; Karimian, N.; Moazallahi, M.; Gazni, R. Evaluation and comparison of ordinary kriging and inverse distance weighting methods for prediction of spatial variability of some soil chemical parameters. *Res. J. Biol. Sci.* **2009**, *4*, 93–102.
14. Wang, Z.; Liu, G.B.; Xu, M.X.; Zhang, J.; Wang, Y.; Tang, L. Temporal and spatial variations in soil organic carbon sequestration following revegetation in the hilly Loess Plateau, China. *Catena* **2012**, *99*, 26–33. [CrossRef]
15. Liu, Z.P.; Shao, M.A.; Wang, Y.Q. Large-scale spatial interpolation of soil pH across the Loess Plateau, China. *Environ. Earth Sci.* **2012**, *69*, 2731–2741. [CrossRef]

16. Kravchenko, A.N. Influence of spatial structure on accuracy of interpolation methods. *Soil Sci. Soc. Am. J.* **2003**, *67*, 1564–1571. [CrossRef]
17. Martín, J.A.R.; Álvaro-Fuentes, J.; Gonzalo, J.; Gil, C.; Ramos-Miras, J.J.; Corbí, J.M.G.; Boluda, R. Assessment of the soil organic carbon stock in Spain. *Geoderma* **2016**, *264*, 117–125. [CrossRef]
18. Zhou, G.M.; Xu, J.M.; Jiang, P.K. Effect of management practices on seasonal dynamics of organic carbon in soils under bamboo plantations. *Pedosphere* **2006**, *16*, 525–531. [CrossRef]
19. Du, M.; Liu, G.; Fan, S.; Feng, H.; Tang, X.; Mao, C. Effects of fertilization on the distribution patterns of biomass and carbon storage in Moso bamboo forest, western Fujian province, China. *Chin. J. Trop. Crops* **2015**, *36*, 1–7. (In Chinese with English Abstract)
20. Liu, J.; Jiang, P.K.; Wang, H.L.; Zhou, G.M.; Wu, J.S.; Yang, F.; Qian, X.B. Seasonal soil CO_2 efflux dynamics after land use change from a natural forest to Moso bamboo plantations in subtropical China. *For. Ecol. Manag.* **2011**, *262*, 1131–1137. [CrossRef]
21. Xu, Q.F.; Jiang, P.K.; Xu, Z.H. Soil microbial functional diversity under intensively managed bamboo plantations in southern China. *J. Soils Sed.* **2008**, *8*, 177–183. [CrossRef]
22. Li, Y.F.; Zhang, J.J.; Chang, S.X.; Jiang, P.K.; Zhou, G.M.; Fu, S.L.; Yan, E.R.; Wu, J.S.; Lin, L. Long-term intensive management effects on soil organic carbon pools and chemical composition in Moso bamboo (*Phyllostachys pubescens*) forests in subtropical China. *For. Ecol. Manag.* **2013**, *303*, 121–130. [CrossRef]
23. Wang, Y.Q.; Zhang, X.C.; Huang, C.Q. Spatial variability of soil total nitrogen and soil total phosphorus under different land uses in a small watershed on the Loess Plateau, China. *Geoderma* **2009**, *150*, 141–149. [CrossRef]
24. Morales, L.A.; Vázquez, E.V.; Paz-Ferreiro, J. Spatial distribution and temporal variability of ammonium-nitrogen, phosphorus, and potassium in a rice field in Corrientes, Argentina. *Sci. World J.* **2014**, *2014*. [CrossRef]
25. Liu, G.L.; Fan, S.H.; Qi, L.H.; Xiao, F.M.; Huang, Y.N. Nutrient distribution and biological cycle characteristics in different types of *Phyllostachys pubescens* forest in Northwest Fujian. *Chin. J. Eco.* **2010**, *29*, 2155–2161. (In Chinese with English Abstract)
26. He, P.; Liu, J.; Yu, K.Y.; Yang, S.P.; Yao, X.; Yu, X.Y.; Deng, Y.B.; Chen, Z.H. Research on spatial heterogeneity of soil organic carbon in the southern bamboo forest. *Chin. J. Soil Sci.* **2016**, *47*, 36–44. (In Chinese with English Abstract)
27. Yong'an Forest Bureau, Forest Management Inventory; Yong'an, Sanming, Fujian, China. Unpublished work, 2014.
28. Tang, X.L.; Fan, S.H.; Qi, L.H.; Liu, G.L.; Guan, F.Y.; Du, M.Y.; Shen, C.X. Effect of different managements on carbon storage and carbon allocation in Moso bamboo forest (*Phyllostachys pubescen*). *Acta Agric. Univ. Jiangxi* **2012**, *34*, 736–742, (In Chinese with English abstract).
29. State Forestry Adimination. *Forestry Standard of People's Republic of China—Methods of Forest Soil Analysis*; Chinese Standard Press: Beijing, China, 1999.
30. Zhang, L.H.; Xie, Z.K.; Zhao, R.F.; Wang, Y.J. The impact of land use change on soil organic carbon and labile organic carbon stocks in the Longzhong region of Loess Plateau. *J. Arid Land* **2012**, *4*, 241–250. [CrossRef]
31. Xie, Z.B.; Zhu, J.G.; Liu, G.; Cadisch, G.; Hasegawa, T.; Chen, C.M.; Sun, H.F.; Tang, H.Y.; Zeng, Q. Soil organic carbon stocks in China and changes from 1980s to 2000s. *Glob. Chang. Biol.* **2007**, *13*, 1989–2007. [CrossRef]
32. Guan, F.Y.; Tang, X.L.; Fan, S.H.; Zhao, J.C.; Peng, C. Changes in soil carbon and nitrogen stocks followed the conversion from secondary forest to Chinese fir and Moso bamboo plantations. *Catena* **2015**, *133*, 455–460. [CrossRef]
33. R Development Core Team. *R: A Language and Environment for Statistical Computing*; R Foundation for Statistical Computing: Vienna, Austria, 2011. Available online: http://www.R-project.Org/ (accessed on 6 March 2015).
34. Gammadesign. Available online: www.gammadesign.com (accessed on 10 January 2015).
35. Esri. Available online: www.esri.com (accessed on 14 June 2013).
36. Krige, D.G. A statistical approach to some basic mine valuation problems on the witwatersrand. *J. Chem. Metall. Min. Soc. S. Afr.* **1951**, *52*, 119–139.
37. Goovaerts, P. Geostatistics in soil science: State-of-the-art and perspectives. *Geoderma* **1999**, *89*, 1–45. [CrossRef]

38. Schöning, I.; Totsche, K.U.; Kögel-Knabner, I. Small scale spatial variability of organic carbon stocks in litter and solum of a forested Luvisol. *Geoderma* **2006**, *136*, 631–642. [CrossRef]

39. Lv, J.S.; Liu, Y.; Zhang, Z.L.; Dai, J.R.; Dai, B.; Zhu, Y.C. Identifying the origins and spatial distributions of heavy metals in soils of Ju country (Eastern China) using multivariate and geostatistical approach. *J. Soil Sed.* **2015**, *15*, 163–178. [CrossRef]

40. Reza, S.K.; Sarkar, D.; Baruah, U.; Das, T.H. Evaluation and comparison of ordinary kriging and inverse distance weighting methods for prediction of spatial variability of some chemical parameters of Dhalai district, Tripura. *Agropedology* **2010**, *20*, 38–48.

41. Yang, Y.H.; Ma, W.H.; Mohammat, A.; Fang, J.Y. Storage, patterns and controls of soil nitrogen in China. *Pedosphere* **2007**, *17*, 776–785. [CrossRef]

42. Liu, J.; Yang, Q.P.; Yu, D.K.; Song, Q.N.; Zhao, G.D.; Wang, B. Contribution of fine root to soil nutrient heterogeneity at two sides of the bamboo and broadleaved forest interface. *Chin. J. Plant Ecol.* **2013**, *37*, 739–749. (In Chinese with English Abstract) [CrossRef]

43. Wang, T.; Yang, Y.H.; Ma, W.H. Storage, patterns and environmental controls of soil phosphorus in China. *Acta Sci. Nat. Univ. Pekin.* **2008**, *44*, 945–951.

44. Chen, J.; Yang, N. Effects of five plantations on soil properties in subtropical red soil hilly region. *J. Northwest A & F Univ. (Nat. Sci. Ed.)* **2013**, *41*, 167–173,178. (In Chinese with English Abstract)

45. Fang, X.; Xue, Z.J.; Li, B.C.; An, S.S. Soil organic carbon distribution in relation to land use and its storage in a small watershed of the Loess Plateau, China. *Catena* **2012**, *88*, 6–13. [CrossRef]

46. Cambardella, C.A.; Moorman, T.B.; Novak, J.M.; Parkin, T.B.; Karlen, D.L.; Turco, R.F.; Konopka, A.E. Field-scale variability of soil properties in central Iowa soils. *Soil Sci. Soc. Am. J.* **1994**, *58*, 1501–1511. [CrossRef]

47. Fu, W.J.; Jiang, P.K.; Zhao, K.L.; Zhou, G.M.; Li, Y.F.; Wu, J.S.; Du, H.Q. The carbon storage in Moso bamboo plantation and its spatial variation in Anji County of southeastern China. *J. Soils Sed.* **2014**, *14*, 320–329. [CrossRef]

48. Veronesi, F.; Corstanje, R.; Mayr, T. Landscape scale estimation of soil carbon stock using 3D modelling. *Sci. Total Environ.* **2014**, *487*, 578–586. [CrossRef]

forests

MDPI

Article

Understanding the Fate of Applied Nitrogen in Pine Plantations of the Southeastern United States Using ^{15}N Enriched Fertilizers

Jay E. Raymond *, Thomas R. Fox and Brian D. Strahm

Department of Forest Resources and Environmental Conservation, Virginia Polytechnic Institution and State University, Blacksburg, VA 24061, USA; trfox@vt.edu (T.R.F.); brian.strahm@vt.edu (B.D.S.)
* Correspondence: jayer11@vt.edu; Tel.: +1-540-267-4991

Academic Editors: Scott X. Chang and Xiangyang Sun
Received: 29 August 2016; Accepted: 5 November 2016; Published: 11 November 2016

Abstract: This study was conducted to determine the efficacy of using enhanced efficiency fertilizer (EEFs) products compared to urea to improve fertilizer nitrogen use efficiency (FNUE) in forest plantations. All fertilizer treatments were labeled with ^{15}N (0.5 atom percent) and applied to 100 m^2 circular plots at 12 loblolly pine stands (*Pinus taeda* L.) across the southeastern United States. Total fertilizer N recovery for fertilizer treatments was determined by sampling all primary ecosystem components and using a mass balance calculation. Significantly more fertilizer N was recovered for all EEFs compared to urea, but there were generally no differences among EEFs. The total fertilizer N ecosystem recovery ranged from 81.9% to 84.2% for EEFs compared to 65.2% for urea. The largest amount of fertilizer N recovered for all treatments was in the loblolly pine trees (EEFs: 38.5%–49.9%, urea: 34.8%) and soil (EEFs: 30.6%–38.8%, urea: 28.4%). This research indicates that a greater ecosystem fertilizer N recovery for EEFs compared to urea in southeastern pine plantations can potentially lead to increased FNUE in these systems.

Keywords: ^{15}N; forest fertilization; nitrogen cycle; plantation forestry; enhanced efficiency fertilizers

1. Introduction

Loblolly pine (*Pinus taeda* L.) is the most widely planted and commercially valuable tree species in the United States [1,2], with large areas in the southeastern United States managed intensively in plantations. Although loblolly pine stemwood can exceed 10 m$^3 \cdot$ha$^{-1} \cdot$year^{-1} in intensively managed plantations [3], growth of many stands is less due to low levels of plant available nitrogen (N) and phosphorous (P) in the soil [4,5]. Nitrogen deficiencies that occur when plant available N in the soil is inadequate to meet tree N demand [6–8] translate to low leaf areas, decreased photosynthetic capacity, and hence reduced growth [9,10]. Temporal patterns in N availability often lead to N deficiencies developing during later parts of the rotation [3,11,12]. Following disturbance, such as harvesting, plant N availability in the soil is high due to N mineralization of organic matter [13,14]. Yet, as the stand develops, plant N availability decreases because of increasing N immobilization in the ecosystem [15], and N fertilization is often required to maintain forest productivity in mid-rotation stands [6]. The average growth response in mid-rotation southeastern pine plantations averages 3 m$^3 \cdot$ha$^{-1} \cdot$year^{-1} over the 8 years following fertilization [3].

However, less than 30% of N applied in fertilizer is taken up by trees [16–20]. The low fertilizer N uptake by trees is likely due to several factors including: (1) N loss from the system; (2) N immobilization; and (3) N application that is asynchronous to seasonal plant N demand [21–26]. Forest soils supporting pine plantations in the southeastern United States contain large natural quantities of N, typically ranging from 2 to 7 Mg$\cdot$ha^{-1} [27–29]. Because of the large amount of N in forest soils, it is difficult to follow the fate, cycling and uptake of fertilizer N which typically adds only 150–250 kg·ha^{-1} of N to the system [28–30].

Fertilizers labeled with ^{15}N can be used to trace the fate of fertilizer N through the ecosystem over time [31–34]. Studies using ^{15}N have improved the understanding of applied N cycling in ecosystems since the 1950s in both agriculture [35] and forested [36] systems. Recent research in forested ecosystems using ^{15}N tracer techniques have focused on understanding N cycling in natural systems [37–39] or the effects of chronic N deposition from industrialization [40,41]. Fertilizers labeled with ^{15}N have also been used to determine fertilizer N uptake and nitrogen use efficiency (FNUE) [21,22,36,42–44].

Urea (46-0-0) is the most commonly used N fertilizer in southern forestry due to its high N content and low cost per unit of applied N [4,5]. However, large N losses following urea fertilization can occur due to ammonia (NH_3) volatilization depending on the interactions of weather and edaphic factors [45–50]. Volatilization, combined with leaching and denitrification, reduce the amount of fertilizer N remaining in the system, and may decrease fertilizer N availability for plant uptake and hence FNUE [51–54].

Enhanced efficiency N fertilizers (EEFs) were developed to reduce N loss and increase N availability [55–65]. The EEFs can be divided into slow release (SRN), controlled release (CRN) and stabilized (SNF) N fertilizers [55–63]. The SRN products slowly release fertilizer N due to microbial decomposition [56]. The CRN products have coatings around the fertilizer N to alter rate, pattern and duration of fertilizer N release [56,57]. The SNF products have compounds to inhibit rapid fertilizer N transformations to less stable forms [61]. The different attributes of the various N containing EEF products increases the flexibility of N fertilization under diverse conditions to optimize plant N uptake and increase FNUE when compared to urea.

Our overall objective was to determine ecosystem uptake of fertilizer N and determine if N uptake was greater for EEFs compared to urea. In this study, we compared fertilizer N uptake in southeastern pine ecosystems following fertilization with urea and three enhanced efficiency fertilizers after a spring application to determine if there were differences among treatments for: (1) total ecosystem fertilizer N recovery; and (2) ecosystem partitioning of fertilizer N.

2. Materials and Methods

2.1. Experimental Design

This study was established as a complete-block design with five fertilizer treatments. Twelve sites were chosen from a network of existing fertilizer and thinning trials, and each site served as a block. At each site, five 100 m^2 circular plots were installed prior to fertilization, and each site was fertilized with a single fertilizer treatment on the same day between 26 March and 8 April 2012.

2.2. Site Description

All sites were considered mid-rotation, with stand ages ranging from 8 to 15 years (Figure 1). The understory of the sites ranged from no understory to encompassing 25% of the plot. Selected climate, physical and stand characteristics are detailed in Table 1.

Table 1. Selected climate, physical and stand characteristics of pine stands in the southeastern United States selected to evaluate ecosystem partitioning of fertilizer N following application of urea or enhanced efficiency N fertilizers enriched with [15]N. MAP = Mean annual precipitation. MAT = Mean annual temperature.

State	Latitude	Longitude	Alt (m)	MAP (cm)	MAT (°C)	Physiographic Region	Soil Taxonomic Class	Drainage Class	Trees plot^{-1}	Trees ha^{-1}	Ht (m)	DBH (cm)
VA	37.445331	78.662917	60	109	13	Piedmont	fine, mixed, subactive, mesic Typic Hapludults	Well	8	880	9.1	15.1
SC	34.450000	80.505383	29	107	16	Sandhills	thermic coated Typic Quartzipsamments	Excessively	15	1560	12.5	16.1
GA1	33.625317	82.801183	35	118	16	Piedmont	fine kaolinitic thermic Rhodic Kandiudults	Well	19	1800	7.2	8.6
GA2	31.339978	81.857283	1	114	18	Atlantic Coastal Plain	sandy siliceous thermic Aeric Alaquods	Poorly	15	1460	14.7	16.5
GA3	31.299333	81.847217	1	114	18	Atlantic Coastal Plain	loamy, siliceous, subactive, thermic Arenic Paleaquults	Somewhat Poorly	13	1340	13.6	15.1
FL	30.205267	83.866817	0.6	142	20	Eastern Gulf Coastal Plain	loamy siliceous superactive thermic Aquic Arenic Hapludalfs	Somewhat Poorly	16	1580	10.2	13.4
MS	31.066717	89.602467	26	152	19	Western Gulf Coastal Plain	coarse loamy siliceous subactive thermic Typic Paleudults	Well	21	2160	12.5	13.8
LA1	31.337017	93.182783	28	147	19	Western Gulf Coastal Plain	fine smectitic thermic Albaquic Hapludalfs	Moderately Well	7	720	14.9	19.3
LA2	31.013333	93.422600	28	147	19	Western Gulf Coastal Plain	fine loamy siliceous subactive thermic Plinthic Paleudults	Well	6	780	14.8	17.5
LA3	30.560533	90.727650	0.9	160	19	Western Gulf Coastal Plain	fine silty mixed active thermic Typic Glossaqualfs	Poorly	23	2380	12.0	13.3
OK	34.029333	94.825017	42	136	16	Western Gulf Coastal Plain	fine silty mixed active thermic Aquic Paleudalfs	Moderately well	15	1580	3.0	4.0
TX	31.13255	94.462533	32	127	20	Western Gulf Coastal Plain	fine loamy siliceous semiacitve thermic Oxyaquic Glossudalfs	Moderately well	13	1360	13.9	11.2

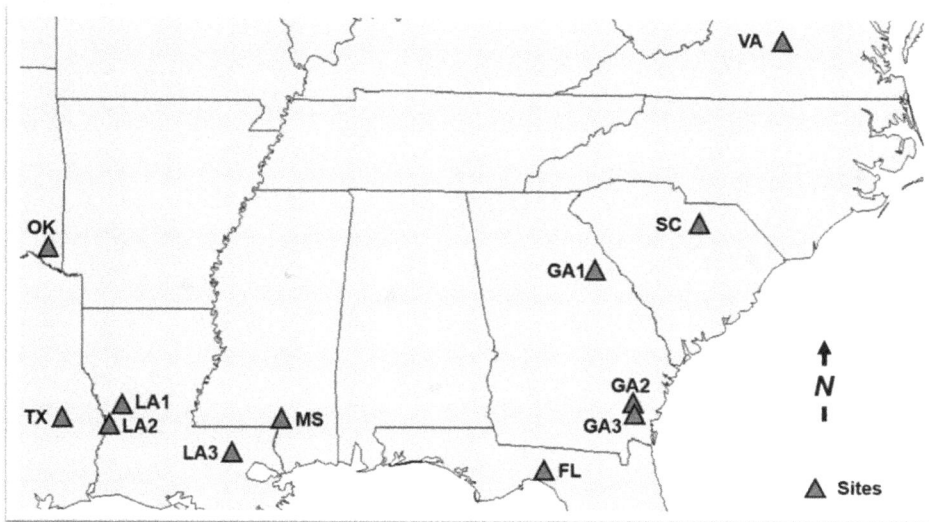

Figure 1. Location map of 12 mid-rotation pine stands across the southeastern United States selected to evaluate ecosystem partitioning of fertilizer N following the application of urea or enhanced efficiency N fertilizers enriched with [15]N.

2.3. Fertilizer Treatments

The five fertilizer treatments used in this study were: (1) urea; (2) urea impregnated with N-(n-Butyl) thiophosphoric triamide (NBPT); (3) urea impregnated with NBPT and coated with monoammonium phosphate (CUF); (4) polymer coated urea (PCU); and (5) a control treatment with no fertilizer added. Urea (46-0-0) was used because it is the most common N fertilizer applied in the southeastern United States. The enhanced efficiency fertilizers (EEFs) tested in this study were developed to reduce NH_3 volatilization and release fertilizer N slowly to the environment. The NBPT treatment (46-0-0) added N-(n-butyl) thiophosphoric triamide at a rate of 26.7% by weight to urea granules to inhibit urease activity. The CUF treatment (39-9-0) also added NBPT to urea granules, which was then coated with an aqueous binder solution of boron and copper sulfate to slow N release. A final coating of monoammonium phosphate was added to provide P. The PCU (44-0-0) treatment encapsulated urea granules with a polymer coating containing pores designed to slowly release N (~80%) over 120 days. All N treatments were applied at an equivalent rate of 224 kg·N·ha^{-1}. Because the CUF treatment had P in a coating, P was applied in the other fertilizer treatments at the equivalent rate of 28 kg·P·ha^{-1} as triple superphosphate (TSP). The urea in all treatments was enriched with the stable isotope [15]N (0.5 atom percent). Each fertilizer treatment was broadcast applied by hand in individual 100 m^2 circular plots at each site. Due to high rates of volatilization and the impact this process has on isotopic fractionation, a fractionation factor of 1.029 was used for each fertilizer treatment as detailed in Högberg [66].

2.4. Field Sampling

The center of each of the 100 m^2 circular plots was located between two co-dominant loblolly pine trees in areas with similar stand, soil and landscape characteristics. Immediately prior to each treatment application, the height and diameter breast height (DBH) of all trees greater than 2.54 cm DBH were measured. The sapling, shrub, vine and herbaceous strata were estimated and individual species in each respective strata was composite sampled. The forest floor (O horizon = Oi + Oe + Oa) was collected with a circular sampler from 4 random locations in the plot and composited. Two mineral soil depth increments (0–15 cm, 15–30 cm) were randomly sampled from 8 locations in the plot with a push tube sampler and composited. Roots were sampled to a depth of 20 cm at 4 random locations in the plot with a bulk density corer and composited. Soil bulk density cores were taken from the 0–15 cm and 15–30 cm depth increments from the center of each plot. After this sampling was completed, two 1 m^2 circular mesh litterfall traps were placed randomly in the plot to sample litterfall from the year after N fertilization.

Fertilizer treatments were randomly applied to the 100 m^2 circular plots (March 26 to April 8). All primary ecosystem components in each plot were resampled at the end of the growing season following fertilization between November 1st and March 31st using similar sampling procedures previously detailed. Litterfall was collected once from the two litterfall traps at the end of the growing season and composited for each individual plot. One of the two central crop trees was selected and felled for sampling. All components of the crop tree sampled were weighed in the field on the same day the tree was felled to obtain field green weights, and subsamples were brought to the laboratory to obtain dry weights. In the field, a 2.54 cm cookie was taken from the tree stem at DBH, height to live crown (HLC), and height to mid crown (HMC). The tree stem was cut into 1.2 m lengths for weighing. The canopy (foliage, fine branches, coarse branches) was randomly placed in 3 piles in the field and each pile was weighed. The foliage, fine branches (branches with foliage attached) and coarse branches (branches with no foliage attached) were separated and also weighed in the field. One of the three canopy piles was randomly selected and returned to the laboratory for analysis. The sapling stratum, if present, was sampled by 2.54 cm diameter classes for individual species categories. Shrubs, vines and herbaceous species were sampled from a randomly selected 3.13 m^2 area of the plot. Individual shrub species were sampled in their entirety, with vines and herbaceous species composite sampled in their respective strata.

2.5. Laboratory Procedures

All samples were dried in a forced air oven at 60 °C. In the laboratory, subsamples of bark, wood from the current year of growth (CGR), and the wood of growth rings prior to fertilizer treatment (PGR) were taken from the stem cookies collected at DBH. Litterfall was separated into pine needles, deciduous leaves, fine branches, coarse branches, bark and unidentifiable litterfall. The forest floor was sieved through a 6 mm sieve and the mineral soil was sieved through a 2 mm sieve. Root samples were elutriated and divided into fine (<2 mm) and coarse (>2 mm) size fractions, dried and weighed.

After drying, all organic material samples were coarse ground in a Wiley Mill to pass a 2 mm sieve. The organic samples were then homogenized to a fine powder with a ball mill (Retsch® Mixer Mill MM 200, Haan, Germany) for 1 min at 25 revolutions per second (rps), whereas all mineral soil samples were ball milled for 2 min at 25 rps. After ball milling, individual homogenized samples were put in separate tin capsules and weighed on a Mettler-Toledo© MX5 microbalance (Mettler-Toledo, Inc., Columbus, OH, USA). These individually weighed samples were analyzed to determine the ^{15}N/^{14}N isotope ratio and total N on a coupled elemental analysis-isotope ratio mass spectrometer (IsoPrime 100 EA-IRMS, Isoprime© Ltd., Manchester, UK) at the Forest Soils and Plant Nutrition Laboratory at Virginia Polytechnic Institute and State University (Virginia Tech). All grinding, ball milling and weighing equipment were cleaned after each sample with ethanol to reduce contamination.

2.6. Calculation of Fertilizer N Recovery

The amount of total fertilizer N recovered in each ecosystem component from the labeled ^{15}N fertilizer was calculated using a mass balance tracer technique that compared individual ecosystem component ^{15}N prior to and 1 year after N fertilization [32,67,68]. Once the fertilizer N recovery for each individual ecosystem component was determined, the fertilizer N recovery for each individual component was summed on a per plot basis to calculate total fertilizer N recovery for the entire plot. The fertilizer N recovery value for the individual loblolly pine sampled in each plot was multiplied by the number of loblolly pine trees in each individual plot to obtain the total fertilizer N recovery for loblolly pine trees on an individual plot basis. The difference between the amount of fertilizer N applied to the plot and the amount of fertilizer N recovered after a single growing season was considered lost from the system.

2.7. Statistical Analysis

Total fertilizer N recovery, expressed as a percentage of fertilizer N applied, was analyzed using a general linear model (GLM) analysis of variance with SAS® 9.4 (SAS Institute Inc., Cary, NC, USA). Percent data was arcsin transformed prior to analysis. Percent fertilizer N recovery (%) was the response variable for the model, fertilizer treatment (CUF, NBPT, PCU, urea, control) was the fixed effects, and site was a random effect. Total fertilizer N recovery, expressed as a percentage of fertilizer N applied, for individual ecosystem components were also analyzed using a GLM analysis of variance, except for analysis of individual mineral soil depth increments (0–15 cm, 15–30 cm) which were analyzed as a repeated measures analysis of variance. Significance levels were set at $\alpha = 0.05$ and the $p > |t|$ values for the treatment means were tested. All post-hoc analysis was conducted with Tukey's HSD.

3. Results

Fertilization increased nitrogen concentrations ($g \cdot kg^{-1}$) in several of the ecosystem components. Foliar N mean ($\pm$ SEM) concentrations increased from 12.5 ± 0.4 $g \cdot kg^{-1}$ in the control to between 13.2 ± 0.5 $g \cdot kg^{-1}$ and 14.2 ± 0.5 $g \cdot kg^{-1}$ for fertilizer treatments, with a significant difference between the control and urea (14.2 ± 0.5 $g \cdot kg^{-1}$) (Table 2). The fine branch N concentrations increased from the control (5.1 ± 0.3 $g \cdot kg^{-1}$) to CUF (7.3 ± 0.6 $g \cdot kg^{-1}$) and NBPT (6.7 ± 0.3 $g \cdot kg^{-1}$), while the coarse branch N concentration increased between the control (2.8 ± 0.2 $g \cdot kg^{-1}$) to urea (3.9 ± 0.4 $g \cdot kg^{-1}$). For the stem, the N concentration of the bark increased between the control (2.1 ± 0.2 $g \cdot kg^{-1}$) and NBPT (2.9 ± 0.1 $g \cdot kg^{-1}$), while the N concentration for the growth ring for the year after fertilization (CGR) increased from 1.8 ± 0.1 $g \cdot kg^{-1}$ for the control to between 2.3 ± 0.1 $g \cdot kg^{-1}$ and 2.7 ± 0.1 $g \cdot kg^{-1}$ for all fertilizer treatments. There were also minor effects of N fertilization on N concentration for fine or coarse roots, litterfall and the mineral soil. The forest floor N concentration was greater for both CUF (8.1 ± 0.6 $g \cdot kg^{-1}$) and PCU (8.7 ± 0.6 $g \cdot kg^{-1}$) compared to the control (6.6 ± 0.6 $g \cdot kg^{-1}$).

Table 2. The mean fertilizer N recovery (% of applied fertilizer N), $\delta^{15}N$ (‰), and N concentration ($g \cdot kg^{-1}$) for individual ecosystem components of pine stands in the southeastern United States selected to evaluate ecosystem partitioning of fertilizer N for urea or enhanced efficiency N containing fertilizers enriched with ^{15}N.

Ecosystem Component	Treatment	Fertilizer N Recovery (% of Applied N)	$\delta^{15}N$ values (‰)	N Concentration ($g \cdot kg^{-1}$)
Foliage	Control	0.0 a	−2.4 (0.5) a	12.5 (0.4) a
	CUF	10.7 (1.1) bc	118.3 (11.2) b	14.1 (0.8) ab
	NBPT	14.8 (1.8) c	126.1 (8.5) b	13.9 (0.6) ab
	PCU	8.1 (1.1) b	101.6 (9.8) b	13.2 (0.5) ab
	Urea	11.0 (1.7) bc	124.5 (11.2) b	14.2 (0.5) b
Fine Branches	Control	0.0 a	−2.0 (0.6) a	5.1 (0.3) a
	CUF	3.2 (0.4) b	110.2 (10.2) b	7.3 (0.6) b
	NBPT	4.1 (0.5) c	111.8 (7.4) b	6.7 (0.3) b
	PCU	2.6 (0.5) b	98.7 (10.5) b	6.1 (0.5) ab
	Urea	2.9 (0.5) b	108.4 (9.67) b	6.2 (0.4) ab

Table 2. Cont.

Ecosystem Component	Treatment	Fertilizer N Recovery (% of Applied N)	δ^{15}N values (‰)	N Concentration (g·kg^{-1})
Coarse Branches	Control	0.0 a	−1.8 (0.4) a	2.8 (0.2) a
	CUF	2.9 (0.4) bc	72.6 (9.3) b	3.7 (0.3) ab
	NBPT	3.0 (0.6) c	78.9 (8.2) b	3.6 (0.4) ab
	PCU	2.6 (0.4) b	73.7 (8.6) b	3.3 (0.3) ab
	Urea	3.2 (1.0) bc	69.3 (6.1) b	3.9 (0.4) b
Bark	Control	0.0 a	−2.1 (0.4) a	2.1 (0.2) a
	CUF	0.6 (0.2) b	22.1 (2.1) b	2.5 (0.2) ab
	NBPT	0.8 (0.2) b	22.5 (2.1) b	2.9 (0.1) b
	PCU	0.5 (0.1) b	20.5 (2.4) b	2.4 (0.1) a
	Urea	0.4 (0.1) ab	20.8 (2.1) b	2.5 (0.2) ab
Current year Growth Ring (CGR)- year of fertilization	Control	0.0 a	−1.9 (0.4) a	1.8 (0.1) a
	CUF	2.2 (0.4) bc	80.0 (8.0) b	2.4 (0.5) bc
	NBPT	3.1 (0.6) c	88.1 (5.8) b	2.7 (0.1) c
	PCU	1.7 (0.3) b	71.9 (7.6) b	2.3 (0.1) b
	Urea	1.9 (0.3) b	78.9 (7.5) b	2.3 (0.1) b
Previous year growth rings (PGR)—growth prior to fertilization	Control	0.0 a	−1.9 (0.4) a	1.4 (0.1) a
	CUF	4.4 (0.8) bc	40.2 (3.9) b	1.6 (0.1) a
	NBPT	5.1 (0.7) c	40.3 (1.3) b	1.6 (0.1) a
	PCU	3.7 (0.4) bc	37.1 (3.9) b	1.6 (0.1) a
	Urea	2.9 (0.5) b	34.8 (5.0) b	1.4 (0.1) a
Fine roots (<2 mm)	Control	0.0 a	−0.3 (0.9) a	9.4 (0.9) a
	CUF	19.2 (1.8) c	36.1 (3.1) b	10.5 (0.6) a
	NBPT	16.2 (1.5) c	33.2 (3.8) b	10.4 (0.4) a
	PCU	16.4 (1.8) c	36.0 (3.5) b	10.3 (0.5) a
	Urea	10.8 (1.4) b	31.2 (5.1) b	9.7 (0.5) a
Coarse roots (>2 mm)	Control	0.0 a	−0.1 (0.37) a	6.0 (0.6) a
	CUF	3.8 (1.4) b	18.3 (1.6) b	6.2 (0.5) a
	NBPT	2.7 (0.6) ab	19.0 (3.2) b	6.3 (0.5) a
	PCU	2.9 (0.9) b	15.4 (2.2) b	6.0 (0.5) a
	Urea	1.7 (0.5) ab	13.9 (3.1) b	6.0 (0.5) a
Litterfall	Control	0.0 a	−3.0 (1.2) a	7.4 (0.3) a
	CUF	1.6 (0.3) b	48.5 (1.4) b	8.0 (0.3) a
	NBPT	1.4 (0.2) b	55.8 (2.1) b	8.3 (0.4) a
	PCU	1.8 (0.3) b	55.6 (1.6) b	8.1 (0.3) a
	Urea	1.7 (0.2) b	55.1 (1.4) b	8.2 (0.2) a
Forest Floor (Organic horizon: Oi + Oe + Oa)	Control	0.0 a	−1.9 (0.4) a	6.6 (0.6) a
	CUF	3.0 (1.4) a	62.9 (6.8) b	8.1 (0.6) b
	NBPT	2.8 (1.3) a	59.3 (5.8) b	8.0 (0.6) ab
	PCU	3.8 (0.9) b	91.1 (9.9) c	8.7 (0.6) b
	Urea	1.2 (0.4) a	55.2 (6.3) b	8.0 (0.7) ab
0–15 cm Mineral Soil	Control	0.0 a	3.3 (0.7) a	0.8 (0.1) a
	CUF	23.1 (3.6) b	21.8 (3.6) b	0.6 (0.1) b
	NBPT	22.4 (2.7) b	22.3 (3.5) b	0.6 (0.1) ab
	PCU	24.2 (5.1) b	21.0 (4.1) b	0.7 (0.1) ab
	Urea	15.9 (2.6) b	15.9 (2.1) b	0.6 (0.1) ab
15–30 cm Mineral Soil	Control	0.0 a	5.4 (0.4) a	0.4 (0.1) a
	CUF	6.7 (1.1) a	13.2 (1.5) b	0.4 (0.0) a
	NBPT	5.4 (0.9) a	11.7 (1.0) b	0.4 (0.1) a
	PCU	10.7 (3.4) b	16.3 (3.0) b	0.4 (0.0) a
	Urea	11.3 (3.1) b	15.2 (2.4) b	0.4 (0.1) a

Different letters represent significant differences at α = 0.05. Numbers in parentheses represent the standard error of the mean (n = 12).

The mean ($\pm$SEM) $\delta^{15}N$ (‰) values at the end of the first growing season after treatment application for all tree and soil ecosystem components were greater for fertilizer treatments compared to the control (Table 2). The mean $\delta^{15}N$ values of loblolly pine trees for fertilizer treatments was greatest in the foliage (101.6‰ $\pm$ 9.8‰ to 126.1‰ $\pm$ 8.5‰), fine branches (98.7‰ $\pm$ 10.5‰ to 111.8‰ $\pm$ 7.4‰), and CGR (71.9 $\pm$ 7.6‰ to 88.1 $\pm$ 5.8‰). The $\delta^{15}N$ values in litterfall (48.5‰ $\pm$ 1.4‰ to 55.8‰ $\pm$ 2.1‰) for the year immediately following fertilization of all fertilized treatments was lower than the foliage mean $\delta^{15}N$ values. The lowest $\delta^{15}N$ values for the loblolly pine components for fertilized treatments were in the coarse roots (13.9‰ $\pm$ 3.1‰ to 19.0‰ $\pm$ 3.2‰), bark (20.5‰ $\pm$ 2.4‰ to 22.5‰ $\pm$ 2.1‰), fine roots (31.2‰ $\pm$ 5.1‰ to 36.1‰ $\pm$ 3.1‰) and stemwood produced in the years prior to fertilization (PGR) (34.8‰ $\pm$ 5.0‰ to 40.3 $\pm$ 1.3‰). The $\delta^{15}N$ value of the soil for the fertilized treatments was greatest in the forest floor (55.2‰ $\pm$ 6.3‰ to 91.1‰ $\pm$ 9.9‰), with PCU significantly greater than other fertilizer treatments. The surface 0–15 cm mineral soil ranged from 15.9‰ $\pm$ 2.1‰ to 22.3‰ $\pm$ 3.5‰ for fertilizer treatments, while the 15–30 cm mineral soil depth increment ranged from 11.7‰ $\pm$ 1.0‰ to 16.3‰ $\pm$ 3.0‰.

There were several significant differences in the mean ($\pm$SEM) fertilizer N recovery for individual ecosystem components (Figure 2, Table 2). The fertilizer N recovered in the loblolly pine trees was greatest in the foliage ranging from 8.1% $\pm$ 1.1% to 14.8% $\pm$ 1.8% of the N applied. More fertilizer N was recovered in the foliage for NBPT (14.8% $\pm$ 1.8%) compared to PCU (8.1% $\pm$ 1.1%). More fertilizer N was also recovered for NBPT (4.1% $\pm$ 0.5%) than the other fertilizer treatments (2.6% $\pm$ 0.5% to 3.2% $\pm$ 0.4%) for fine branches, and coarse branches for NBPT (3.0% $\pm$ 0.6%) compared to PCU (2.6% $\pm$ 0.5%). Fertilizer N recovery was greater for NBPT (3.1% $\pm$ 0.6%) for CGR than PCU (1.7% $\pm$ 0.3%) and urea (1.9 $\pm$ 0.3%), and the PGR for NBPT (5.1% $\pm$ 0.7%) compared to urea (2.9% $\pm$ 0.5%). For belowground loblolly pine biomass, more fertilizer N was recovered for EEFs (16.2% $\pm$ 1.5% to 19.2% $\pm$ 1.8%) compared to urea (10.8% $\pm$ 1.4%) in fine roots. In the soil, there was greater fertilizer N recovery for PCU (3.8% $\pm$ 0.9%) compared to the other fertilizer treatments (1.2% $\pm$ 0.4% to 3.0% $\pm$ 1.4%) in the forest floor, and for both PCU (10.7% $\pm$ 3.4%) and urea (11.3% $\pm$ 3.1%) compared to CUF (6.7% $\pm$ 1.1%) and NBPT (5.4% $\pm$ 0.9%) in the 15–30 cm mineral soil.

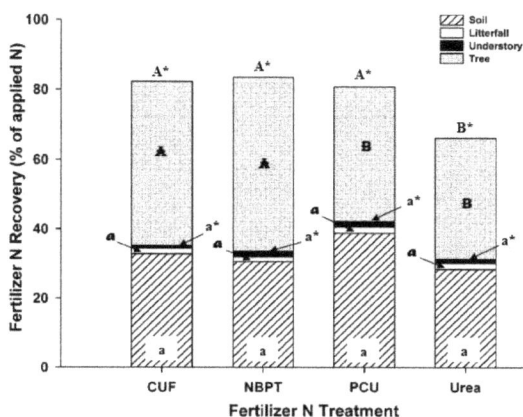

Figure 2. The total fertilizer N recovery (% of fertilizer N applied) of the major ecosystem components (Tree—loblolly pine aboveground + belowground biomass, understory, litterfall, and soil (O horizon + mineral soil- 0 to 30 cm) for pine stands in the southeastern United States selected to evaluate ecosystem partitioning of fertilizer N of urea or enhanced efficiency N fertilizers enriched with ^{15}N. Data represents fertilizer application dates for spring (March–April 2012). Different letters represent significant differences at α = 0.05. Different letter fonts represent comparisons among treatments between same ecosystem components (soil, litterfall, understory, tree). n = 12.

Differences between fertilizer treatments also occurred when individual ecosystem components were combined into primary components (tree, soil) (Figure 2). For the loblolly pine canopy (foliage, fine branches, coarse branches), fertilizer N recovery for NBPT (22.0%) was greater than PCU (13.3%). For the stem (bark, CGR, PGR), NBPT (9.1%) was greater than both PCU (5.9%) and urea (5.2%). When all aboveground biomass components are combined (canopy, stem), NBPT (31.0%) was greater than both PCU (19.2%) and urea (22.3%). For belowground biomass (total roots), all EEFs (CUF = 23.0%, NBPT = 18.9%, PCU = 19.3%) were greater than urea (12.5%). When all pine components were combined (canopy, stem, roots), both CUF (46.9%) and NBPT (49.9%) had a greater fertilizer N recovery compared to PCU (38.5%) and urea (34.8%). For the entire soil (forest floor, 0–30 cm mineral soil), the percentage of recovery was similar for EEFs (30.6% to 38.8%) and urea (28.4%). When all ecosystem components were combined to determine the total ecosystem recovery of fertilizer N, all the EEFs (CUF = 81.9%, NBPT = 84.2%, PCU = 79.8%) had a greater recovery total fertilizer N recovery compared to urea (65.2%).

4. Discussion

This study evaluated differences in the ecosystem retention and crop tree uptake of applied fertilizer N between urea and three enhanced efficiency fertilizers in mid-rotation pine plantations of the southeastern United States. The sites in this study covered the entire southeastern United States region where loblolly pine is planted to improve the understanding of the ultimate fate of fertilizer N in these systems to augment the results of numerous site specific studies. The primary hypotheses tested in this study were: (1) whether there were differences in the total fertilizer N recovery among conventional and enhanced efficiency fertilizers; and (2) if there were differences in the ecosystem partitioning of fertilizer N among conventional and enhanced efficiency N fertilizers. The overall primary objective was to improve fertilizer N use efficiency in southeastern pine plantations to increase the productivity and efficiency of these systems in a sustainable, environmentally responsible approach.

There were significant differences in the total ecosystem fertilizer N recovery among individual fertilizer treatments (CUF, NBPT, PCU, urea). The total ecosystem fertilizer N recovery was greater for all enhanced efficiency fertilizers (CUF, NBPT, PCU) compared to urea, with no differences between individual EEFs. The primary reason for lower fertilizer N recovery for urea compared to EEFs was likely due to initially large ammonia (NH_3) volatilization losses from the urea treatment compared to the EEFs immediately following fertilization. Raymond et al. [49] compared NH_3 volatilization losses of the same fertilizer treatments used in this study and found higher losses with urea (26%–49%) compared to all EEFs (4%–26%). Other studies in pine plantations in the southeastern United States have also shown fertilizer N losses through the NH_3 volatilization pathway after urea fertilization exceeding 25%, with lower losses using various EEF products [47,48,64–66,69,70]. Interestingly, when the results for NH_3 volatilization reported by Raymond et al. [49] 15 days following fertilization for the same fertilizer treatments used in this study are included in the treatments in respect to mass balance accounting of ecosystem fertilizer N recovery on a per plot basis, total ecosystem recovery of fertilizer N for each treatment exceeds 90%. This evidence indicates that greater fertilizer N loss via NH_3 volatilization immediately following fertilization, combined with other minor initial fertilizer N loss pathways (denitrification, leaching) for urea compared to EEFs, translates to less fertilizer N remaining in the ecosystem for urea. Although the results from this study are similar to those of other tracer studies in a variety of ecosystems [39,40,71,72], results from this study indicated a generally higher fertilizer N uptake by trees for the EEFs.

Results from numerous studies specific to loblolly pine plantations and fertilizer N show generally lower fertilizer N uptake (<30%) by loblolly pine trees [21–25]. Several potential explanations for this low fertilizer N uptake by loblolly pine plantations exist. First, many studies in loblolly plantations have used urea as the fertilizer N source. Because high fertilizer N losses can occur immediately following fertilization with urea [46–49], less fertilizer N remains in the ecosystem and hence less fertilizer N is available for plant uptake. As previously stated, this explanation partially explains the results for the fertilizer N recovery in the loblolly pine trees in this study (35%) which is near the upper end reported in the literature. The fertilizer N recovery for the EEF treatments used in this study had higher fertilizer N recovery for the entire tree compared to urea, ranging from 39% to 50%. Clearly, the lower amount of fertilizer N remaining in the system when urea is used as a fertilizer source affects the quantity, and hence intensity, of fertilizer N in the soil over the growing season that becomes available for uptake by the desired crop trees. Second, the application of fertilizer N for this study in the spring (March-April) may have been more synchronous to the seasonal growth patterns of desired crop trees and could explain higher recovery rates of fertilizer N for all treatments found in this study. For example, results from Blazier et al. [25] had higher fertilizer N recovery after a spring and summer fertilization compared to a winter application, when operational N fertilization is traditionally conducted in southeastern pine plantations in an effort to minimize high NH_3 volatilization losses. Although losses were high for urea compared to the EEFs, recovery for the entire tree for fertilizer N was greater compared to most fertilizer studies. This result may indicate that although losses can be high in the spring, conditions for N demand and uptake by the crop tree for a readily available N source may still be high. Low fertilizer N uptake by trees after fertilization during the dormant season (winter) in the south may cause other loss pathways, such as denitrification [51] and leaching [53] to become more important than NH_3 volatilization and also contribute to lower fertilizer N recovery in studies. Third, many studies only account for N uptake in the foliage. Although foliage in loblolly pine trees has a high amount of N in the tree, other portions of the tree also contain N. If only N in the foliage is measured, N uptake may be underestimated.

There were also significant differences in ecosystem partitioning of fertilizer N among treatments. Analysis of the percent of fertilizer N recovered in individual ecosystem components (foliage, fine branches, soil, etc.) showed numerous differences among fertilizer treatments. Despite differences among fertilizer treatments in the percentage of fertilizer N recovered in different ecosystem components, most fertilizer N for all treatments was recovered in either the loblolly pine trees or soil. There was a difference for the fertilizer N recovered in the entire tree between NBPT and CUF and urea and PCU, yet no significant differences occurred in the soil. Although there were no significant differences in the percentage of fertilizer N in the soil, there was 10% more fertilizer N recovered (NS) for PCU compared to urea. This difference may have important implications for the long term fate and cycling of fertilizer N for PCU, which provides a more gradual constant release compared to the other fertilizers. Additionally, significantly higher amounts of fertilizer N were found in the fine roots of all EEFs compared to urea. This additional amount of fertilizer N was likely the result of a higher amount of fertilizer N in the soil and potentially greater root uptake for EEFs compared to urea.

Although results from this research showed a greater ecosystem recovery for each EEF compared to urea, several primary questions remain concerning the long-term fate of each of these fertilizer treatments. For example, significantly more fertilizer N was recovered in the forest floor for PCU than other treatments. Additionally, although not significant, there was a high proportion of fertilizer N recovered for all EEFs in the 0–15 cm mineral soil compared to urea. A shift in the ecosystem recovery of fertilizer N compared to urea in the ecosystem raises the question of whether additional gains in fertilizer nitrogen use efficiency for the system for EEFs will increase if it becomes bioavailable in the future of the stand or becomes immobilized in the system for the stand rotation.

Results from agroecosystems have shown that the majority of fertilizer N that is not incorporated into plant biomass during the initial growing season after fertilization becomes immobilized in the soil [73]. Similar results specific to fertilizer N immobilization in the soil have also been shown in forested systems [74–78]. Despite these results, laboratory experiments using a bioassay approach with soil collected from long term ^{15}N tracer experiments show a measureable uptake of ^{15}N labeled fertilizers by seedlings from both forest floor and mineral soil sources a decade after fertilization [79]. The mechanisms governing this disconnect between field and laboratory studies will require additional examination. Yet, if it is found that the additional fertilizer N becomes bioavailable to seedlings after the harvesting of the remaining stand, this finding could translate to additional gains in fertilizer nitrogen use efficiency in these managed pine ecosystems. At certain managed pine ecosystem sites, especially those which have been thinned, fertilizer management may include an initial fertilizer N application at stand establishment or shortly after. Yet, if the increase in existing fertilizer N remaining in the soil is found to be bioavailable, a management system may be altered to forgo the initial N fertilization while still maintaining high levels of productivity.

Extending research for this study to monitor the long term fate of the ^{15}N enriched fertilizer treatments will be needed to answer the questions specific to the bioavailability of the fertilizer N remaining in the soil. Answering whether the increase in fertilizer N in the soil with EEFs will translate to a higher N availability for trees in these intensively managed pine ecosystems in the future or become immobilized and/or leached from the system is an important question to continue improving the economic viability and environmental stewardship of fertilization in these systems. If future research for the ^{15}N plots used in this study find a majority of the fertilizer N immobilized in the soil for EEFs, as has been found with other studies using traditional N fertilizers in forest ecosystems, the primary question becomes whether fertilizer N application rates can be reduced by a percentage when compared to the traditional rates used in forestry while maintaining productivity. Results indicating either a higher fertilizer N availability for extended periods from the soil using EEFs and/or a reduction in fertilizer N application rates while maintaining productivity would assist in improving the FNUE efficiency of these pine plantation systems.

This research has continued to refine our understanding of the differences in the ecosystem fate and cycling of fertilizer N between enhanced efficiency and urea fertilizers through the use of stable isotopes in southeastern pine plantations. Results from this study are from a broad geographic region, and indicate similar cycling patterns and differences between enhanced efficiency N containing fertilizers compared to urea after a spring application. The results from this study provide forest managers a level of confidence that the enhanced efficiency products used in this research after a spring fertilization will have a generally higher fertilizer N ecosystem retention when compared with urea across the southeastern United States where pine plantations are intensively managed. The results from this study will continue to assist forest managers in improving fertilizer nitrogen use efficiency in pine plantations across the southeastern United States.

5. Conclusions

The principal results of this study were: (1) there was a greater amount of fertilizer N recovered from the ecosystem for all EEFs compared to urea; and (2) there were differences in the ecosystem partitioning of fertilizer N among treatments. The reduced ecosystem loss of fertilizer N for all EEFs compared to urea may translate to an increased quantity of fertilizer N remaining in the system that could be available for crop tree N uptake and increase the FNUE for the EEFs used in this study compared to urea. The increased fertilizer N retention in the soil for all EEFs, although not significant, may also contribute to increasing FNUE over the course of the stand rotation if the increased quantity of soil N translates to an increase in intensity of N supply. Additional research will be required to determine the long term fate of EEFs in forest plantations, and whether this added quantity of soil N from fertilization using EEFs translates to increased productivity. The results from this study will continue to improve site specific management and increase the efficiency of southeastern pine plantations. The use of enhanced efficiency fertilizers, if found to be economically viable, could provide managers the flexibility to apply fertilizer N under a variety of conditions to improve the synchronicity of fertilizer application and plant N demand.

Acknowledgments: This research was primarily supported by the Forest Productivity Cooperative (FPC), the National Science Foundation Center for Advanced Forest Systems (CAFS) and the United States Department of Agriculture National Institute of Food and Agriculture McIntire-Stennis Capacity Grant. We thank David Mitchem for assistance in the laboratory, Andy Laviner and Eric Carbaugh for field assistance and the numerous work study and seasonal student employees for both field and laboratory assistance.

Author Contributions: "Thomas Fox, Jay Raymond and Brian Strahm conceived and designed the experiments. Jay Raymond performed the experiments. Jay Raymond analyzed the data; Thomas Fox and Brian Strahm contributed reagents/materials/analysis tools; Jay Raymond wrote the paper.

Conflicts of Interest: The authors declare no conflict of interest.

References

1. Schultz, R.P. *The Ecology and Culture of Loblolly Pine (Pinus taeda L.)*; Agriculture Handbook #713; USDA Forest Service: Washington DC, USA, 1997; p. 493.
2. US Department of Agriculture (USDA), Forest Service. *Future of America's Forest and Rangelands: Forest Service 2010 Resources Planning Act Assessment*; US Department of Agriculture (USDA), Forest Service: Washington DC, USA, 2012; p. 198. Available online: http://www.fs.fed.us/research/publications/gtr/gtr_wo87.pdf (accessed on 5 February 2016).
3. Fox, T.R.; Jokela, E.J.; Allen, H.L. The development of pine plantation silviculture in the southern United States. *J. For.* **2007**, *105*, 337–347.
4. Allen, H.L. Forest fertilizers: Nutrient amendment, and productivity, and environmental impact. *J. For.* **1987**, *85*, 37–46.
5. Fox, T.R.; Allen, H.L.; Albaugh, T.J.; Rubilar, R.; Carlson, C.A. Tree nutrition and forest fertilization of pine plantations in the southern United States. *South. J. Appl. For.* **2007**, *31*, 5–11.
6. Miller, H.G. Forest fertilization: Some guiding concepts. *Forestry* **1981**, *54*, 157–167. [CrossRef]
7. Chapin, F.S.; Vitousek, P.M.; Van Cleve, K. The nature of nutrient limitation in plant-communities. *Am. Nat.* **1986**, *127*, 48–58. [CrossRef]
8. Vitousek, P.M.; Howarth, R.W. Nitrogen limitation on land and in the sea—How can it occur? *Biogeochemistry* **1991**, *13*, 87–115. [CrossRef]
9. Linder, S. Responses to water and nutrients in coniferous ecosystems. In *Potentials and Limitations of Ecosystems Analysis*; Schulze, E.D., Wolfer, H.Z., Eds.; Springer-Verlag: New York, NY, USA, 1987; pp. 180–202.
10. LeBauer, D.S.; Treseder, K.K. Nitrogen limitation of net primary productivity in terrestrial ecosystems is globally distributed. *Ecology* **2008**, *89*, 371–379. [CrossRef] [PubMed]
11. Switzer, C.L.; Nelson, L.E. Nutrient accumulation and cycling in loblolly pine (*Pinus taeda* L.) plantation ecosystems: The first 20 years. *Soil Sci. Soc. Am. Proc.* **1972**, *36*, 143–147. [CrossRef]
12. Wells, C.G.; Jorgensen, J.R. Effect of intensive harvesting on nutrient supply and sustained productivity. In Proceedings of the Impacts of Intensive Harvesting on Forest Nutrient Cycling, Syracuse, NY, USA, 1979; pp. 212–230.
13. Vitousek, P.M.; Matson, P.A. Intensive harvesting and site preparation decrease soil nitrogen availability in young plantations. *South. J. Appl. For.* **1985**, *9*, 120–125.
14. Fox, T.R.; Burger, J.A.; Kreh, R.E. Effects of site preparation on nitrogen dynamics in the southern Piedmont. *For. Ecol. Manag.* **1986**, *15*, 241–256. [CrossRef]
15. Birk, E.M.; Vitousek, P.M. Nitrogen availability and nitrogen use efficiency in loblolly pine stands. *Ecology* **1986**, *67*, 69–79. [CrossRef]
16. Pritchett, W.L.; Smith, W.H. Fertilizer response in young slash pine stands. *Soil Sci. Soc. Am. J.* **1972**, *36*, 660–663. [CrossRef]
17. Martin, S.W.; Bailey, R.L.; Jokela, E.J. Growth and yield predictions for lower Coastal Plain slash pine plantations fertilized at mid-rotation. *South. J. Appl. For.* **1999**, *23*, 39–45.
18. Amateis, R.L.; Liu, J.; Ducey, M.J.; Allen, H.L. Modeling response to midrotation nitrogen and phosphorus fertilization in loblolly pine plantations. *South. J. Appl. For.* **2000**, *24*, 207–212.
19. Carlson, C.A.; Fox, T.R.; Allen, H.L.; Albaugh, T.J.; Rubilar, R.A.; Stape, J.L. Growth responses of loblolly pine in the southeast united states to midrotation applications of nitrogen, phosphorus, potassium, and micronutrients. *For. Sci.* **2014**, *60*, 157–169. [CrossRef]
20. Rojas, J.C. Factors influencing responses of loblolly pine stands to fertilization. Ph.D. Dissertation, North Carolina State University, Raleigh, NC, USA, 2005.
21. Mead, D.J.; Pritchett, W.L. Fertilizer movement in a slash pine ecosystem. *Plant Soil* **1975**, *43*, 451–465. [CrossRef]

22. Melin, J.; Nommik, H.; Lohm, U.; Flower-Ellis, J. Fertilizer nitrogen budget in a Scots pine ecosystem attained by using root-isolated plots and ^{15}N technique. *Plant Soil* **1983**, *74*, 249–263. [CrossRef]

23. Johnson, D.W.; Todd, D.E. Nitrogen fertilization of young yellow poplar and loblolly pine plantations at differing frequencies. *Soil Sci. Soc. Am. J.* **1988**, *52*, 1468–1477. [CrossRef]

24. Albaugh, T.J.; Allen, H.L.; Dougherty, P.M.; Johnsen, K.H. Long term growth responses of loblolly pine to optimal nutrient and water resource availability. *For. Ecol. Manag.* **2004**, *192*, 3–19. [CrossRef]

25. Blazier, M.A.; Hennessey, T.C.; Dougherty, P.; Campbell, R. Nitrogen accumulation and use by a young loblolly pine plantation in southeast Oklahoma: Effects of fertilizer formulation and date of application. *South. J. Appl. For.* **2006**, *30*, 66–78.

26. Raison, R.J.; Khanna, P.K.; Connell, M.J.; Falkiner, R.A. Effects of water availability and fertilization on nitrogen cycling in a stand of Pinus radiate. *For. Ecol. Manag.* **1990**, *30*, 31–43. [CrossRef]

27. Kiser, L.C.; Fox, T.R. Soil accumulation of nitrogen and phosphorus following annual fertilization of loblolly pine and sweetgum on sandy sites. *Soil Sci. Soc. Am. J.* **2012**, *76*, 2278–2288. [CrossRef]

28. Villanueva, A.T. Impacts of Fertilization on Soil Properties in Loblolly Pine Plantations in the Southeastern United States. Master's Thesis, Virginia Polytechnic Institute and State University, Blacksburg, VA, USA, 2015.

29. Fisher, R.F.; Binkley, D. *Ecology and Management of Forest Soils*; John Wiley & Sons: New York, NY, USA, 2000.

30. Flint, C.M. Leaching of Nitrogen from the Rooting Zone of Douglas-Fir Forests Following Urea Fertilization and Potential Impacts on the Water Quality of Hood Canal. Master's Thesis, University of Washington, Seattle, WA, USA, 2007.

31. Knowles, R.; Blackburn, T.H. *Nitrogen Isotope Techniques*; Academic Press, Inc.: New York, NY, USA, 1993.

32. Nadelhoffer, K.J.; Fry, B. Nitrogen isotope studies in forest ecosystems. In *Stable Isotopes in Ecology and Environmental Science*; Lajtha, K., Michener, R.H., Eds.; Blackwell Scientific Publications: New York, NY, USA, 1994; pp. 22–44.

33. Robinson, D. δ15N as an integrator of the nitrogen cycle. *Trends Ecol. Evol.* **2001**, *16*, 153–162. [CrossRef]

34. Dawson, T.E.; Mambelli, S.; Plamboeck, A.H.; Templer, P.H.; TuSource, K.P. Stable isotopes in plant ecology. *Annu. Rev. Ecol. Syst.* **2002**, *33*, 507–559. [CrossRef]

35. Hauck, R.D.; Bystrom, M. ^{15}N-A Selected Bibliography for Agricultural Scientists; The Iowa State University Press: Ames, IA, USA, 1971; p. 206.

36. Hauck, R.D. Nitrogen source requirements in different soil-plant systems. In *Forest Fertilization: Theory and Practice*; TVA National Fertilizer Development Center: Muscle Shoals, AL, USA, 1968; pp. 47–57.

37. Currie, W.S.; Nadelhoffer, K.J. Dynamic redistribution of isotopically labeled cohorts of nitrogen inputs in two temperate forests. *Ecosystems* **1999**, *2*, 4–18. [CrossRef]

38. Dinkelmeyer, H.; Lehmann, J.; Renck, A.; Trujillo, L.; Pereira da Silva, J., Jr.; Gebauer, G.; Kaiser, K. Nitrogen uptake from ^{15}N-enriched fertilizer by four tree crops in an Amazonian agroforest. *Agrofor. Syst.* **2003**, *57*, 213–224. [CrossRef]

39. Nadelhoffer, K.J.; Colman, B.P.; Currie, W.S.; Magill, A.; Aber, J.D. Decadal-scale fates of ^{15}N tracers added to oak and pine stands under ambient and elevated N inputs at the Harvard Forest (USA). *For. Ecol. Manag.* **2004**, *196*, 89–107. [CrossRef]

40. Tietema, A.; Emmett, B.A.; Gundersen, P.; Kjonaas, O.J.; Koopmans, C.J. The fate of ^{15}N-labelled nitrogen deposition in coniferous forest ecosystems. *For. Ecol. Manag.* **1998**, *101*, 19–27. [CrossRef]

41. Templer, P.H.; Mack, M.C.; Chapin, F.S., III; Christenson, L.M.; Compton, J.E.; Crook, H.D.; Currie, W.S.; Curtis, C.J.; Dail, D.B.; D'Antonio, C.M.; et al. Sinks for nitrogen inputs in terrestrial ecosystems: A meta-analysis of ^{15}N tracer field studies. *Ecology* **2012**, *93*, 1816–1829. [CrossRef] [PubMed]

42. Chang, S.X.; Preston, C.M.; McCullough, K.; Weetman, G.; Barker, J. Effect of understory competition on distribution and recovery of ^{15}N applied to a western red cedar—Western hemlock clear-cut site. *Can. J. For. Res.* **1996**, *26*, 313–321. [CrossRef]

43. Bubb, K.A.; Xu, Z.H.; Simpson, J.A.; Saffigna, P.G. Growth response to fertilization and recovery of ^{15}N-labelled fertilizer by young hoop pine plantations of subtropical Australia. *Nutr. Cycling Agroecosystems* **1999**, *54*, 81–92. [CrossRef]

44. Werner, A.T. Nitrogen Release, Tree Uptake, and Ecosystem Retention in a Mid-Rotation Loblolly Pine Plantation Following Fertilization with 15N-Enriched Enhanced Efficiency Fertilizers. Master's Thesis, Virginia Tech, Blacksburg, VA, USA, 2013.

45. Zhang, X.; Mauzerall, D.L.; Davidson, E.A.; Kanter, D.R.; Cai, R. The economic and environmental consequences of implementing nitrogen-efficient technologies and management practices in agriculture. *J. Environ. Qual.* **2015**, *44*, 312–324. [CrossRef] [PubMed]

46. Cabrera, M.L.; Kissel, D.E.; Vaio, N.; Craig, J.R.; Rema, J.A.; Morris, L.A. Loblolly pine needles retain urea fertilizer that can be lost as ammonia. *Soil Sci. Soc. Am. J.* **2005**, *69*, 1525–1531. [CrossRef]

47. Zerpa, J.L.; Fox, T.R. Controls on volatile NH_3 losses from loblolly pine plantations fertilized with urea in the southeast USA. *Soil Sci. Soc. Am. J.* **2011**, *75*, 257–266. [CrossRef]

48. Elliot, J.R.; Fox, T.R. Ammonia volatilization following fertilization with urea or ureaform in a thinned loblolly pine plantation. *Soil Sci. Soc. Am. J.* **2014**, *78*, 1469–1473. [CrossRef]

49. Raymond, J.E.; Fox, T.R.; Strahm, B.; Zerpa, J. Ammonia Volatilization following Nitrogen Fertilization with Enhanced Efficiency Fertilizers and Urea in Loblolly Pine Plantations of the Southern United States. *For. Ecol. Manag.* **2016**, *376*, 247–255. [CrossRef]

50. Engel, R.; Jones, C.; Wallander, R. Ammonia volatilization from urea and mitigation by by NBPT following surface application to cold soils. *Soil Sci. Soc. Am. J.* **2011**, *75*, 2348–2357. [CrossRef]

51. Shrestha, R.; Strahm, B.D.; Sucre, E.B. Nitrous oxide fluxes in fertilized *Pinus taeda* plantations across a gradient of soil drainage classes. *J. Environ. Qual.* **2014**, *43*, 1823–1832. [CrossRef] [PubMed]

52. Binkley, D.; Carter, R.; Allen, H.L. Nitrogen fertilization practices in forestry. In *Nitrogen Fertilization and the Environment*; Bacon, P., Ed.; Marcel Dekker: New York, NY, USA, 1995; pp. 421–441.

53. Aust, W.M.; Blinn, C.R. Forestry best management practices for timber harvesting and site preparation in the eastern United States: An overview of water quality and productivity research during past 20 years. *Water Air Soil Poll. Focus* **2004**, *4*, 5–36. [CrossRef]

54. Galloway, J.N.; Aber, J.D.; Erisman, J.W.; Seitzinger, S.P.; Howarth, R.W.; Cowling, E.B.; Cosby, B.J. The Nitrogen Cascade. *Bioscience* **2003**, *53*, 341–356. [CrossRef]

55. Chien, S.H.; Prochnow, L.I.; Cantarella, H. Recent developments of Fertilizer Production and Use to Improve Nutrient Efficiency and Minimize Environmental Impacts. In *Advances in Agronomy 102*; Elsevier Inc.: Newark, NJ, USA, 2009; pp. 267–322.

56. Trenkel, M.E. *Slow-and Controlled-Release and Stabilized Fertilizers: An option for Enhancing Nutrient Use Efficiency in Agriculture*; IFA, International Fertilizer Industry Association: Paris, France, 2010.

57. Azeem, B.; KuShaari, K.; Man, Z.B.; Basit, A.; Thanh, T.H. Review on materials & methods to produce controlled release coated urea fertilizer. *J. Control. Release* **2014**, *181*, 11–21. [PubMed]

58. Fertilizer Institute. Available online: https://www.tfi.org/introduction-fertilizer/nutrient-science/enhanced-efficiency-fertilizers (accessed on 11 October 2015).

59. AAPFCO. Association of American Plant Food Control Officials. 2011. Available online: http://www.aapfco.org/index.html (accessed on 1 October 2015).

60. Shaviv, A. *Controlled Release Fertilizers. IFA International Workshop on Enhanced-Efficiency Fertilizers, Frankfurt*; International Fertilizer Industry Association: Paris, France, 2005.

61. Hauck, R.D. Slow release and bio-inhibitor-amended nitrogen fertilizers. In *Fertilizer Technology and Use*, 3rd ed.; Engelstad, O.P., Ed.; SSSA: Madison, WI, USA, 1985; pp. 293–322.

62. Goertz, H.M. Controlled Release Technology. In *Kirk-Othmer Encyclopedia of Chemical Technology*; Controlled Release Technology (Agricultural): New York, NY, USA, 1993; pp. 251–274.

63. Gowariker, V.; Krishnamurthy, V.N.; Gowariker, S.; Dhanorkar, M.; Paranjape, K. *The Fertilizer Encyclopedia*; John Wiley & Sons, Inc.: Hoboken, NJ, USA, 2009.

64. Bremner, J.M.; Douglas, L.A. Inhibition of urease activity in soils. *Soil Biol. Biochem.* **1971**, *3*, 297–307. [CrossRef]

65. Bremner, J.M.; Chai, H.S. Evaluation of N-butyl phosphorothioic triamide for retardation of urea hydrolysis in soil. *Commun. Soil Sci. Plant Anal.* **1986**, *17*, 337–351. [CrossRef]

66. Högberg, P. ^{15}N natural abundance in soil-plant systems. *New Phytol.* **1997**, *137*, 179–203. [CrossRef]

67. Powlson, D.S.; Barraclough, D. Mineralization and assimilation in soil-plant systems. In *Nitrogen Isotope Techniques*; Academic Press, Inc.: New York, NY, USA, 1993; pp. 209–242.

68. Nadelhoffer, K.J.; Downs, M.R.; Fry, B.; Aber, J.D.; Magill, A.H.; Melillo, J.M. The Fate of ^{15}N-Labelled Nitrate Additions to a Northern Hardwood Forest in Eastern Maine, USA. *Oecologia* **1995**, *103*, 292–301. [CrossRef]

69. Kissel, D.; Cabrera, M.; Vaio, N.; Craig, J.; Rema, J.; Morris, L. Rainfall timing and ammonia loss from urea in a loblolly pine plantation. *Soil Sci. Soc. Am. J.* **2004**, *68*, 1744–1750. [CrossRef]

70. Kissel, D.; Cabrera, M.L.; Vaio, N.; Craig, J.R.; Rema, J.A.; Morris, L.A. Forest floor composition and ammonia loss from urea in a loblolly pine plantation. *Soil Sci. Soc. Am. J.* **2009**, *73*, 630–637. [CrossRef]

71. Burke, I.C.; Lauenroth, W.K. Ecosystem ecology at regional scales. *Ecology* **2002**, *83*, 305–306. [CrossRef]

72. Blanes, M.C.; Emmett, B.A.; Viñegla, B.; Carreira, J.A. Alleviation of P limitation makes tree roots competitive for N against microbes in a N-saturated conifer forest: A test through P fertilization and ^{15}N labelling. *Soil Biol. Biochem.* **2012**, *48*, 51–59. [CrossRef]

73. Preston, C.M. The availability of residual fertilizer nitrogen immobilized as clay-fixed ammonium and organic N. *Can. J. Soil Sci.* **1982**, *62*, 479–486. [CrossRef]

74. Foster, N.W.; Beauchamp, E.G.; Corke, C.T. Reactions of ^{15}N-labelled urea with Jack pine forest floor materials. *Soil Biol. Biochem.* **1985**, *17*, 699–703. [CrossRef]

75. Foster, N.W.; Beauchamp, E.G.; Corke, C.T. Immobilization of ^{15}N-labelled urea in a Jack pine forest floor. *Soil Sci. Soc. Am. J.* **1985**, *49*, 448–452. [CrossRef]

76. Hulm, S.C.; Killham, K. Response over two growing seasons of a Sitka spruce stand to ^{15}N-urea fertilizer. *Plant Soil.* **1990**, *124*, 65–72. [CrossRef]

77. NÔmmik, H.; Larsson, K. Effects of nitrogen source and placement on fertilizer ^{15}N enrichment in *Pinus sylvestris* foliage. *Scand. J. For. Res.* **1992**, *7*, 155–163. [CrossRef]

78. Preston, C.M.; Mead, D.J. Growth response and recovery of ^{15}N-fertilizer one and eight growing season after application to lodgepole pine in British Columbia. *For. Ecol. Manag.* **1994**, *65*, 219–229. [CrossRef]

79. Swanston, C.W.; Preston, C.M. Availability of residual fertilizer ^{15}N from forest floor and mineral soil to Douglas-fir seedlings ten years after fertilization. *Plant Soil* **2014**, *381*, 381–394. [CrossRef]

forests

MDPI

Article

Resource Utilization by Native and Invasive Earthworms and Their Effects on Soil Carbon and Nitrogen Dynamics in Puerto Rican Soils

Ching-Yu Huang [1,2,*], Grizelle González [3] and Paul F. Hendrix [1]

[1] Odum School of Ecology, University of Georgia, Athens, GA 30602, USA; hendrixp@uga.edu
[2] Department of Biology, University of North Georgia, Dahlonega, GA 30597, USA
[3] USDA Forest Service, International Institute of Tropical Forestry, Jardín Botánico Sur, 1201 Ceiba St., Río Piedras 00926, Puerto Rico; ggonzalez@fs.fed.us
* Correspondence: ching-yu.huang@ung.edu; Tel.: +1-706-867-2952

Academic Editors: Scott X. Chang and Xiangyang Sun
Received: 22 September 2016; Accepted: 6 November 2016; Published: 15 November 2016

Abstract: Resource utilization by earthworms affects soil C and N dynamics and further colonization of invasive earthworms. By applying ^{13}C-labeled *Tabebuia heterophylla* leaves and ^{15}N-labeled *Andropogon glomeratus* grass, we investigated resource utilization by three earthworm species (invasive endogeic *Pontoscolex corethrurus*, native anecic *Estherella* sp., and native endogeic *Onychochaeta borincana*) and their effects on soil C and N dynamics in Puerto Rican soils in a 22-day laboratory experiment. Changes of ^{13}C/C and ^{15}N/N in soils, earthworms, and microbial populations were analyzed to evaluate resource utilization by earthworms and their influences on C and N dynamics. *Estherella* spp. utilized the ^{13}C-labeled litter; however, its utilization on the ^{13}C-labeled litter reduced when cultivated with *P. corethrurus* and *O. borincana*. Both *P. corethrurus* and *O. borincana* utilized the ^{13}C-labeled litter and [15]C-labeled grass roots and root exudates. *Pontoscolex corethrurus* facilitated soil respiration by stimulating ^{13}C-labeled microbial activity; however, this effect was suppressed possibly due to the changes in the microbial activities or community when coexisting with *O. borincana*. Increased soil N mineralization by individual *Estherella* spp. and *O. borincana* was reduced in the mixed-species treatments. The rapid population growth of *P. corethrurus* may increase competition pressure on food resources on the local earthworm community. The relevance of resource availability to the population growth of *P. corethrurus* and its significance as an invasive species is a topic in need of future research.

Keywords: carbon and nitrogen mineralization; invasive earthworms; Luquillo mountains; microbial respiration; Puerto Rico; stable isotope; tropics

1. Introduction

Invasive earthworms have caused significant effects on local biota and ecosystem processes (such as nutrient dynamics) in the invaded areas, e.g., European Lumbricids in North America [1–3]. Population declines of native earthworms, particularly in remote and non-fragmented forests, have contributed to a result of competitive exclusion by expanding invasive earthworm populations [2,4,5]. Lachnicht et al. [6] observed that invasive *Pontoscolex corethrurus* (Müller, 1856) earthworms, when incubated with native *Estherella* sp., utilized different N resources, possibly avoiding direct competition on food resource. Winsome et al. [7] found that invasive *Aporrectodea trapezoides* (Dugès, 1828) lost its competition advantage when co-existing with native *Argilophilus marmoratus* (Eisen, 1893) in the resource-poor habitat of a Californian grassland. Interactions between native and invasive earthworms varied with resource utilization of earthworm species and resource availability [6,7]. Earthworms are categorized into three ecological groups, epigeic, endogeic, and anecic, based on their preferences on space and food resources [8]. Epigeic earthworms mainly consume leaf litter (and microbial populations colonizing on it) and inhabit the litter layer, while endogeic earthworms occupy mineral soils and use soil organic matter as their main food resources. Anecic earthworms utilize mainly leaf litter but with the ability to build burrows deep in the soil [8]. Earthworms with same feeding strategies are expected to evolve stronger competitive interactions because they share the same food resources [2,9,10]. Hence, resource utilization of earthworms could serve as a determinant for the success of earthworm invasions and its effects on the native earthworm community [7].

Earthworm invasions have significantly altered nutrient dynamics (e.g., carbon (C) and nitrogen (N)) in invaded soils [1,11,12]. A mixed-species of European Lumbricid earthworm assemblage has been documented to lessen organic layers and relocate leaf litter and humus fragments (C) into the deeper mineral soils, as well as to cause an increase of N loss in the soil adjacent to plant roots in the temperate forests of North America [1]. The effects of earthworms on soil C and N dynamics may vary with the feeding strategies of earthworms and composition of earthworm assemblages [13]. For example, epigeic earthworms may have stronger effects on nutrient fluxes between leaf litter layers and microbial populations that colonized on it (detritusphere) from their comminution and digestion of the leaf litter substrate [1,11,12]. Endogeic/anecic earthworms, on the other hand, may play a significant role in regulating nutrient dynamics in mineral soil and plant root zones (rhizosphere) by their consumption of soil organic matter and root exudates (and depositions) and their active burrowing activity [14–16]. In an area inhabited by a mixture of earthworms (either different feeding strategies or native co-existing with invasive worms), whether earthworm effects on soil nutrient dynamics can be explained by a summation of individual earthworm effects or disproportionally dominated by one aggressive earthworm species is a topic of interest, yet still in need of more research.

Stable isotope ^{13}C and ^{15}N techniques, including ^{13}C- and ^{15}N-labeled plant materials and a natural abundance of ^{13}C and ^{15}N isotopes, have recently provided invaluable information for studying earthworm feeding strategies and their effects on soil C and N dynamics [6,17–20]. For example, Hendrix et al. [17] suggested an inter-specific competition for N resources based on their observation of overlapped natural abundance ^{15}N in both *Estherella* sp. and *P. corethrurus* in a lower altitude tabonuco forest, Puerto Rico. Neilson et al. [18] found that a natural abundance of ^{13}C and ^{15}N in earthworms can be used to assess the availability and diversity of food resources in the environment. With the application of ^{13}C- and ^{15}N-enriched plant materials, how earthworms utilize different type of food resources and the corresponding effects on soil C and N dynamics can be evaluated by tracking changes of δ^{13}C and δ^{15}N associated with ^{13}C and ^{15}N-labeled plant materials in soils, earthworms, and the microbial populations. In this study, we applied ^{13}C-labeled *Tabebuia heterophylla* (DC.) Britton leaves and ^{15}N-labeled *Andropogon glomeratus* (Walter) Britton, Sterns, & Poggenb. grass to investigate resource utilization of three earthworm species from Puerto Rico (invasive *Pontoscolex corethrurus*, native *Estherella* spp., and native *Onychochaeta borincana* (Borges, 1994) and their effects on soil C and N dynamics in Puerto Rican soils.

Pontoscolex corethrurus has invaded multiple habitats in Puerto Rico, in contrast to the restricted distribution of the native earthworms in mature forests [21,22]. Competition pressure from invasive *P. corethrurus* to native earthworms has been suggested to be responsible for the absence of native earthworms in most disturbed areas, i.e., pasture and young forests [22–24]. Lachnicht et al. [6] observed that endogeic *P. corethrurus* and anecic *Estherella* sp. showed resource partitioning (in terms of space and food) to avoiding direct competition in a 19-day laboratory experiment. The interactions observed between *P. corethrurus* and *Estherella* sp. have also caused differential influences on soil C and N mineralization [6]. In this study, we investigated feeding strategies of endogeic *P. corethrurus*, anecic *Estherella* sp., and endogeic *Onychochaeta borincana* (single-species earthworm treatments) on ^{13}C-labeled *Tabebuia* leaves and ^{15}N-labeled *Andropogon* grass. Changes in resource utilization of individual earthworm species would be evaluated by comparing earthworm tissue ^{13}C and ^{15}N of single-species earthworm treatments to those of mixed-species earthworm treatments (co-existed with other anecic/endogeic earthworms). Influences of individual earthworm species and inter-specific earthworm interactions on soil C and N dynamics would be assessed by tracking the changes of ^{13}C and ^{15}N in soils, earthworms, and microbial populations in single- and mixed-species earthworm treatments. Anecic *Estherella* spp. was expected to utilize more ^{13}C-labeled *Tabebuia* leaves, as compared with endogeic *O. borincana* and endogeic *P. corethrurus*. Given that *P. corethrurus* is believed to exhibit flexible feeding behaviors and enhance soil mineralization [6,17], we expected that *P. corethrurus* would utilize more leaf litter (detritusphere) than plant roots (rhizosphere) resources, when incubated with endogeic *O. borincana*, to avoid competition with *O. borincana*. Higher population growth would be observed in a *P. corethrurus* population, which would enhance soil C and N mineralization. However, the presence of anecic *Estherella* sp. and endogeic *O. borincana* would weaken enhanced soil mineralization caused by *P. corethrurus*.

2. Materials and Methods

2.1. Experiment Design and Setup

The experiment was conducted at Sabana Field Research Station in Luquillo, Puerto Rico, from November to December 2006. A total of 60 soil mesocosms (Polyvinyl Chloride (PVC) material, 11 cm in diameter and 20 cm in depth) were set up with 15-cm-deep field soils with the bottoms sealed with a 1 mm mesh fiberglass window screen. Experimental treatments included (1) control mesocosms ($n = 4$, no earthworms; Control) with isotope-labeled *Tabebuia* litter and *Andropogon glomeratus* grass; and (2) seven earthworm treatments (each treatment: $n = 4$) with isotope-labeled *Tabebuia* litter and *A. glomeratus* grass: single and mixed earthworm treatments (two- and three-species earthworm combination; see below). Four soil mesocosms with no isotope-labeled plant materials and no earthworms (Soil; $n = 4$), four soil mesocosms with ^{15}N-labeled grass plants (Grass; $n = 4$), and four soil mesocosms with ^{13}C-labeled leaf litter (Litter; $n = 4$) were also analyzed as reference data to evaluate the efficiency of ^{13}C- and ^{15}N-labeled methods.

Experimental soil was collected from the forest at the Bisley Experimental Watersheds (BEW) in the Luquillo Mountains (18°18′ N; 65°50′ W). The forest at BEW is mostly dominated by a secondary growth of tabonuco trees, and its soils are clayey and well weathered Ultisols. Detailed description of BEW can be found in Scatena [25]. The collected soils were separated by three depths of 0–5, 5–10, and 10–15 cm to air-dry for 48 h and sieved through a 5 mm mesh size sieve to exclude plant roots, rocks, cocoons, and earthworms. Three depths of air-dry soils were used to set up the 0–5, 5–10, and 10–15 cm depth in the mesocosms. Total soil C and N in 0–5 cm were 3.96 ± 0.05% and 0.37%, respectively. Three *Andropogon glomeratus* seedlings (ca. 8 cm tall), the common grass species in Puerto Rico, were transplanted into each control and earthworm mesocosm a week before the beginning of the experiment. The *Andropogon* grass leaves were brushed with 2 atom % ^{15}N-urea solution every day to establish ^{15}N-labeled plant roots and root-derived substrates (the rhizosphere) during the experiment [26]. Seedlings of *Tabebuia heterophylla*, one of the common, native woody species (Family: Bignoniaceae) in Puerto Rico, were incubated in a growth chamber with pulse injection of 99 atom % ^{13}CO$_2$ to acquire ^{13}C-labeled *Tabebuia* leaves through photosynthesis cycles during June–July 2006. After labeling procedures, *Tabebuia* senescent leaves were collected, air-dried for 48 hours, and then shredded into 1 cm^2 pieces (δ^{13}C varied from 385‰ to 804‰). A total of 3.7 g of dry ^{13}C-labeled *Tabebuia* litter (calculated based on field litterfall data) was applied to the soil surface of each control and earthworm mesocosm to establish ^{13}C-labeled litter and related microbial populations (detritusphere).

2.2. Earthworm Species and Collection

Three earthworm species from Puerto Rico were chosen for this experiment. Two native species, *Estherella* spp. and *Onychochaeta borincana*, were collected from the BEW forests (18.5°18′51.893″ N, 65.5°44′41.694″ W) and a riparian forest in Almirante Norte (18°41′ N, 65°38′ W; alluvial soil) in Puerto Rico [27], respectively; while *Pontoscolex corethrurus* was collected from the pasture at the Sabana Field Research Station (18°18′ N, 65°50′ W) in the town of Luquillo, Puerto Rico. Anecic *Estherella* spp. has dark pigmentation on the dorsal side and stays in leaf litter and upper soil layers. Endogeic *O. borincana* has pale coloration and stays in the subsoil layer. The invasive earthworm species, *P. corethrurus*, as an endogeic species, is the dominant peregrine earthworm that has colonized most habitats of Puerto Rico [6,17]. Before introducing into the earthworm mesocosms, gut contents of all earthworms were voided for 24 h, and their fresh biomass was recorded as the initial biomass data at the beginning of the experiment. Earthworms were introduced to assigned single or mixed earthworm treatments as followed: single species treatments—*O. borincana* only (O; 4 worms), *Estherella* spp. only (E; 4 worms), and *P. corethrurus* only (P; 4–5 worms); two-species mixed treatments—*Estherella* and *P. corethrurus* (E + P; 3 worms from each species), *Estherella* and *O. borincana* (E + O; 4 *Estherella* worms and 3 *O. borincana* worms), and *O. borincana* and *P. corethrurus* (O + P; 3 worms from each species); and three-species mixed treatments—*Estherella*, *P. corethrurus*, and *O. borincana* (E + P + O; 2 *Estherella*, 2 *O. borincana*, and 3 *P. corethrurus*). Four soil mesocosms were assigned to the control and each earthworm treatment as experimental replicates. The earthworm species were introduced into the experimental mesocosms following the order of *O. borincana*, *Estherella* spp., and *P. corethrurus*. Average fresh biomass of earthworms for each earthworm treatment is listed in Table 1. Each mesocosm was watered with 35 mL of water every day to maintain soil moisture during the 22-day experiment. The mesocosms were rotated randomly every week during the experiment.

Table 1. Average fresh biomass of *Estherella* spp. (E), *Onychochaeta borincana* (O), and *Pontoscolex corethrurus* (P) earthworms introduced into different earthworm mesocosm (g per mesocosm).

Variables	Single species (E, O, P)	E + O	E + P	O + P	E + O + P
Earthworm Treatments					
Estherella spp.					
Fresh weight (before)	5.3 (0.5)	4.2 (0.6)	3.2 (0.4)	n/a	2.2 (0.2)
Fresh weight (after)	4.6 (1.2)	3.9 (1.1)	2.7 (0.6)	n/a	2.3 (0.4)
Onychochaeta borincana					
Fresh weight (before)	4.9 (0.6)	3.6 (0.5)	n/a	2.7 (0.6)	2.2 (0.4)
Fresh weight (after)	2.9 (1.8)	2.4 (0.6)	n/a	2.3 (0.7)	1.6 (0.3)
Pontoscolex corethrurus					
Fresh weight (before)	2.0 (0.3)	n/a	1.5 (0.3)	1.4 (0.1)	1.2 (0.1)
Fresh weight (after)	1.8 (0.4)	n/a	1.5 (0.1)	1.7 (0.2)	1.4 (0.2)

Capital letters (E, O, and P) represent treatments with different earthworm assemblages. Single-species: E = *Estherella* spp.; O = *Onychochaeta borincana*; P = *Pontoscolex corethrurus*. Two-species: E + O = *Estherella* spp. and *O. borincana* assemblage; E + P = *Estherella* spp. and *P. corethrurus* assemblage; O + P = *O. borincana* and *P. corethrurus* assemblage. Three-species: E + O + P = *Estherella* spp., *O. borincana*, and *P. corethrurus* assemblage. Value is shown as mean (S.D.) (*n* = 4) at the beginning of the experiment (before) and after the 22-day experiment (after). "n/a" indicates the particular earthworm species was not introduced into the corresponding experimental mesocosm.

2.3. Experiment Responding Variables

2.3.1. Soil CO_2 and ^{13}C-CO_2

At Day 21 of the experiment, soil carbon dioxide (CO_2) evolution was collected using the alkali absorption technique [28]. At each sampling, a circular area (5 cm in diameter) in between the center and the edge of the mesocosm was randomly chosen for each mesocosm, and the *Tabebuia* litter within was gently removed to the side. A PVC chamber (10 cm tall and 5 cm in diameter) was inserted 1 cm into the soil surface of each mesocosm with a scintillation vial containing 10 mL of a 1 mol/L NaOH solution placed inside each PVC chamber. The chamber was sealed with plastic wrap and aluminum foil on the top for soil CO_2 absorption. Five NaOH solution vials (control) were kept closed during the 24 h absorption, except to open only at the beginning and the end of absorption to assess sampling contamination. Twenty-four hours later, each alkali solution was removed from the chamber, and 2 mL of 1 mol/L $BaCl_2$ was added to form $BaCO_3$ precipitate. Total CO_2 trapped by alkali solution was determined by titration with 1 mol/L HCl to reach a pH neutral point (phenolphthalein endpoint) [28]. BaCO$_3$ precipitate from each sample was air dried and packed in tin capsules for ^{13}C-CO$_2$ analysis.

2.3.2. The Remaining Mass of the *Tabebuia* Litter

Soil mesocosms were deconstructed at Day 22 to collect final data of the experiment. *Tabebuia* litter was carefully picked up and oven-dried at 60 °C for 48 h. The litter samples were ground, and a subsample of 0.5 g litter was burned at 550 °C for 4 h to obtain ash-free dry matter (AFDM) data. The data were used to calculate the remaining litter mass at the end of the experiment.

2.3.3. Survivorship, Growth, and the ^{13}C and ^{15}N Composition of Earthworms

The number of earthworms that survived at the end of the experiment was used to determine earthworm survivorship. All earthworms were put into separate containers to void their gut contents for 24 h. Final fresh biomass was recorded after gut-voiding. Earthworms were killed by dipping in boiling water for 3 seconds. One-third of the earthworm body (tail part) was cut and rinsed with deionized water with the gut content removed. Earthworm tissue was then freeze-dried and ground. Two milligrams of earthworm tissue was packed into a tin capsule and analyzed by dry combustion on a Carlo Erba NA1500 CN analyzer (Thermo Scientific, Waltham, MA, USA) for earthworm total C, N, and ^{13}C and ^{15}N.

2.3.4. Soil and Soil Microbes

Soil was separated into three soil depths, 0–5, 5–10, and 10–15 cm. Ten grams of soil from each depth was oven-dried at 105 °C for 48 h to calculate soil moisture. Subsamples of soils were ground and packed into tin capsules (ca. 20 mg) for total soil carbon (C) and nitrogen (N) and isotopic analysis (^{13}C and ^{15}N) by dry combustion on a Carlo Erba NA1500 CN analyzer. Two sets of 20 g 0–5 cm soils were extracted with 60 mL of a 0.5 mol/L potassium sulfate (K$_2$SO$_4$) solution (3:1 solution to soil mass ratio) for soil microbial biomass analysis by using the fumigation–extraction method [29,30]. Total microbial biomass C and ^{13}C was analyzed from K$_2$SO$_4$-extracted samples using an OI analytical TIC/TOC analyzer (Shimaduz, Kyoto, Japan) coupled with a Thermo-Finnigan Delta Plus Isotope Ratio Mass Spectrometer (IRMS) (Thermo Scientific, Waltham, MA, USA). The persulfate digestion method was adapted to obtain microbial N data [31]. The K$_2$SO$_4$-extracted samples and persulfate digestion samples were analyzed with an Alpkem nitrogen autoanalyzer (OI analytical, College Station, TX, USA). Dissolved inorganic N (DIN; NH$_4^+$-N and NO$_3^-$-N) was calculated from a non-fumigated K$_2$SO$_4$ extract. Microbial biomass N (MBN) was calculated from the difference between total persulfate nitrogen from fumigated and non-fumigated samples. Total persulfate nitrogen from fumigated samples was used to determine total dissolved nitrogen (TDN).

Delta ^{15}N data for each portion (DIN, MBN, and TDN) were obtained by running the samples through the isotope diffusion method [32]. The δ^{13}C/ δ^{15}N value is calculated based on the measure isotope ratios between the samples and the standard:

$$\delta\,^{13}\text{C}\,(\text{\textperthousand}) = \left((R_{\text{sample}} - R_{\text{standard}}) \,/\, R_{\text{standard}} \right) \times 10^3 \tag{1}$$

$$\delta\,^{15}\text{N}\,(\text{\textperthousand}) = \left((R_{\text{sample}} - R_{\text{standard}}) \,/\, R_{\text{standard}} \right) \times 10^3 \tag{2}$$

where δ^{13}C (δ^{15}N) unit is the parts per thousand and R is the mass ratio of ^{13}C/^{12}C (^{15}N/^{14}N) in the sample and standard [33].

For DIN (K$_2$SO$_4$) extracts, KCl was added along with MgO and Devarda's alloy to increase the ionic strength of the solution. For microbial N and TDN (persulfate digests) samples, 10 M NaOH was added to raise pH (>13) of the solution instead. Pairs of glass filter disks (Whatman GF/D) were prepared by baking in a muffle furnace at 500 °C for 4 h. They were acidified with 35 μL of 2M H$_2$SO$_4$

and then wrapped with Teflon tape. The Teflon-filter packages were incubated in the solutions for 6 days. After the incubation, the packages were dried over concentrated H_2SO_4 for at least 48 h, then packed in silver capsules for dry combustion on a Carlo Erba NA1500 CN analyzer and IRMS for total N and ^{15}N data.

2.4. Statistic Analysis

The differences of litter remaining mass (data transformed), soil respiration (C-CO$_2$, ^{13}C-CO$_2$, and δ^{13}C), total C/N concentration, atom percentage of ^{13}C/^{15}N, and δ^{13}C/ δ^{15}N in soil, microbial biomass, and earthworm tissue, dissolved inorganic nitrogen (DIN), DIN-^{15}N, and total dissolved nitrogen (TDN) between control and earthworm treatments were analyzed by a one-way ANOVA procedure (a generalized linear model (GLM) was used if data were not balanced) in SAS statistical software [34]. A GLM was also used to compare the differences of earthworm biomass and survivorship (data transformed) between earthworm treatments. If significantly different, Tukey's HSD method was applied for the comparisons between treatments. Student's *t*-test and GLM were applied to compare worm ^{13}C and ^{15}N differences between earthworm species in two-species and three-species mixed earthworm treatments, respectively. The significance level was set as $\alpha = 0.05$.

3. Results

3.1. Litter Mass Loss and Soil C and N

The remaining mass of the *Tabebuia* litter (ash-free dry weight), ranging from 21.6% in the control treatment to 45.3% in the E + O earthworm mesocosms, was not significantly different between control and earthworm treatments (data transformed; GLM, $F_{7, 31} = 2.1$, $p = 0.08$). At the end of the experiment, soil total C and total N concentrations were not significantly different between the initial soil, the soil samples (with no worms), and earthworm treatments (soil C: $F_{8, 27} = 1.0$, $p = 0.43$; soil N: $F_{8, 27} = 0.2$, $p = 0.9$; Table 2). Soil carbon in the O + P earthworm mesocosms showed a significantly higher soil ^{13}C percentage (1.0786 ± 0.002%) and soil ^{15}N percentage (0.36915 ± 0.00072%), as compared with those in the initial soil (soil ^{13}C = 1.0753 ± 0.0001% and soil ^{15}N = 0.36815 ± 0.00005%) and the control soil (soil ^{13}C = 1.0752 ± 0.0002% and soil ^{15}N = 0.36810 ± 0.00005%) (both $p < 0.01$; Table 2). Soil C and N from the earthworm treatments showed stronger δ^{13}C (average = −25.9 ± 0.9‰) and δ^{15}N (average = 6.5 ± 1.0‰) signatures as compared with the control soil (δ^{13}C= −27.9 ± 0.2‰ and δ^{15}N = 4.5 ± 0.1‰).

Table 2. Total soil carbon (mg C/ g soil) and nitrogen (µg N/ g soil), atom percentages of ^{13}C and ^{15}N (%) and delta ^{13}C ($\delta^{13}C$; ‰) and delta ^{15}N ($\delta^{15}N$; ‰) from the initial soil samples (no isotope-labeled materials and no worms at Day 0; Initial), control treatment (no isotope-labeled materials and no worms at Day 22; Soil) and earthworm treatments at the end of the 22-day mesocosm experiment with Puerto Rican soils.

Variables	Initial	Soil	E	O	P	E + O	E + P	O + P	E + O + P	Statistics
Soil Carbon										
Total C	39.6 (0.5)	43.5 (2.5) a	42.5 (3.7)	43.8 (2.0)	41.0 (1.6)	42.9 (3.8)	42.1 (2.7)	42.0 (1.5)	42.1 (2.2)	$F_{8,27} = 1.0; p = 0.43$
Atom ^{13}C (%)	1.0753 a (0.0001)	1.0752 a (0.0002)	1.0767 bc (0.0003)	1.0771 abc (0.0009)	1.0767 ac (0.0004)	1.0768 abc (0.0002)	1.0776 bc (0.0005)	1.0786 b (0.0020)	1.0769 abc (0.0007)	$F_{8,27} = 7.2; p < 0.0001$
$\delta^{13}C$	−27.8 a (0.1)	−27.9 a (0.2)	−25.8 bc (0.3)	−26.1 abc (0.9)	−26.5 ab (0.4)	−26.4 abc (0.2)	−25.6 bc (0.4)	−24.8 c (1.8)	−26.3 abc (0.6)	$F_{8,27} = 7.2; p < 0.0001$
Soil Nitrogen										
Total N	371.5 (2.3)	367.9 (19.0) a	375.0 (19.5) ab	370.8 (21.4) ab	364.3 (11.0) ab	369.1 (19.6) ab	362.3 (19.7) ab	367.4 (11.4) b	365.6 (8.3) ab	$F_{8,27} = 0.2; p = 0.9$
Atom ^{15}N (%)	0.36815 a (0.00005)	0.36810 a (0.00005)	0.36885 ab (0.00042)	0.36858 ab (0.00012)	0.36866 ab (0.00036)	0.36889 ab (0.00025)	0.36885 ab (0.00004)	0.36915 b (0.00072)	0.36882 ab (0.00033)	$F_{8,27} = 4.3; p = 0.002$
$\delta^{15}N$	4.6 (0.1) a	4.5 (0.2) a	6.5 ab (1.1)	5.8 ab (0.3)	6.0 ab (1.0)	6.6 ab (0.7)	6.5 ab (0.1)	7.3 b (2.0)	6.4 ab (0.9)	$F_{8,27} = 4.3; p = 0.002$

Capital letters (E, O, and P) represent treatments with different earthworm assemblages. Single-species: E = Estherella spp.; O = Onychochaeta borincana; P = Pontoscolex corethrurus. Two-species: E + O = Estherella spp. and O. borincana assemblage; E + P = Estherella spp. and P. corethrurus assemblage; O + P = O. borincana and P. corethrurus assemblage. Three-species: E + O + P = Estherella spp., O. borincana and P. corethrurus assemblage. Value is shown as mean (S.D.) (n = 4). Statistics shows the statistical results (F ratios and p values) from one-way ANOVA (GLM for unbalanced data). Different letters indicate significant difference among earthworm treatments (Tukey's HSD, p < 0.05).

3.2. Earthworm Populations

3.2.1. Earthworm Biomass and Survivorship

Average fresh biomass of the surviving earthworms for each earthworm treatment at the end of the 22-day mesocosm experiment is listed in Table 1. The endogeic earthworm, *Onychochaeta borincana*, showed significantly lower survivorship (71.8 ± 25.0%) than the other two earthworm species (epi-endogeic *Pontoscolex corethrurus*: 96.9 ± 8.3%; anecic *Estherella* spp.: 93.8 ± 13.0%) (data-transformed, GLM, $F_{2,47} = 9.56$, $p = 0.0003$). However, the survivorship of individual earthworm species did not significantly differ between the single or the mixed-earthworm treatments (GLM, *Estherella*: $F_{3,15} = 1.6$, $p = 0.2$; *O. borincana*: $F_{3,15} = 0.4$, $p = 0.8$; *P. corethrurus*: $F_{3,15} = 0.7$, $p = 0.6$), nor did the biomass changes (%) of individual earthworm species (*Estherella*: $F_{3,15} = 0.6$, $p = 0.7$; *O. borincana*: $F_{3,15} = 1.1$, $p = 0.4$; *P. corethrurus*: $F_{3,15} = 2.1$, $p = 0.2$). A total of eight *P. corethrurus* were reproduced during the 22-day mesocosm experiment.

3.2.2. Tissue C/^{13}C and N/^{15}N in Native *Estherella* spp.

Percentage of tissue biomass C of anecic *Estherella* spp. showed no significant difference between its single species treatment (cultivated alone; tissue C = 46.3 ± 0.3%) and the mixed-species treatments (cultivated with *O. borincana* and/or *P. corethrurus*; tissue C (%) = 45.7%–47.2%) ($F_{3,39} = 2.2$, $p = 0.1$; Table 3). However, *Estherella* spp. when cultivated alone was found to have significantly higher ^{13}C enrichment (as in δ^{13}C and atom percentage of ^{13}C) as compared with the mixed-species treatments (for δ^{13}C: E + P and E + O + P mesocosms; $F_{3,39} = 2.0$, $p = 0.04$) (for tissue ^{13}C (%): E + P mesocosms; $F_{3,39} = 2.9$, $p = 0.047$) (Table 3). *Estherella* spp. did not show a significant difference in worm tissue N (%), δ^{15}N, and ^{15}N (%) between its single species and the mixed-species mesocosms (all $p > 0.4$; Table 3).

Table 3. Earthworm tissue total carbon (C) and nitrogen (N) percentages (%), atom percentages of ^{13}C and ^{15}N (%), and delta ^{13}C (δ^{13}C; ‰) and delta ^{15}N (δ^{15}N; ‰) in native earthworms *Estherella* spp. (E) and *Onychochaeta borincana* (O) at each earthworm mesocosm from different earthworm treatments at the end of the 22-day experiment with Puerto Rican soils.

Variables	Single species (E or O)	E + O	E + P	O + P	E + O + P	Statistics
Estherella spp.						
Total C (%)	46.3 (0.8)	45.7 (1.3)	46.2 (1.1)	n/a	47.2 (1.8)	$F_{3,39} = 2.2$; $p = 0.10$
Atom ^{13}C (%)	1.0805 a (0.0039)	1.0785 ab (0.0004)	1.0781 b (0.0006)	n/a	1.0788 ab (0.0007)	$F_{3,39} = 2.9$; $p = 0.047$
δ^{13}C	−23.0 a (3.5)	−24.8 ab (0.4)	−25.2 b (0.5)	n/a	−24.6 b (0.7)	$F_{3,39} = 2.9$; $p = 0.040$
Total N (%)	12.4 (0.5)	12.3 (0.8)	12.2 (1.0)	n/a	12.7 (0.4)	$F_{3,39} = 0.8$; $p = 0.5$
Atom ^{15}N (%)	0.3690 (0.0002)	0.3688 (0.0003)	0.3689 (0.0004)	n/a	0.3688 (0.0002)	$F_{3,39} = 1.0$; $p = 0.4$
δ^{15}N	6.8 (0.6)	6.2 (0.9)	6.6 (1.0)	n/a	6.5 (0.6)	$F_{3,39} = 1.0$; $p = 0.4$
Onychochaeta borincana						
Total C (%)	46.0 (1.2)	46.6 (1.5)	n/a	46.6 (1.3)	46.5 (1.2)	$F_{3,25} = 0.4$; $p = 0.8$
Atom ^{13}C (%)	1.0823 (0.0046)	1.0812 (0.0016)	n/a	1.0845 (0.0102)	1.0812 (0.0006)	$F_{3,25} = 0.5$; $p = 0.7$
δ^{13}C	−21.4 (4.2)	−22.4 (1.5)	n/a	−19.3 (9.4)	−22.3 (0.5)	$F_{3,25} = 0.5$; $p = 0.7$
Total N (%)	11.8 (0.8)	12.5 (0.7)	n/a	12.3 (0.7)	12.4 (0.5)	$F_{3,25} = 1.8$; $p = 0.2$

Table 3. *Cont.*

Variables	Single species (E or O)	E + O	E + P	O + P	E + O + P	Statistics
		Earthworm Treatments				
Atom^{15}N (%)	0.3693 (0.0013)	0.3694 (0.0006)	n/a	0.3705 (0.0035)	0.3697 (0.0004)	$F_{3,25} = 0.4; p = 0.7$
δ^{15}N	8.69 (3.6)	8.2 (1.6)	n/a	11.0 (9.5)	8.9 (1.0)	$F_{3,25} = 0.4; p = 0.7$

Capital letters (E, O, and P) represent treatments with different earthworm assemblages. Single-species: E = *Estherella* spp.; O = *Onychochaeta borincana*; P = *Pontoscolex corethrurus*. Two-species: E + O = *Estherella* spp. and *O. borincana* assemblage; E + P = *Estherella* spp. and *P. corethrurus* assemblage; O + P = *O. borincana* and *P. corethrurus* assemblage. Three-species: E + O + P = *Estherella* spp., *O. borincana* and *P. corethrurus* assemblage. Value is shown as mean (S.D.). Statistics shows the statistical results (*F* ratios and *p* values) from one-way ANOVA (GLM for unbalanced data). Different letters indicate significant difference among earthworm treatments (Tukey's HSD, $p < 0.05$).

3.2.3. Tissue C/^{13}C and N/^{15}N in Native *O. Borincana*.

For endogeic *O. borincana*, there was no significant difference in worm tissue C and N (%), δ^{13}C and δ^{15}N signatures, and tissue ^{13}C and ^{15}N (%) between its own single species and the mixed-species mesocosms (all $p > 0.2$; see Table 3).

3.2.4. Tissue C/^{13}C and N/^{15}N in Invasive *P. corethrurus*

Invasive *P. corethrurus* earthworms did not show significant differences in worm tissue C and N (%), δ^{13}C and δ^{15}N enrichments, and tissue ^{13}C and ^{15}N (%) between its single species and the mixed-species mesocosms (all $p > 0.2$; Table 4). However, the juvenile *P. corethrurus* reproduced during this 22-day mesocosm experiment did show significant lower tissue C (41.0 ± 4.7%) and N percentages (9.2 ± 2.0%), as compared with the adult *P. corethrurus* worms (tissue C (%): $F_{4,52} = 3.9$, $p = 0.007$; tissue N (%): $F_{4,52} = 6.0$, $p < 0.001$) (Table 4). Juvenile *P. corethrurus* worms also showed lower enrichment of δ^{13}C ($-24.3 ± 1.4‰$) and tissue ^{13}C (1.0791 ± 0.0011%), as compared with the adult *P. corethrurus* worms (δ^{13}C: $F_{4,52} = 4.7$, $p = 0.002$; tissue ^{13}C (%): $F_{4,52} = 4.8$, $p = 0.002$) (Table 4). There was a significantly higher enrichment of ^{15}N (as in δ^{15}N = 7.6 ± 0.9‰ and an atom percentage of ^{15}N = 0.3692 ± 0.0003%), as compared with the adult *P. corethrurus* worms (δ^{15}N: $F_{4,52} = 7.2$, $p < 0.001$; atom percentage ^{15}N: $F_{4,52} = 7.2$, $p < 0.001$) (Table 4).

Table 4. Earthworm tissue total carbon (C) and nitrogen (N) percentages (%), atom percentages of ^{13}C and ^{15}N (%) and delta ^{13}C (δ^{13}C; ‰) and delta ^{15}N (δ^{15}N; ‰) in native earthworms *Estherella* spp. (E) and *Onychochaeta borincana* (O) at each earthworm mesocosm from different earthworm treatments at the end of the 22-day experiment with Puerto Rican soils.

Variables	Single Species (P)	PJ	E + P	O + P	E + O + P	Statistics
		Earthworm Treatments				
Pontoscolex corethrurus						
Total C (%)	47.0 (1.0) a	41.0 (4.7) b	44.7 (6.1) ab	46.3 (1.6) a	46.9 (0.3) a	$F_{4,52} = 3.9; p = 0.007$
Atom^{13}C (%)	1.0809 a (0.0008)	1.0791 b (0.0011)	1.0812 a (0.0009)	1.0813 a (0.0012)	1.0812 a (0.0005)	$F_{4,52} = 4.8; p = 0.002$
δ^{13}C	-22.6 a (0.8)	-24.3 b (1.4)	-22.4 a (0.8)	-22.2 a (1.1)	-22.3 a (0.5)	$F_{4,52} = 4.7; p = 0.002$
Total N (%)	11.9 (0.5) a	9.2 (2.0) b	11.3 (1.5) a	11.4 (1.1) a	11.9 (0.4) a	$F_{4,52} = 6.0; p < 0.001$
Atom^{15}N (%)	0.3686 a (0.0001)	0.3692 b (0.0003)	0.3686 a (0.0002)	0.3687 a (0.0003)	0.3686 a (0.0001)	$F_{4,52} = 7.2; p < 0.001$
δ^{15}N	5.9 (0.4) a	7.6 (0.9) b	5.9 (0.5) a	6.1 (0.9) a	5.9 (0.3) a	$F_{4,52} = 7.2; p < 0.001$

Capital letters (E, O, and P) represent treatments with different earthworm assemblages. Single-species: E = *Estherella* spp.; O = *Onychochaeta borincana*; P = *Pontoscolex corethrurus*. Two-species: E + O = *Estherella* spp. and *O. borincana* assemblage; E + P = *Estherella* spp. and *P. corethrurus* assemblage; O + P = *O. borincana* and *P. corethrurus* assemblage. Three-species: E + O + P = *Estherella* spp., *O. borincana* and *P. corethrurus* assemblage. Value is shown as mean (S.D.). Statistics shows the statistical results (*F* ratios and *p* values) from one-way ANOVA (GLM for unbalanced data). Different letters indicate significant difference among earthworm treatments (Tukey's HSD, $p < 0.05$).

3.3. Microbial Biomass Carbon and Soil Respiration

There was no significant difference in microbial biomass carbon (MBC) and MBC-^{13}C between the soil (MBC = 741.3 $\pm$ 103.0 ug C g^{-1} soil; MBC-^{13}C = 8.0 $\pm$ 1.1 ug C g^{-1} soil), control (MBC = 332.1 $\pm$ 183.2 ug C g^{-1} soil; MBC-^{13}C = 3.6 $\pm$ 1.3 ug C g^{-1} soil), and earthworm treatments (MBC ranged from 340.1 to 532.1 ug C g^{-1} soil; MBC-^{13}C from 3.7 to 7.2 ug C g^{-1} soil) (MBC: $F_{10, 29}$ = 1.3, p = 0.26; MBC-^{13}C: $F_{10, 29}$ = 1.3, p = 0.27; Table 5). Microbial biomass ^{13}C (%) was significantly higher in the control treatments, as compared with those in the Soil Only or Grass mesocosms ($F_{10, 29}$ = 2.8, p = 0.015; Table 5), which suggested the microbes utilized and incorporated the ^{13}C-labeled litter into their biomass. The microbial biomass δ^{13}C enrichment from P (-28.8 ± 3.4‰), O + P (-28.8 ± 3.4‰), and E + O + P (-29.1 ± 2.5‰) treatments were significantly higher than the Soil Only treatment (-36.1 ± 1.4‰) ($F_{10, 29}$ = 3.3, p = 0.006; Table 5).

At the end of the experiment (Day 21), soil respiration C-CO$_2$ and ^{13}C-CO$_2$ (%) from the control (soil with both ^{13}C- and ^{15}N-labeled materials but no worms) and the earthworm treatments were significantly higher than the Soil Only and Grass treatments (both p < 0.0001; Table 5). This suggested that the input of ^{13}C-labeled leaf-litter and earthworms facilitated microbial respiration. However, different earthworm treatments showed differential effects on ^{13}C-CO$_2$ (%) evolved in the microbial respiration. The *P. corethrurus* earthworm treatment showed higher ^{13}C evolved from the microbial respiration (as in ^{13}C-CO$_2$ and δ^{13}C; Table 5) as compared with that from the O + P earthworm treatment at the end of the experiment (both p < 0.0001; Table 5).

Table 5. Microbial biomass total carbon (MBC, µg C / g soil), carbon-^{13}C (MBC-^{13}C; µg ^{13}C / g soil), atom percentage of ^{13}C (%), and soil delta ^{13}C (δ^{13}C; ‰), soil respiration C-CO$_2$ (µg C per day), atom percentage of ^{13}C-CO$_2$ (%), and delta ^{13}C-CO$_2$ (δ^{13}C; ‰) from the control treatments (Soil Only, soil with ^{15}N-labeled grass (Grass), soil with ^{13}C-labeled leaf litter (Litter), and Control (soil with both grass and leaf litter but no worms)) and earthworm treatments at the end of the 22-day mesocosm experiment with Puerto Rican soils.

Variables	Soil Only	Grass	Litter	Control	E	O	P	E + O	E + P	O + P	E + O + P	Statistics
Microbial biomass												
MBC	741.3 (103.0)	570.2 (167.9)	568.3 (166.3)	332.1 (183.2)	659.8 (119.2)	340.1 (115.3)	528.8 (91.4)	479.4 (119.2)	448.1 (110.3)	483.4 (199.5)	431.7 (224.4)	$F_{10,29} = 1.3; p = 0.26$
MBC-^{13}C	8.0 (1.1)	6.2 (1.3)	6.2 (1.1)	3.6 (1.3)	7.2 (1.3)	3.7 (1.2)	5.7 (1.0)	5.2 (1.3)	4.9 (1.1)	5.3 (2.2)	4.7 (1.1)	$F_{10,29} = 1.3; p = 0.27$
Atom^{13}C (%)	1.078 a (0.002)	1.079 a (0.0004)	1.083 ab (0.004)	1.09 b (0.009)	1.083 ab (0.003)	1.082 ab (0.003)	1.086 ab (0.004)	1.085 ab (0.002)	1.083 ab (0.002)	1.086 ab (0.004)	1.086 ab (0.003)	$F_{10,29} = 2.8; p = 0.015$
δ^{13}C	−36.1 a (1.4)	−34.9 ab (0.4)	−31.1 ab (3.5)	−30.6 ab (1.4)	−31.1 ab (2.3)	−32.0 ab (2.9)	−28.8 b (3.4)	−29.4 ab (1.8)	−31.5 ab (1.6)	−28.8 ab (3.4)	−29.1 b (2.5)	$F_{10,29} = 3.3; p = 0.006$
Variables	Soil Only	Grass	Litter	Control	E	O	P	E + O	E + P	O + P	E + O + P	Statistics
Soil respiration at day 21												
C-CO$_2$	1.73 a (0.79)	3.87 ab (0.91)	5.24 abc (0.92)	9.51 c (4.95)	8.01 bc (2.05)	6.35 abc (1.58)	9.32 bc (1.50)	9.50 c (0.99)	8.90 bc (1.09)	9.99 c (3.17)	7.88 bc (2.23)	$F_{10,32} = 5.2; p < 0.001$
^{13}C- CO$_2$ (%)	1.085 a (0.002)	1.088 a (0.004)	1.228 b (0.014)	1.223 b (0.012)	1.206 bc (0.020)	1.209 bc (0.004)	1.220 b (0.023)	1.205 bc (0.007)	1.215 bc (0.018)	1.183 c (0.008)	1.195 bc (0.010)	$F_{10,32} = 53.6; p < 0.0001$
δ^{13}C	−18.8 a (2.1)	−16.5 a (3.6)	111.9 b (13.1)	107.5 b (11.3)	91.7 bc (18.7)	94.3 bc (3.3)	104.9 b (21.2)	91.3 bc (6.9)	100.2 bc (16.9)	71.1 c (7.1)	81.8 bc (8.9)	$F_{10,32} = 53.5; p < 0.0001$

Capital letters (E, O, and P) represent treatments with different earthworm assemblages. Single-species: E = *Estherella* spp.; O = *Onychochaeta borincana*; P = *Pontoscolex corethrurus*. Two-species: E + O = *Estherella* spp. and *O. borincana* assemblage; E + P = *Estherella* spp. and *P. corethrurus* assemblage; O + P = *O. borincana* and *P. corethrurus* assemblage. Three-species: E + O + P = *Estherella* spp., *O. borincana* and *P. corethrurus* assemblage. Value is shown as mean (S.D.) (*n* = 4, except data with: *n* = 3). Statistics shows the statistical results (*F* ratios and *p* values) from one-way ANOVA (GLM for unbalanced data). Different letters indicate significant difference among earthworm treatments (Tukey's HSD, *p* < 0.05).

Table 6. Soil microbial total nitrogen (MBN; μg N/g soil), atom percentage of ^{15}N (MBN-^{15}N; %), and delta^{15}N (δ^{15}N; ‰) signature and dissolved inorganic nitrogen (DIN; μg N/g soil), and atom percentage of ^{15}N (DIN-^{15}N; %) in DIN from the control treatments (Soil Only, soil with ^{15}N-labeled grass (Grass), soil with ^{13}C-labeled leaf litter (Litter), and Control (soil with grass and leaf litter but no worms)) and earthworm treatments at the end of the 22-day mesocosm experiment with Puerto Rican soils. See Table 5 for definitions of abbreviations.

Variables	Soil Only	Grass	Litter	Control	Earthworm Treatments E	O	P	E + O	E + P	O + P	E + O + P	Statistics
Microbial biomass												
MBN	124.6 (30.1)	96.8 (26.8)	110.2 (25.5)	129.1 (54.5)	190.4 (110.2)	114.5 (31.0)	162.1 (65.0)	92.3 (27.9)	111.0 (39.0)	90.4 (21.9)	136.9 (74.3)	$F_{10,31} = 1.2$; $p = 0.3$
MBN-^{15}N (%)	0.3691 a (0.0005)	0.3747 b (0.0017)	0.3693 a (0.0007)	0.3708 a (0.0019)	0.3709 a (0.0019)	0.3694 a (0.0012)	0.3711 a (0.0017)	0.3721 ab (0.0015)	0.3709 a (0.0004)	0.3711 a (0.0008)	0.3698 a (0.0001)	$F_{10,31} = 6.0$; $p < 0.0001$
δ^{15}N	7.5 a (1.3)	23.0 b (4.8)	8.1 a (1.9)	12.3 a (5.3)	12.7 a (5.3)	8.5 a (3.4)	12.0 a (4.7)	15.7 ab (4.1)	12.5 a (1.2)	13.2 a (2.3)	9.4 a (0.1)	$F_{10,31} = 6.0$; $p < 0.0001$
Dissolved inorganic N												
DIN	62.9 a (12.8)	37.0 b (8.9)	22.6 b (8.4)	18.4 b (4.1)	40.4 ab (8.6)	38.8 ab (20.2)	23.8 b (7.7)	31.1 b (6.7)	25.7 b (6.4)	28.8 b (7.1)	33.0 b (6.2)	$F_{10,31} = 5.7$; $p < 0.0001$
DIN-^{15}N (%)	0.3692 a (0.0008)	0.3958 b (0.0149)	0.3687 a (0.0003)	0.3740 a (0.0003)	0.3770 a (0.0041)	0.3749 a (0.0025)	0.3751 a (0.0087)	0.3784 a (0.0046)	0.3749 a (0.0025)	0.3813 ab (0.0076)	0.3776 a (0.0043)	$F_{10,31} = 6.4$; $p < 0.0001$

Value is shown as mean (S.D.) ($n = 4$, except data with: $n = 3$). Statistics shows the statistical results (F ratios and p values) from one-way ANOVA (GLM for unbalanced data; significant level $\alpha = 0.05$).

3.4. Soil and Microbial Nitrogen Dynamics

There was no significant difference in microbial biomass nitrogen (MBN) between the control and earthworm treatments. However, the higher MBN-^{15}N and microbial $\delta^{15}N$ signature from the Grass treatment, compared with those in the control and earthworm treatments (except E + O treatment), indicated that the microbes did utilize and incorporate the ^{15}N-labeled grass resources (plant roots or root exudates) into the microbial biomass (both $p < 0.0001$; Table 6).

At the end of experiment (Day 21), lower soil dissolved inorganic nitrogen (DIN) was found in the control and the earthworm treatments, except native *Estherella* spp. (E) and *O. borincana* (O) treatments, as compared with the Soil Only mesocosms ($F_{10, 31} = 5.7$, $p < 0.0001$; Table 6). Earthworms not only reduced the DIN in the experimental soil but also reduced the ^{15}N percentage in DIN (except O + P treatment) ($F_{10, 31} = 6.4$, $p < 0.0001$; Table 6). There was no significant difference total dissolved nitrogen (TDN) between the control (10.8 ± 1.5 µg N/g soil) and the earthworm treatments (ranged from 12.0–17.1 µg N/g soil) ($F_{10, 31} = 1.3$, $p = 0.3$).

4. Discussion

In this study, newly added ^{13}C-labeled leaf litter and ^{15}N-labeled grass were sufficiently incorporated into 10 cm of top soil, soil microbial biomass, and earthworm tissue. Natural abundance of $\delta^{13}C$ in earthworms was suggested to be 1−3‰ heavier than its dietary sources (such as leaf litter, root exudates, and microbial populations in the soil) [18,35]. In this study, earthworm $\delta^{13}C$ showed on average 1.4‰ heavier in *Estherella* spp., 3.5‰ heavier in *P. corethrurus*, and 5‰ heavier in *O. borincana*, with respect to the soil $\delta^{13}C$, while earthworm tissue showed on average 5.9‰ heavier $\delta^{13}C$ in *Estherella* spp., 7.2‰ heavier in *P. corethrurus*, and 8.5‰ heavier in *O. borincana* than the microbial biomass $\delta^{13}C$ in which they inhabited (Tables 2–5).

We did not observe any competition exclusion among three earthworm species based on the survivorship and biomass gain among the single-species and the mixed-species treatments for each individual species. However, anecic *Estherella* spp., when cultivated alone, did show higher tissue—[13]C (%) and $\delta^{13}C$—compared with when it was cultivated with other earthworm species. This suggested that *Estherella* spp. might change its feeding strategy by reducing its utilization of ^{13}C-labeled litter materials and/or the microbial community that was related to the ^{13}C-labeled litter when cultivated with *P. corethrurus* or both *P. corethrurus* and *O. borincana*. Lachnicht et al. [6] observed that *P. corethrurus* and *Estherella* spp., while cultivated together, excluded each other in the bottom and upper layers of soil, respectively, in a 19-day laboratory experiment in Puerto Rican soils. The authors also found that *P. corethrurus* acquired more ^{15}N-labeled leaf litter when co-occurring with *Estherella* spp. [6]. We did not find that *P. corethrurus* changed its feeding preference in this 22-day experiment based on worm tissue ^{13}C and $\delta^{13}C$ as well as tissue ^{15}N and $\delta^{15}N$ between the single-species and mixed-species earthworm treatments, nor did *O. borincana*. In this study, cultivating live *A. glomeratus* grass plants could provide a steady, continuous supply of root exudates and rhizodeposits for soil microbes and earthworms, as compared to the one-time application of ^{13}C-labeled glucose and ^{15}N-labeled leaf litter adopted by Lachnicht et al. [6]. Such a continuous supply of food resources might relieve potential inter-specific competitive pressure derived from limited food resources in short-term experiments, especially for endogeic earthworms like *O. borincana* and *P. corethrurus* that strongly rely on rhizosphere resources.

Both endogeic *O. borincana* and *P. corethrurus* showed 5‰ or higher $\delta^{13}C$ signature than their food resources (soil organic matter and soil microbial biomass). Higher $\delta^{13}C$ signature in both endogeic earthworms could be explained by their utilization on soil microbial populations (i.e., bacteria and fungi) as food resources. Fungal species (such as mycorrhizal and saprotrophic fungi) have been reported to have a higher ^{13}C enrichment than plant foliage, fine roots, and soils because of fungal biochemical synthesis and transport between plant parts [36]. Microbial activity releases the lighter ^{12}C in respiration and gradually results in an increase of ^{13}C concentration in humified residues and its own population [37,38]. As a result, endogeic earthworms (active in rhizosphere and the mineral soils), *P. corethrurus* and *O. borincana* in this study, showed higher $\delta^{13}C$ signature and tissue [13]C (%)

than anecic *Estherella* spp. due to their preferential consumption of ^{13}C-enriched decayed/humified debris in the mineral soil layer, to a significant portion of ^{13}C-enriched microbial (higher microbial δ^{13}C observed in P, O + P and E + O + P earthworm treatments; Table 5) and fungal populations, or to both [6,36,37]. The possibility that both endogeic *O. borincana* and *P. corethrurus* consumed the microbial populations in the mineral soil, the rhizosphere, or both is also confirmed by their heavier δ^{15}N signatures (0.6‰ and 2.7‰ δ^{15}N heavier, respectively) compared with the soil δ^{15}N (Tables 2–4).

We found that soil microbial-δ^{15}N was on average 6.1‰ heavier than *Estherella* spp., 5.8‰ heavier than *P. corethrurus*, and 2.6‰ heavier than *O. borincana* (Tables 3, 4 and 6). The stronger ^{15}N enrichment in endogeic *O. borincana* could be derived from its utilization of ^{15}N-labeled rhizosphere (plant roots, root exudates, and rhizosphere-related microbes). Even though no study has yet investigated the feeding behavior of *O. borincana*, some endogeic earthworms (e.g., *P. corethrurus*) are often found aggregated in the root zones utilizing living root fragments and dead root cells, or as response to enhanced microbial activities in the rhizosphere [35,39]. In this study, the presence of *O. borincana* seemed to relate to higher microbial biomass ^{15}N and δ^{15}N (in the E + O earthworm mesocosms) and higher DIN and higher ^{15}N-DIN (%) in the O + P treatment (although not statistically significant), as compared with other earthworm treatments (Table 6). The potential effect of endogeic *O. borincana* on rhizospheric microbial populations and activities is a topic of interest, yet in need for further research.

Pontoscolex corethrurus showed a prolific reproduction (a total of eight juvenile *P. corethrurus*) within the 22-day soil mesocosm experiment. The stronger δ^{15}N signal observed in juvenile *P. corethrurus*, as compared with the adults, might be explained by (1) the possibility that adult *P. corethrurus* allocated its assimilated ^{15}N into cocoon reproduction, which later integrated into the tissue of juvenile *P. corethrurus*, and (2) a higher soil consumption and biomass increase in relation to overall biomass by juvenile worms than the adult worms [6]. *Pontoscolex corethrurus* has been described as one of the cosmopolitan earthworm species that has aggressively invaded many regions in the tropics, including Puerto Rico, Central Amazonian, and Peruvian soils [40–43]. Exceptional reproductive strategies of *P. corethrurus*, such as a high rate of cocoon production and hatching success, a short development time, and the ability of parthenogenesis, critically influence the local native earthworm community in the invaded soils [2]. The rapid population growth of *P. corethrurus* may increase competition pressure on food resources to the local native earthworm community [22]. The relevance of resource availability to the population growth of *P. corethrurus* and its significance in a *P. corethrurus* invasion is certainly a topic in need of future research.

Earthworms showed differential effects on soil mineralization processes in this study. All earthworm treatments along with the control (no worms) had higher soil respiration C-CO_2 at Day 21, especially in the P, E + O, and O + P treatments, as compared with other control treatments (Soil Only, Grass, and Litter mesocosms). There were higher ^{13}C-CO_2 (%) and δ^{13}C from the P mesocosms (Tukey's HSD, $p < 0.001$) and the mixed E + O mesocosms (marginally significant; $p = 0.06$) compared with those from the O + P treatments. The effects on soil microbial activities by earthworms could be explained by earthworms' direct grazing behavior on soil microbial community or indirect burrowing and casting activities [11,14]. Whether the higher soil respiration C-CO_2 from the control (no worms) mesocosms was due to the release from earthworms' grazing activity is uncertain. However, the significantly higher soil respiration ^{13}C-CO_2 (%) and δ^{13}C from *P. corethrurus* (includes P and E + P) were an indicator of facilitating effects of earthworms on the enriched soil microbial biomass δ^{13}C from the same mesocosms. *Pontoscolex corethrurus* might cause an increase in soil respiration via its simulation on the activity of the ^{13}C-labeled microbial population. However, the lower soil respiration ^{13}C-CO_2 (%) and δ^{13}C in the mixed *P. corethrurus* and *O. borincana* treatments (i.e., O + P) suggested that the presence of *O. borincana* and its interaction with *P. corethrurus* might have a negative effect on the ^{13}C-labeled microbial community and facilitate the ^{15}N-labeled microbial communities in the rhizosphere. Such a possibility is supported by the observation of the slightly increased [15]N (%) in the soil DIN from the increased microbial activity related to the ^{15}N-labeled rhizosphere in the O + P treatment (Table 6).

The individual stimulation on soil N mineralization by *Estherella* spp. and *O. borincana* was slightly reduced when they were incubated with other earthworm species (mixed-species earthworm treatments; Table 6). No significant change was observed in microbial biomass (C and N) between treatments, thus the changes shown in soil respiration $\delta^{13}C$ and DIN could be explained by the changed activities from the microbial population or possibly alternation of microbial community induced by the inter-specific earthworm interactions from the mixed earthworm treatments. Studies have suggested that the preference of earthworms on utilizing different food resources can reshape microbial communities in the detritusphere and the rhizosphere [44,45]. Native *Estherella* spp. and *O. borincana* may individually sustain a microbial community that specialized on N mineralization in the rhizopshere, yet the microbes switched to those which utilized a labile, newly added ^{13}C-labeled resource when sharing resources with the other species. Earthworm effect on either microbial activities or microbial community by individual species is confounded when inter-specific interactions are considered, and the individual effect on microbial activities and communities was not additive. Furthermore, changes in microbial activities and alterations to the microbial community by earthworms could gradually alter soil nutrient dynamics and availability of labile C and N over time [46], which later has an effect on habitat suitability for other species. For example, invasive *Amynthas agrestis* (Goto and Hatai, 1899) was documented to change soil microbial communities, which positively affected the habitat invasibility for another invasive species, *Lumbricus rubellus* (Hoffmeister, 1843) [47]. Many studies have focused on the earthworm effects on soil microbial biomass and soil mineralization [11,47–53]; however, research investigating the effects of earthworms with different feeding strategies (i.e., epigeic, anecic, and endogeic) on soil microbial activities and communities in terms of functional groups is still limited.

5. Conclusions

In this study, anecic *Estherella* spp. was observed to reduce its utilization on ^{13}C-labeled litter or ^{13}C-related microbial community when cultivated with *P. corethrurus* or both *P. corethrurus* and *O. borincana*. Resource utilization by different earthworms changed the activities and composition of soil microbial community and further affected soil respiration and nitrogen mineralization processes. However, the individual species effect on soil C and N dynamics was altered with mixed earthworm assemblages. *Pontoscolex corethrurus* was found to stimulate soil respiration by facilitating the activity of the ^{13}C-labeled microbial activity; however, the positive effect was suppressed when it coexisted with *O. borincana*. The stimulated N mineralization process by native *Estherella* spp. and *O. borincana* individually were reduced when each of them cultivated with other earthworm species. We concluded that the earthworm effect on soil microbial community and activity varies by species, and the individual species effect is not additive when considering multiple earthworm species assemblages. Regulation on soil nutrient dynamics by native *Estherella* sp. and *O. borincana* may potentially affect habitat suitability (e.g., resource availability) to invasive *P. corethrurus* during colonization. However, the rapid population growth of *P. corethrurus* may increase competition pressure on food resources to the local earthworm community. The relevance of resource availability to the population growth of *P. corethrurus* and its significance as an invasive species is a topic in need of future research.

Acknowledgments: We thank the staff at the Sabana Field Research Station and the International Institute of Tropical Forestry (IITF) for their help in the lab and in the field. We also thank David Kissel, the Soil Plant and Water Laboratory and the University of Georgia, for helping us with the soil transportation and sterilization process. This research was supported by the National Science Foundation grant number 0236276 to the University of Georgia Research Foundation, Inc.; the Graduate Student Research Prize, Center for Biodiversity and Ecosystem Processes, University of Georgia; grant DEB-0218039 from the NSF to the Institute of Tropical Ecosystem Studies, University of Puerto Rico, and the USDA Forest Service-IITF as part of the LTER program in the Luquillo Experimental Forest. All research at the USDA Forest Service International Institute of Tropical Forestry was performed in collaboration with the University of Puerto Rico. We also thank Ariel E. Lugo, Frank Wadsworth, and two anonymous reviewers, for their comments and suggestions on an earlier version of this manuscript.

Author Contributions: Ching-Yu Huang and Paul F. Hendrix conceived the study and designed the experiment. Grizelle González provided field site information and aided in field site selection and field sample collection.

Ching-Yu Huang processed the samples, analyzed the data, and prepared the manuscript. Paul F. Hendrix and Grizelle González provided suggestions and reviews at various stages of the manuscript.

Conflicts of Interest: The authors declare no conflict of interest.

References

1. Frelich, L.E.; Hale, C.M.; Scheu, S.; Holdsworth, A.R.; Heneghan, L.; Bohlen, P.J.; Reich, P.B. Earthworm invasion into previously earthworm-free temperate and boreal forests. *Biol. Invasions* **2006**, *8*, 1235–1245. [CrossRef]

2. Hendrix, P.F.; Baker, G.H.; Callaham, M.A., Jr.; Damoff, G.A.; Fragoso, C.; González, G.; Winsome, T.; Zou, X. Invasion of exotic earthworms into ecosystems inhabited by native earthworms. *Biol. Invasions* **2006**, *8*, 1287–1300. [CrossRef]

3. Hendrix, P.F.; Callaham, M.A., Jr.; Drake, J.M.; Huang, C.-Y.; James, S.W.; Snyder, B.A.; Zhang, W. Pandora's box contained bait: The global problem of introduced earthworms. *Annu. Rev. Ecol. Evol. Syst.* **2008**, *39*, 593–613. [CrossRef]

4. Callaham, M.A., Jr.; Hendrix, P.F.; Phillips, R.J. Occurrence of an exotic earthworm (*Amynthas agrestis*) in undisturbed soils of the southern Appalachian mountains, USA. *Pedobiologia* **2003**, *47*, 466–470. [CrossRef]

5. Kalisz, P.J.; Wood, H.B. Native and exotic earthworms in wildland ecosystems. In *Earthworm Ecology and Biogeography in North America*, 1st ed.; Hendrix, P.F., Ed.; Lewis Publishers: Boca Raton, FL, USA, 1995; pp. 117–126.

6. Lachnicht, S.L.; Hendrix, P.F.; Zou, X. Interactive effects of native and exotic earthworms on resource use and nutrient mineralization in a tropical wet forest soil of Puerto Rico. *Biol. Fertil. Soils* **2002**, *36*, 43–52. [CrossRef]

7. Winsome, T.; Epstein, L.; Hendrix, P.F.; Horwath, W.R. Competitive interactions between native and exotic earthworm species as influenced by habitat quality in a California grassland. *Appl. Soil. Ecol.* **2006**, *32*, 38–53. [CrossRef]

8. Bouché, M.B. Strategies Lombriciennes. In *Soil Organisms as Components of Ecosystems: Proceedings of the VI International Soil Zoology Colloquium of the International Society of Soil Science (ISSS)*; Lohm, U., Persson, T., Eds.; Swedish Natural Science Research Council: Stockholm, Sweden, 1977; pp. 122–132.

9. Lavelle, P.; Barois, I.; Cruz, I.; Fragoso, C.; Hernandez, A.; Pineda, A.; Rangel, P. Adaptive strategies of Pontoscolex corethrurus (Glossoscolecidae, Oligochaeta), a peregrine geophagous earthworm of the humid tropics. *Biol. Fertil. Soils* **1987**, *5*, 188–194. [CrossRef]

10. Lavelle, P.; Lapied, E. Endangered earthworms of Amazonia: An homage to Gilberto Righi. *Pedobiologia* **2003**, *47*, 419–417. [CrossRef]

11. Groffman, P.M.; Bohlen, P.J.; Fisk, M.C.; Fahey, T.J. Exotic earthworm invasion and microbial biomass in temperate forest soils. *Ecosystems* **2004**, *7*, 43–54. [CrossRef]

12. Hale, C.M.; Frelich, L.E.; Reich, P.B.; Pastor, J. Effects of European earthworm invasion on soil characteristics in Northern hardwood forests of Minnesota, USA. *Ecosystems* **2005**, *8*, 911–927. [CrossRef]

13. Huang, C.-Y.; Hendrix, P.F.; Fahey, T.J.; Bohlen, P.J.; Groffman, P.M. A simulation model to evaluate the impacts of invasive earthworms on soil carbon dynamics. *Ecol. Model* **2010**, *20*, 2447–2457. [CrossRef]

14. Bossuyt, H.; Six, J.; Hendrix, P.F. Rapid incorporation of carbon from fresh residues into newly formed stable microaggregates within earthworm casts. *Eur. J. Soil Sci.* **2004**, *55*, 393–399. [CrossRef]

15. Curry, J.P.; Schmidt, O. The feeding ecology of earthworms—A review. *Pedobiologia* **2007**, *50*, 463–477. [CrossRef]

16. Mummey, D.L.; Rillig, M.C.; Six, J. Endogeic earthworms differentially influence bacterial communities associated with different soil aggregate size fractions. *Soil Biol. Biochem.* **2006**, *38*, 1608–1614. [CrossRef]

17. Hendrix, P.F.; Lachnicht, S.L.; Callaham, M.A., Jr.; Zou, X. Stable isotopic studies of earthworm feeding ecology in tropical ecosystems of Puerto Rico. *Rapid Commun. Mass Sp.* **1999**, *13*, 1295–1299. [CrossRef]

18. Neilson, R.; Boag, B.; Simth, M. Earthworm $\delta^{13}C$ and $\delta^{15}C$ analyses suggest that putative functional classifications of earthworms are site-specific and may also indicate habitat diversity. *Soil Biol. Biochem.* **2000**, *32*, 1053–1061. [CrossRef]

19. Schmidt, O.; Scrimgeour, C.M.; Handley, L.L. Natural abundance of ^{15}N and ^{13}C in earthworms from a wheat and a wheat-clover field. *Soil Biol. Biochem.* **1997**, *29*, 1301–1308. [CrossRef]

20. Zhang, W.; Hendrix, P.F.; Snyder, B.A.; Molina, M.; Li, J.; Rao, X.; Siemann, E.; Fu, S. Dietary flexibility aids Asian earthworm invasion in North American forests. *Ecology* **2010**, *91*, 2070–2079. [CrossRef] [PubMed]

21. González, G.; Zou, X.; Borges, S. Earthworm abundance and species composition in abandoned tropical croplands: Comparison of tree plantations and secondary forests. *Pedobiologia* **1996**, *40*, 385–391.

22. Sánchez-de León, Y.; Zou, X.; Borges, S.; Ruan, H. Recovery of native earthworms in abandoned tropical pastures. *Conserv. Biol.* **2003**, *17*, 999–1006. [CrossRef]

23. González, G.; Zou, X.; Sabat, A.; Fetcher, N. Earthworm abundance and distribution pattern in contrasting plant communities within a tropical wet forest in Puerto Rico. *Caribb. J. Sci.* **1999**, *35*, 93–100.

24. Huang, C.-Y.; González, G.; Hendrix, P.F. The re-colonization ability of native earthworm, Estherella spp., in Puerto Rican forests and pastures. *Caribb. J. Sci.* **2006**, *42*, 386–396.

25. Scatena, F.N. An Introduction to the Physiography and History of the Bisley Experimental Watersheds in the Luquillo Mountains of Puerto Rico. Available online: www.srs.fs.usda.gov/pubs/gtr/gtr_so072.pdf (accessed on 19 September 2016).

26. Schmidt, O.; Scrimgeour, C.M. A simple urea leaf-feeding method for the production of ^{13}C and ^{15}N labelled plant material. *Plant Soil* **2001**, *229*, 197–202. [CrossRef]

27. Abelleira, O.J. Ecology of Novel Forests Dominated by the African Tulip Tree (*Spathodea campanulata* Beauv.) in Northcentral Puerto Rico. Master's Thesis, University of Puerto Rico, Rio Piedras, Puerto Rico, 2009.

28. Liu, Z.G.; Zou, X.M. Exotic earthworms accelerate plant litter decomposition in a Puerto Rican pasture and a wet forest. *Ecol. Appl.* **2002**, *12*, 1406–1417. [CrossRef]

29. Joergensen, R.G. The fumigation-extraction method to estimate soil microbial biomass: Calibration of the k_{EC} value. *Soil Biol. Biochem.* **1996**, *28*, 25–31. [CrossRef]

30. Sparling, G.P.; West, A.W. A direct extraction method to estimate soil microbial C: Calibration *in situ* using microbial respiration and ^{14}C labelled cells. *Soil Biol. Biochem.* **1988**, *20*, 337–343. [CrossRef]

31. Cabrera, M.L.; Beare, M.H. Alkaline persulfate oxidation for determining total nitrogen in microbial biomass extracts. *Soil Sci. Soc. Am. J.* **1993**, *57*, 1007–1012. [CrossRef]

32. Stark, J.M.; Hart, S.C. Diffusion technique for preparing salt solutions, Kjeldahl digests, and persulfate digests for nitorgen-15 analysis. *Soil Sci. Soc. Am. J.* **1996**, *60*, 1846–1855. [CrossRef]

33. Coleman, D.C.; Fry, B. *Carbon Isotope Techniques*, 1st ed.; Academic Press, Inc.: San Diego, CA, USA, 1991.

34. SAS Institute Inc. *SAS Technical Report, SAS/STAT Software: The GLM Procedure*; Version 6; SAS Institute Inc.: Cary, NC, USA, 1991; p. 217.

35. Spain, A.V.; Saffigna, P.G.; Wood, A.W. Tissue carbon sources for *Pontoscolex corethrurus* (Oligochaeta: Glossoscolecidae) in a sugarcane ecosystem. *Soil Biol. Biochem.* **1990**, *22*, 703–706. [CrossRef]

36. Hobbie, E.A.; Macko, S.A.; Shugart, H.H. Insights into nitrogen and carbon dynamics of ectomycorrhizal and saprotrophic fungi from isotopic evidence. *Oecologia* **1999**, *118*, 353–360. [CrossRef]

37. Pollierer, M.M.; Langel, R.; Scheu, S.; Maraun, M. Compartmentalization of the soil animal food web as indicated by dual analysis of stable isotope ratios ($^{15}N/^{14}N$ and $^{13}C/^{12}C$). *Soil Biol. Biochem.* **2009**, *41*, 1221–1226. [CrossRef]

38. Dijkstra, P.; Ishizu, A.; Doucett, R.; Hart, S.C.; Schwartz, E.; Menyailo, O.V.; Hungate, B.A. ^{13}C and ^{15}N natural abundance of the soil microbial biomass. *Soil Biol. Biochem.* **2006**, *38*, 3257–3266. [CrossRef]

39. Binet, F.; Hallaire, V.; Curmi, P. Agricultural practices and the spatial distribution of earthworms in maize fields. Relationships between earthworm abundance, maize plants and soil compaction. *Soil Biol. Biochem.* **1997**, *29*, 577–583. [CrossRef]

40. Chauvel, A.; Grimaldi, M.; Barros, E.; Blanchart, E.; Desjardins, T.; Sarrazin, M.; Lavelle, P. Pasture damage by an Amazonian earthworm. *Nature* **1999**, *398*, 32–33. [CrossRef]

41. Fragoso, C.; Kanyonyo, J.; Moreno, A.; Senapati, B.K.; Blanchart, E.; Rodríguez, C. A survey of tropical earthworms: Taxonomy, biogeography and environmental plasticity. In *Earthworm Management in Tropical Agroecosystems*; Lavelle, P., Brussaard, L., Hendrix, P., Eds.; CABI: New York, NY, USA, 1999; pp. 1–26.

42. González, G.; Huang, C.-Y.; Zou, X.; Rodriguez, C. Earthworm invasions in the tropics. *Biol. Invasions* **2006**, *8*, 1247–1256. [CrossRef]

43. Hallaire, V.; Curmi, P.; Duboisset, A.; Lavelle, P.; Pashanasi, B. Soil structure changes induced by the tropical earthworm *Pontoscolex corethrurus* and organic inputs in a Peruvian ultisol. *Euro. J. Soil Biol.* **2000**, *36*, 35–44. [CrossRef]

44. Butenschoen, O.; Marhan, S.; Scheu, S. Response of soil microorganisms and endogeic earthworms to cutting of grassland plants in a laboratory experiment. *Appl. Soil Ecol.* **2008**, *38*, 152–160. [CrossRef]

45. Sheehan, C.; Kirwan, L.; Connolly, J.; Bolger, T. The effects of earthworm functional diversity of microbial biomass and the microbial community level physiological profile of soils. *Euro. J. Soil Biol.* **2008**, *44*, 65–70. [CrossRef]

46. Bohlen, P.J.; Edwards, C.A.; Zhang, Q.; Parmelee, R.W.; Allen, M. Indirect effects of earthworms on microbial assimilation of labile carbon. *App. Soil Ecol.* **2002**, *20*, 255–261. [CrossRef]

47. Zhang, B.-G.; Li, G.-T.; Shen, T.-S.; Wang, J.-K.; Sun, Z. Changes in microbial C, N, and P and enzyme activities in soil incubated with the earthworms *Metaphire guillelmi* or *Eisenia fetida*. *Soil Biol. Biochem.* **2000**, *32*, 2055–2062. [CrossRef]

48. Bohlen, P.J.; Scheu, S.; Hale, C.M.; McLean, M.A.; Migge, S.; Groffman, P.M.; Parkinson, D. Non-native invasive earthworms as agents of change in northern temperate forests. *Front. Ecol. Environ.* **2004**, *2*, 427–435. [CrossRef]

49. Eisenhauer, N.; Partsch, S.; Parkinson, D.; Scheu, S. Invasion of a deciduous forest by earthworms: Changes in soil chemistry, microflora, microarthopods and vegetation. *Soil Biol. Biochem.* **2007**, *39*, 1099–1110. [CrossRef]

50. Fisk, M.C.; Fahey, T.J.; Groffman, P.M.; Bohlen, P.J. Earthworm invasion, fine-root distributions, and soil respiration in North temperate forests. *Ecosystems* **2004**, *7*, 55–62. [CrossRef]

51. Lachnicht, S.L.; Hendrix, P.F. Interaction of the earthworm *Diplocardia mississippiensis* (Megascolecidae) with microbial and nutrient dynamics in a subtropical Spodosol. *Soil Biol. Biochem.* **2001**, *33*, 1411–1417. [CrossRef]

52. Li, X.; Fisk, M.C.; Fahey, T.J.; Bohlen, P.J. Influence of earthworm invasion on soil microbial biomass and activity in a northern hardwood forest. *Soil Biol. Biochem.* **2002**, *34*, 1929–1937. [CrossRef]

53. Wolters, V.; Joergensen, R.G. Microbial carbon turnover in beech forest soils worked by *Aporrectodea caliginosa* (Savigny) (Oligochaeta: Lumbricidae). *Soil Biol. Biochem.* **1992**, *24*, 171–177. [CrossRef]

![forests logo] *forests*

MDPI

Article

Surface CO$_2$ Exchange Dynamics across a Climatic Gradient in McKenzie Valley: Effect of Landforms, Climate and Permafrost

Natalia Startsev [1], Jagtar S. Bhatti [1,*] and Rachhpal S. Jassal [2]

[1]   Natalia Startsev. Canadian Forest Service, Northern Forestry Centre, 5320 122 Street, Edmonton, AB T6H 3S5, Canada; natalia.startsev@canada.ca

[2]   Biometeorology and Soil Physics Group, University of British Columbia, Vancouver, BC V6T 1Z4, Canada; rachhpal.jassal@ubc.ca

*   Correspondence: jagtar.bhatti@canada.ca; Tel.: +1-780-435-7241; Fax: +1-780-435-7359

Academic Editors: Scott X. Chang and Xiangyang Sun
Received: 26 July 2016; Accepted: 5 November 2016; Published: 15 November 2016

Abstract: Northern regions are experiencing considerable climate change affecting the state of permafrost, peat accumulation rates, and the large pool of carbon (C) stored in soil, thereby emphasizing the importance of monitoring surface C fluxes in different landform sites along a climate gradient. We studied surface net C exchange (NCE) and ecosystem respiration (ER) across different landforms (upland, peat plateau, collapse scar) in mid-boreal to high subarctic ecoregions in the Mackenzie Valley of northwestern Canada for three years. NCE and ER were measured using automatic CO$_2$ chambers (ADC, Bioscientific LTD., Herts, England), and soil respiration (SR) was measured with solid state infrared CO$_2$ sensors (Carbocaps, Vaisala, Vantaa, Finland) using the concentration gradient technique. Both NCE and ER were primarily controlled by soil temperature in the upper horizons. In upland forest locations, ER varied from 583 to 214 g C·m^{-2}·year^{-1} from mid-boreal to high subarctic zones, respectively. For the bog and peat plateau areas, ER was less than half that at the upland locations. Of SR, nearly 75% was generated in the upper 5 cm layer composed of live bryophytes and actively decomposing fibric material. Our results suggest that for the upland and bog locations, ER significantly exceeded NCE. Bryophyte NCE was greatest in continuously waterlogged collapsed areas and was negligible in other locations. Overall, upland forest sites were sources of CO$_2$ (from 64 g·C·m^{-2}·year^{-1} in the high subarctic to 588 g C·m^{-2}·year^{-1} in mid-boreal zone); collapsed areas were sinks of C, especially in high subarctic (from 27 g· C· m^{-2} year^{-1} in mid-boreal to 86 g·C·m^{-2}·year^{-1} in high subarctic) and peat plateaus were minor sources (from 153 g· C· m^{-2}· year^{-1} in mid-boreal to 6 g·C·m^{-2}·year^{-1} in high subarctic). The results are important in understanding how different landforms are responding to climate change and would be useful in modeling the effect of future climate change on the soil C balance in the northern regions.

Keywords: net carbon exchange; ecosystem respiration; upland forest; bogs; collapse scar; permafrost

1. Introduction

Northern regions have experienced considerable climate change during the last few decades, affecting the substantial pool of soil carbon stored in the permafrost [1–3]. Regions exposed to these changes are in many respects fragile and may undergo significant ecological alterations including permafrost thawing, potential release of fossil methane, and decomposition of previously frozen peat. More than one third of the planetary soil organic C is stored in the permafrost area [4,5], therefore assessing and forecasting of the carbon (C) storage dynamics in this pool has a major impact on the understanding of global C balance. Particular interest in this case lies in the monitoring of the changes in the surface CO_2 flux, and climate-driven changes in the permafrost status and peat accumulation rates. These changes are sensitive to local landforms and soil microclimate variations.

Effects that warming, and lengthening of the growing season, have on the C pool in the northern areas are multifaceted. On one hand, increases in temperature, depth of active soil layer, and length of growing season increase the rate of C loss from the soil; on the other hand, the same factors, aided by potential CO_2 fertilization effects, can increase the assimilation rate of the plants and hence biomass production [6]. A large amount of C stored in the soil in northern regions is evidence that assimilation rates continuously exceed of ecosystem respiration. The primary reason for this positive net C exchange (NCE) is lower decomposition rates of soil organic C due to low soil temperatures, soil waterlogging, and permafrost. The increase in mean annual air temperature affected soil temperatures, increasing the depth of the active layer, and escalating organic matter decomposition and thus ecosystem respiration, thereby shifting the NCE balance from carbon accumulation to degradation of stored carbon [7]. The implications of this shift include a release of vast amounts of carbon to the atmosphere in response to a comparatively minor increase in mean annual soil temperature [1,8].

Soil microbial activity response to changes in soil temperature varies depending on the temperature range. In the frozen soil layers at the temperature well below zero and with biological activity nearly non-existent [5,9], an increase of a few degrees Celsius would not change the carbon balance significantly as long as the soil stays frozen. However, in soil layers above freezing point, similar increase in soil temperature would cause an exponential increase in decomposition and ecosystem respiration [7,10]. The most appreciable effect of the climate change on the carbon balance can be expected in the areas of the zonal boundaries where relatively small changes in the soil temperature signify the difference between the frozen and thawed state influencing soil water regime, permafrost degradation, and ground subsidence, thereby resulting in a nearly instantaneous rise in CO_2 production.

Thawing of the permafrost follows a complex pattern of more intricate localized conditions than ecoclimatic gradient [11]. Areas affected by the permafrost frequently have a mosaic of landforms with locations of distinctly different soil moisture and temperature regimes in close proximity. Some of the most common landforms in the permafrost-affected areas include forested uplands, lichen-dominated peat plateaus, and collapse features of different origins. They differ in vegetation, allocation of C storage, soil types, hydrological regime, and consequently in the C balance. The effect of the landform and depth of the saturated soil layer on CO_2 fluxes is not fully understood and existing studies offer conflicting information. For example, Waddington and Roulet [12] suggested that CO_2 fluxes from the collapsed areas exceeded those from peat plateaus, while Mitsch and Gosselink [13] showed an opposite relationship.

Uplands generally have better drainage, mineral horizons are present within the active layer, and most of the CO_2 assimilation is performed by the tree crowns, while shaded ground covering the bryophyte layer plays only a minor role in total ecosystem assimilation [14]. Organic layers are frequently coarse mor to moder, and although comparatively thick, contain a limited and relatively constant amount of C [15]. This suggests that these areas are historically C neutral, neither sinks nor sources of CO_2, but that the assimilation and respiration functions of the ecosystem are largely stratified: tree crowns responsible for the bulk of total assimilation, and decomposition taking place at the ground surface, making the soil surface a strong source of CO_2 while the ecosystem as a whole remains carbon neutral [1]. Climate warming might increase or decrease total pools of stored carbon as higher temperatures would promote faster decomposition of soil C, greater tree growth rate would provide more litter fall and more carbon stored in standing biomass [3]. It is reasonable to expect that in either case, these ecosystems would reach another state of equilibrium with new, somewhat higher, decomposition rates of organic matter on the ground compensating for the greater assimilation rate of the crowns and undergrowth.

Peat plateaus landforms are composed of perennially frozen peat and appear to have developed under non-permafrost conditions which subsequently became elevated and permanently frozen [16].

These areas contain a large amount of carbon stored in the peat, indicating that historically they underwent a period of C accumulation [10]. However, measurements indicate that peat plateaus at present are either minor C sources or C neutral, or alternate between minor C sources and minor C sinks from year to year [17]. Most studies suggest that C balance of peat plateaus can be positive or negative varying from year to year [12,18]. This indicates that either warming has already shifted the NCE or that present peat plateaus have reached a stable stage of ecosystem development. Further warming is expected to increase respiration [5], and the depth of active layer thereby exposing stored C to degradation, or destabilization and collapse of the permafrost to create collapsed areas in place of peat plateaus [10,19].

Collapsed areas are also known as thermokarst form where permafrost thawed and ground surface subsided. These are soils composed of water-saturated peat with the water table close to the surface or above it, initially forming fast growing sphagnum communities [16]. Thermokarst features of the landform are predominantly sources of carbon emissions in the form of methane [12]. There are indications that high assimilation rates and low soil respiration due to waterlogging create conditions for net C sequestration [13]. Several studies suggested that collapsed areas are the greatest sinks of CO_2 in northern regions due to active sphagnum growth under conditions of constant supply of water, sunlight, and warmer conditions, while at the same time producing the greatest amounts of methane [20–23]. Carbon accumulation in these areas is conditioned by the presence of the high water table, which depends on the presence of permafrost in the surrounding areas that prevents water from draining away from the collapsed areas [24]. Further permafrost thaw would cause drainage of collapsed areas thereby affecting conservation of soil organic matter and supporting sphagnum growth, thus converting them from sinks to major sources of CO_2.

In Canada, the Mackenzie valley region in the northwest has undergone the most warming (1.7 °C) over the last century [25]. As a result, changes in the permafrost distribution have affected forest and peatland ecosystems in the Mackenzie valley region. It is uncertain how the above mentioned changes will affect the distribution, composition and C source/sink capacity of forests and peatlands, i.e., balance between organic matter production, heterotrophic mineralization, and greenhouse gases (GHG) emissions [2]. Limited data and understanding of how changing environmental conditions and permafrost thawing influences the C cycle of forests and peatlands over short timescales poses an even greater challenge to predict the changes in the C sink/source relationships and GHG dynamics under a changing climate. The goal is to advance and further develop C monitoring and assessment along diverse landscape (forest–peatland) and ecoregion (boreal–subarctic–arctic) gradients across the Mackenzie valley. This study was conducted with the following objectives: (1) To measure soil CO_2 fluxes, including NCE, along a climatic gradient and across different landforms; (2) To assess effects of permafrost, temperature and solar radiation on NCE and soil CO_2 production; and (3) To develop a statistical relationship to estimate ecosystem respiration (ER) along latitudinal gradient. Considering the difference in the soil surface carbon balance along the climatic gradient and between most common landforms, spanning from strong C sources to sinks, the task of evaluating and modeling of CO_2 emissions and assimilation in changing climate requires an approach that would take all of these variables into account.

2. Methods

2.1. Site Description

The study was conducted at four locations along a north-south climatic gradient in the Mackenzie Valley, Western Canada (Figure 1), from mid boreal with isolated patches of permafrost near Anzac, northern Alberta (AZ), to boreal forest with sporadic discontinuous permafrost zone in the area near Fort Simpson (FS), low subarctic with extensive discontinuous permafrost in the proximity of Norman Wells (NW), and high subarctic with extensive continuous permafrost in the Inuvik (IN) area. The mean values of the main climatic parameters for the sites (Table 1) indicate mean annual temperature and growing degree days decreasing but mean annual snow pack and precipitation increasing with latitude. Each study site was chosen to include a gradient from upland to peatland conditions and to a collapsed scar (Figure 2). Individual study plots were located within the upland forest (UL), peat plateau (PP; in Anzac site, a lichen-populated peat bog was used as ecologically comparable substitute), and a collapse feature (CS; in Anzac site, collapse feature was represented by an internal lawn; in Fort Simpson and Norman Wells, collapse scars; and in Inuvik a polygonal trench). Internal lawns are characteristically less than 50 cm lower than the surrounding bog and contain dead stands of *Picea mariana* and a *Sphagnum* ground cover [16].

Figure 1. Map of the site locations.

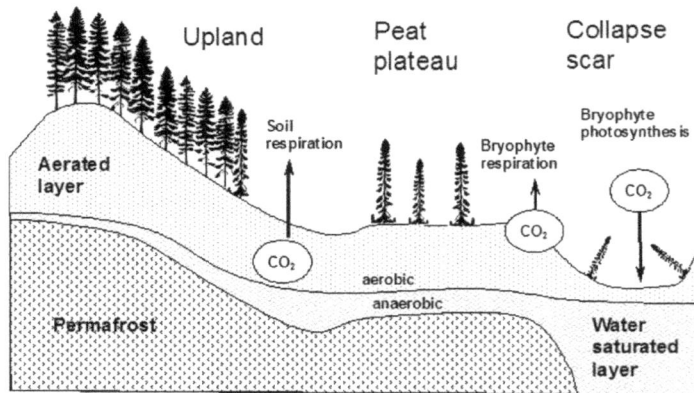

Figure 2. Schematic of the site landform features and key fluxes of CO_2. The arrows only represent the major fluxes not their magnitude at any point in time.

Table 1. Locations, ecozone permafrost status, and climatic description of the study sites.

	Latitude (°)	Longitude (°)	Ecozone	Permafrost status	Mean Annual Air Temperature (°C)	Growing Degree Days above 0 °C, (Days)	Mean Annual Snow Pack, (cm)	Mean Annual Precipitation, (mm)
Anzac (AZ)	56.40	−111.03	Mid-boreal	Isolated patches (0%–10%)	1.49	2251	32	464
Fort Simpson (FS)	61.63	−121.40	High-boreal	Sporadic discontinuous (10%–50%)	−2.17	2064	54	361
Norman Wells (NW)	65.21	−127.01	Low Subarctic	Extensive discontinuous (50%–90%)	−4.89	1817	48	318
Inuvik (IN)	68.32	−133.43	High Subarctic	Continuous (90%–100%)	−6.9	1295	60	259

Study Plots Establishment

At each site and each landform location, a study plot was established to facilitate measurements and minimize site disturbance. Board walks were laid in a cross pattern for easy reach to permanent sampling points; 10 cm in diameter plastic tubes (10 cm in length) were driven into the soil to serve as permanent collars for CO_2 measurements. In addition, 10 cm in diameter and 40 cm long plastic tubes were installed for the measurements of the water table.

2.2. Vegetation Description

In Anzac and Fort Simpson, uplands were characterized by a dense *Picea mariana* stand with a feathermoss understorey consisting primarily of *Hylocomium splendens* and *Pleurozium schreberi*. Peat plateaus had a sparse *Picea mariana* canopy with *Ledum groenlandicum* and *Vaccinium vitis-idaea* dominating the shrub layer, and a ground layer of lichen species, including *Cladina mitis*, *C. rangiferina*, and *C. stellaris*. The Norman Wells upland had a sparse *Picea mariana* canopy with *Ledum groenlandicum*, *Arctostaphylos rubra*, and *Vaccinium uliginosum* dominating the shrub layer, and a ground layer of primarily moss species, including *Tomenthypnum nitens*, *Hylocomium splendens*, and *Aulacomnium palustre*. The Inuvik upland plot was dominated by an open canopy of *Picea mariana* with a shrub layer of *Vaccinium vitis-idaea* and *Ledum decumbens*. *Petasites* sp. and *Rubus chamaemorus* were predominant in the herbaceous layer, while *Cladina mitis*, *Sphagnum angustifolium*, *Peltigera* spp. and *Cladonia* spp. comprised the ground layer.

Both Norman wells and Inuvik peat plateaus were treeless, with vegetation in Norman Wells dominated by a ground layer of *Cladina mitis* with patches of *Sphagnum fuscum*. The vascular vegetation consisted of a mix of *Ledum decumbens* with *Vaccinium vitis-idaea*, *Andromeda polifolia*, *Myrica gale*, and *Vaccinium uliginosum* in the shrub layer, and an herbaceous layer of scattered *Carex* sp. and *Rubus chamaemorus*. The Inuvik peat plateau was vegetated by *Ledum decumbens*, *Andromeda polifolia*, *Betula glandulosa*, and *Vaccinium vitis-idaea* dominating the shrub layer, a herbaceous layer of *Rubus chamaemorus* and *Carex* spp., and a ground layer comprised primarily of the lichen species *Flavocetraria nivalis* and *Cladina mitis*. Collapse scars were treeless at all sites, with vegetation dominated by a moss layer of *Sphagnum* sp. and *S. fuscum* with shrub cover including *Chamaedaphne calyculata* and *Andromeda polifolia*, and a herbaceous layer of *Smilacina trifolia*, *Eriophorum* sp. and *Carex* sp.

2.3. Soil Description

Soils at the study sites differed most notably with landforms and the presence of permafrost in the soil profile. Upland soils had comparatively better drainage and varied from moderately to imperfectly drained. The water table was within the upper 40 cm of the soil profile at the peat plateaus. Collapse features of the landforms did not have permafrost with the exception of the Inuvik site where the collapsed area was represented by a polygonal trough with ice-wedge at the depth of 50 cm. Upland soils were represented by Gleyed Dystric Brunisol in Anzac, Orthic Dystric Brunisol in Fort Simpson, Orthic Static Cryosol in Norman Wells, and Gleysolic Turbic Cryosol in Inuvik. Peat plateaus at all sites had a cover of Fibric organic cryosol except for Anzac where it was Typic Mesic Fibrosol developed in the absence of permafrost. The soil in collapse scars was classified as Hydric Fibrisol at all sites except Inuvik, where Mesic Organic Cryosol was found in the polygonal trench.

2.4. CO_2 Flux Measurements

2.4.1. Intra-Site Spatial Variability of CO_2 Fluxes

To determine spatial variability within a site, we measured NCE and ER using the PP system (EGM-4 portable CO_2 Gas Analyzer, PP System Inc., Amesbury, MA, USA) at three different locations, namely upland, peat plateau, and collapse scars. At each study location, six permanent gas flux sampling collars were installed within easy reach of the boardwalk in such a way that they were not shaded by the boardwalk. In order to improve the representation of spatial variability in hummocky areas, both hummocks and hollows were included as sample points. The measurements began in the fall of 2007 and continued throughout the growing seasons of 2008 and 2009. Surface CO_2 fluxes were measured using the PP system. Surface NCE and ER were measured using a clear and an opaque cover, respectively. The rate of assimilation of the soil cover vegetation was obtained from the difference between NCE and ER.

Spatially explicit measurements were taken during daytime, usually in the early afternoon, during the warmest part of the day when temperature-driven microbiological activity would have reached its maximum. In order to calculate daily flux values, a relationship was established between hourly fluxes and soil temperature (Table 2) and used to extrapolate flux values for all twenty-four hours of the day. To estimate net carbon exchange (NCE) at the soil surface, exponential regression (1):

$$NCE = ae^{(bT)} \tag{1}$$

was used where NCE is the net carbon exchange ($g \cdot CO_2 \cdot m^{-2} \cdot h^{-1}$) and T is the soil temperature at the 5-cm depth ($°C$), a is the reference NCE at zero $°C$, and coefficient b represents the temperature sensitivity. For the purpose of modeling the temperature effect on ER, an exponential regression (2):

$$ER = ae^{(bT)} \tag{2}$$

was used where ER is the ecosystem respiration ($g \cdot CO_2 \cdot m^{-2} \cdot h^{-1}$) and T is soil temperature at the 5-cm depth ($°C$), a is reference respiration at zero $°C$, and coefficient b represents the temperature sensitivity. Soil moisture has insignificant effect on the soil CO2 fluxes. Values of Q_{10}, which is the relative increase in respiration for a 10 $°C$ increase in temperature, were obtained as Equation (3):

$$Q_{10} = b(1 + 10 \ T^{-1}) \tag{3}$$

Similarly, the effect of solar irradiation on the net assimilation rate (NAR) was evaluated using a log function (Equation (4)):

$$NAR = a'\ln(Q) - b' \tag{4}$$

where NAR is surface CO_2 net assimilation rate ($g \cdot CO_2 \cdot m^{-2} \cdot h^{-1}$) and Q is PAR in $\mu mol \cdot photons \cdot m^{-2} \cdot s^{-1}$, and a' and b' are the coefficients.

Table 2. Parameter values to estimate the effect of soil temperature T ($°C$) at the 5-cm depth on net carbon exchange (NCE) ($g \cdot CO_2 \cdot m^{-2} \cdot h^{-1}$) using equation $NCE = ae^{(bT)}$ and coefficient of determination for landforms across different ecoregions (sites).

Site	Location	a ($g \cdot CO_2 \cdot m^{-2} \cdot h^{-1}$)	b ($°C^{-1}$)	R^2
	Upland	0.128	0.17	0.91
Mid-boreal	Peat Plateau	0.018	0.28	0.75
	Collapse scar	0.072	0.08	0.79
	Upland	0.151	0.09	0.84
High-boreal	Peat Plateau	0.048	0.11	0.91
	Collapse scar	0.012	0.15	0.82
	Upland	0.035	0.15	0.82
Low Subarctic	Peat Plateau	0.034	0.11	0.76
	Collapse scar	0.047	0.11	0.68
	Upland	0.023	0.16	0.97
High Subarctic	Peat Plateau	0.010	0.27	0.87
	Collapse scar	0.011	0.356	0.82

2.4.2. Temporal Variation in CO_2 Fluxes

Surface NCE was also measured for extended periods of time (up to 45 days) with clear chambers using ACE automatic CO_2 exchange systems (ADC BioScientific Ltd., Hertfordshire, UK). ACE units were equipped with a built-in photosynthetically active radiation (PAR) silicone photocell sensor and a soil temperature probe, and were powered by gel-cell batteries and solar panels. The ACE units were deployed simultaneously in upland, and peat plateau locations with one unit in each plot for comparison of local CO_2 fluxes, surface temperatures, and illumination intensities between landform positions.

We also measured soil CO_2 concentrations using Vaisala solid-state infrared CO_2 sensors (Vaisala Oyj, Helsinki, Finland) at three depths (5, 20, and 40 cm) with Vaisala model GMP222 (measurement range 0–10,000 $\mu mol \cdot CO_2 \cdot mol^{-1}$) at the 5 cm depth, and model GMP221 (measurement range 0–20,000 $\mu mol \cdot CO_2 \cdot mol^{-1}$) at the 20 and 40 cm depths. The sensors were protected with Teflon socks and installed at different depths as explained in Jassal et al. [26], and connected to a data logger (Campbell Scientific CR1000) recording soil CO_2 concentrations every two hours. Soil respiration was obtained with the gradient technique using Equation (5) (see below). To obtain diffusivity values, soil physical properties, such as bulk density and porosity, and variations in soil temperature and moisture content were also measured. The gradient approach allowed for the calculation of the contribution of different soil layers to the total surface flux.

Soil temperature and volumetric water content (θ_v) were measured continuously starting spring 2008 to fall 2010 using HOBO U12 units equipped with temperature sensors (Model TMC6-HD, Onset Computer Corp., Bourne, MA, USA) and soil moisture sensors (Watermark Sensor type WMSM, Delta-T Device Ltd., Cambridge, UK), respectively. The temperature sensors were installed at the 5, 25, and 50 cm depths while the moisture sensors were installed at the 5 cm depth. Soil temperature and moisture readings were also taken every two hours.

2.4.3. Calculating Soil CO_2 Fluxes

CO_2 flux was calculated from the Fick's first law of diffusion (Equation (5)):

$$F = -D_s \frac{dC}{dz} \tag{5}$$

where F is CO_2 flux in $\mu mol \cdot m^{-1} \cdot s^{-1}$, D_s is diffusion coefficient in the soil ($m^2 \cdot s^{-1}$), C is soil CO_2 concentration at the given depth and dC/dz is vertical gradient of CO_2 concentrations in the soil between two measurement depths. The value of D_s was calculated as $D_s = \zeta D_a$ where ζ is soil gas tortuosity factor and D_a is CO_2 diffusivity in the air. The latter is calculated from the standard diffusivity value (D_{a0}) at 20 °C (293.15 K) and 101.3 kPa using the following equation (Equation (6)):

$$D_a = D_{a0} \left(\frac{T}{293.15}\right)^{1.75} \left(\frac{P}{101.3}\right) \tag{6}$$

where T is temperature (K) and P is pressure (kPa). The value of D_{a0} is empirically given as $14.7 \ m^2 \cdot s^{-1}$. We used the Millington-Quirk model [27] for calculation of tortuosity (ζ) as it was proven to provide the best correlation to the measured tortuosity [28] (see Equation (7)).

$$\zeta = \frac{\alpha^{\frac{10}{3}}}{\phi^2} \tag{7}$$

where α is air-filled porosity and ϕ is total porosity. The calculated soil CO_2 fluxes, using the gradient technique, for the uppermost soil layer (0–5 cm) were validated by plotting them against soil respiration values measured using the PP system on bare soil.

2.5. Statistical Analysis

Annual NCE and ER values were calculated for each site and landform position for the growing season (1 June to 30 September) of 2008, 2009, and 2010 by summing all daily values. As outlined in Section 2.4.1, collars were installed along three arms of the boardwalk, with two pseudo-replications on each arm. Arms were treated as replications. Re-measurements of the collars each month were initially treated as repeated measures. However, that did not permit for the account of variations in the temperature and light. The latest calculation treats re-measurements as replicated measurements using ANCOVA testing for month $\times$ temperature covariance. Barlett's test was used to test homoscedasticity and Pearson's chi-square test for normality for comparison of the groups by site and by plot. In case, the assimilation in the two plots variability ratio did not meet the criteria for homoscedasticity, the weighed variables method in analysis was employed. Comparison of temporal variations of NCE and ER between different sites and landform positions was conducted using analysis of Variance (ANOVA) with repeated measures coupled with Tukey's Studentized Range Test in SAS (SAS Institute Inc., Cary, NC, USA, 2004) [29]. Interaction effects were elucidated using the *pdiff* option in SAS. To determine the significance of trends in NCE, ER, and assimilation rate and landforms variables, Pearson product-moment correlations were carried out in SAS. Significance was tested using *p*-value = 0.05.

3. Results

3.1. Weather Monitoring

During three years of the study, weather at all four sites was close to their climatic normals (Table 1). The temperature and precipitation at different sites are presented in Figure 3. Year 2007 was somewhat warmer at all sites except for Norman Wells where the annual temperature was the same as the long-term average. The annual mean temperature in 2007 in Anzac and Inuvik was warmer than their long-term means by 0.2 and 1.2 °C, respectively. 2008 was also somewhat warmer at all sites except Norman Wells. Year 2009 was slightly cooler at all sites except Inuvik where the mean annual temperature was about 1 °C higher than normal. Though the length of the warm (above freezing) season was significantly longer in the southern sites with mild winters, there were no significant differences in the mid-summer temperatures between the sites. Soil temperature generally followed air temperature with a maximum in July, though it did not warm up to the same degree in the north as in the south during shorter summers. Precipitation along the valley typically is between 300 to 400 mm, tapering off along the Mackenzie Delta to 220 mm at Inuvik. Individual daily rainfalls are typically light with few exceeding 5 mm; however, heavy rains can happen during summer months with rainfall occasionally exceeding 30 mm for sites in Fort Simpson and Newman Wells. After October, most of the precipitation falls as snow.

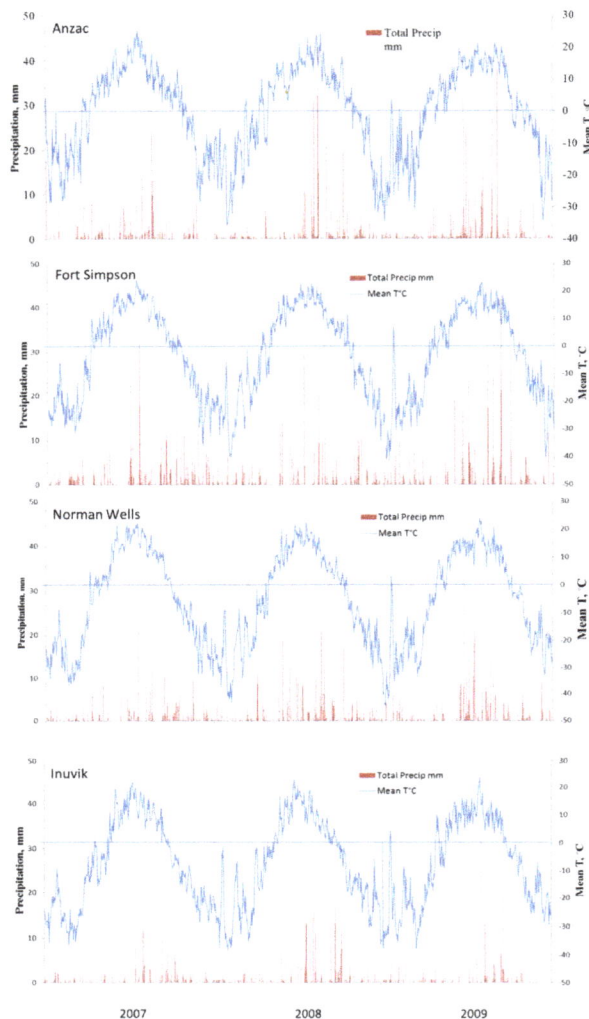

Figure 3. Seasonal patterns of soil temperature (T_s) and precipitation (Ppt) during the three year period at Anzac (AB), Fort Simpson (NWT), Norman Wells (NWT) and Inuvik (NWT).

3.2. Diurnal Variations in NCE

Diurnal measurements of surface NCE indicated the surface was a source of CO_2 with respiration dominating the gas exchange balance at most of the study sites. Hourly measurement of NCE corresponded closely to diurnal temperature variations. Figure 4 shows the distinct diurnal pattern in measured NCE in collapse scars, peat plateau, and upland areas at the Fort Simpson site with upland showing maximum (9.0 $\mu mole \cdot m^{-2} \cdot s^{-1}$) NCE followed by peat plateau (3.9 $\mu mole \cdot m^{-2} \cdot s^{-1}$) and collapse scars (0.3 $\mu mole \cdot m^{-2} \cdot s^{-1}$), also suggesting that photosynthetic activity of the ground cover bryophytes played a notable role only in collapsed areas. Both upland and tree-covered peat plateau areas had higher surface CO_2 fluxes (emissions) during the daytime, but collapse scar exhibited a double-peak pattern with higher respiration during nighttime, which began to decrease with sunrise and reached negative values (becoming carbon sink) during morning hours. During mid-day when the temperature was at the peak, collapse scar NCE turned positive again followed by another period of negative NCE showing the prevalence of assimilation during late afternoon until sunset.

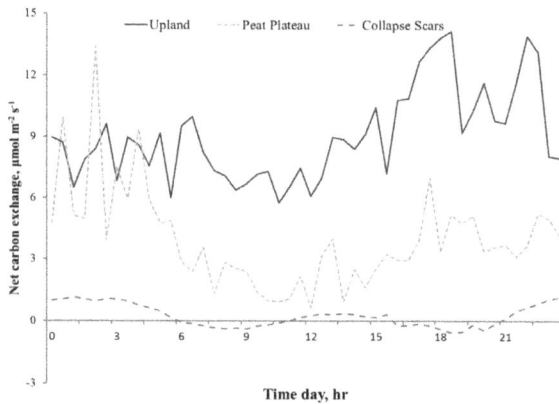

Figure 4. Role of landform on diurnal variation in NCE in upland, peat plateau, and collapse scar on 25 June 2009, at Fort Simpson (NWT) Canada.

3.3. Statistical Model of Surface CO_2 Fluxes against Temperature

Net carbon exchange (NCE) was primarily controlled by soil temperature at the 5 cm depth as the exponential model (Equation (1), NCE = $ae^{(bT)}$) explained between 68% to 97% variation in observed fluxes for upland, peat plateau, and collapse scar sites along the latitudinal gradient (Table 2). Adding soil moisture factor into the model did not improve it any further. The model showed variation in coefficients a and b, and Q_{10} values, based on landform along the north-south gradient. At peat plateau and collapsed areas, the effect of temperature was less obvious (Tables 3 and 4) where soil moisture would have a dominant control. There was also a good agreement between ER measurements and soil temperature at the 5 cm depth at different sites and landforms. The values of the parameters in Equation (2) (ER = $ae^{(bT)}$) are presented in Table 3. The value of the a coefficient was higher for all the upland sites as compared to peat plateau. The ER-temperature and NCE temperature responses did not differ dramatically among landform units suggesting that ER was a major component of the NCE. The $Q_{10\ values}$ estimated for different landform positions along the latitudinal gradient are also shown in Table 4. The response of ER to the temperature varied between ecoregions and landform positions with Q_{10} values being the greatest in the peat plateau of south most sites (5.6–11.5) followed by uplands (2.9–5.4) and collapsing scar (1.2–4.0), and decreasing with increases in latitude (Table 4). Furthermore, we also found that ER was negatively affected by water table and permafrost depths.

Table 3. Parameter values to estimate the effect of soil temperature T ($^\circ$C) at the 5-cm depth on surface respiration (g·CO$_2$·m^{-2}·h^{-1}) using equation ER = $ae^{(bT)}$ and coefficient of determination for landforms across different ecoregions (sites).

Site	Location	a (g·CO$_2$·m^{-2}·h^{-1})	b ($^\circ$C^{-1})	R^2
Mid-boreal	Upland	0.061	0.251	0.68
	Peat Plateau	0.010	0.339	0.83
High-boreal	Upland	0.146	0.080	0.63
	Peat Plateau	0.001	0.216	0.91
Low Subarctic	Upland	0.008	0.208	0.65
	Peat Plateau	0.001	0.519	0.58
High Subarctic	Upland	0.010	0.183	0.56
	Peat Plateau	0.002	0.299	0.46

Table 4. Q_{10} values for surface respiration in upland (UL), peat plateau (PP) and collapse scars (CS) across different ecoregions (sites).

Site	Uplands	Peat Plateau	Collapse Scar
Mid-boreal	5.38	11.49	3.97
High-boreal	3.41	11.36	3.28
Low Subarctic	3.16	5.45	2.18
High Subarctic	2.86	5.53	1.22

3.4. Time Series of NCE and ER

The temporal variations in surface NCE and ER are shown in Figures 5 and 6, respectively, depicting most pronounced differences among sites. Surface NCE and ER followed a bell-shaped curve during the year closely corresponding to the seasonal variation in soil temperature. The peak NCE and ER values were observed during the summer months of July and August (Figure 5). The greatest amount of CO_2 was released in the southernmost site; however, from mid-boreal to low-arctic, the rate of CO_2 release from uplands dropped sharply due to a lower seasonal temperature in the north. A comparison of Figures 5 and 6 show that in upland locations, ER was the major component of NCE, exceeding assimilation rate by ground cover vegetation three-fold. Therefore, the response of NCE to ecoclimatic factors was, to a large extent, dictated by the response of ER. NCE did not vary much between peat plateaus with extensive and continuous permafrost in low or high subarctic regions. Results show that total CO_2 emissions were greatest in the upland location, least in the collapse scar with peat plateau occupying the intermediate position (Figure 5).

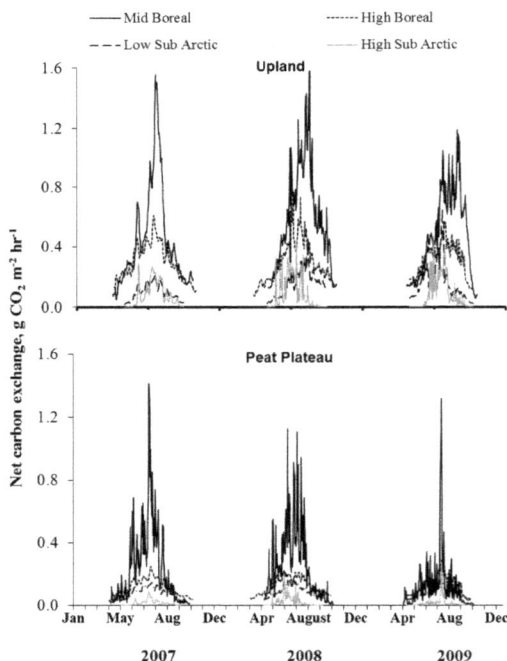

Figure 5. Temporal variation in net carbon exchange (NCE) for different landforms at sites within different ecoregions.

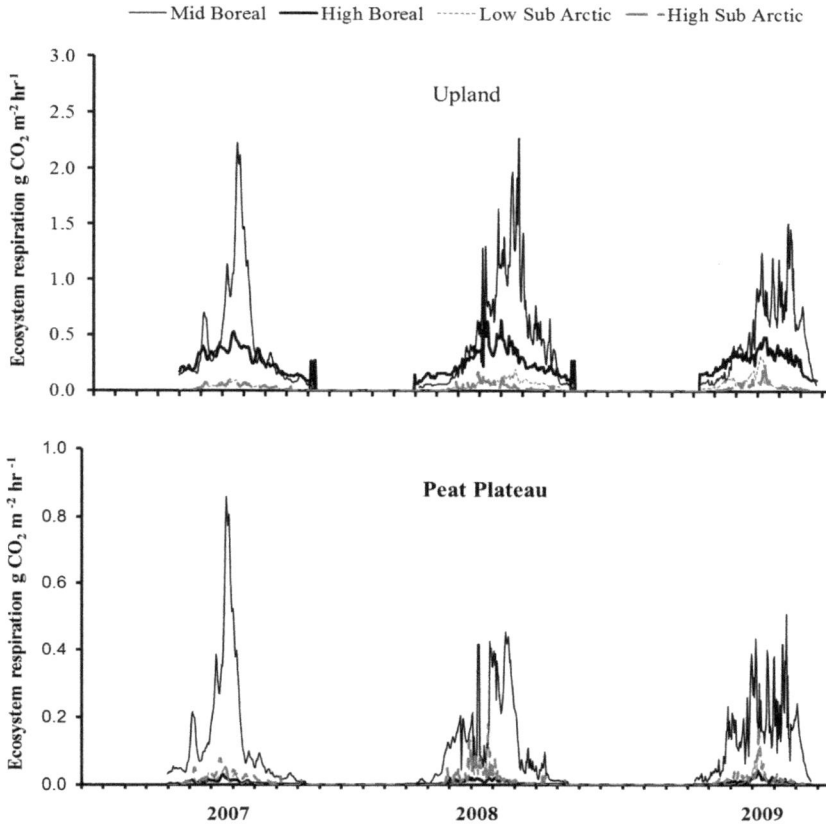

Figure 6. Temporal variation in ecosystem respiration (ER) at upland and peat plateau landforms at sites within different ecoregions.

Calculated flux contributions from different soil layers using the CO_2 gradient technique showed that in both peat plateau (PP) and upland (UL) areas, the greatest contribution (about 75%) of soil respiration (SR) originated in the upper 5 cm layer (Figure 7). Combined contribution of the lower horizons did not exceed 25% of the total flux. Both in upland and peat plateau, soil temperature was a strong driver of CO_2 release in SR. In peat plateau, the lower layer (21 to 40 cm) contributed about only 5% of SR, which could be due to the presence of the water table. Peat plateau areas exhibited greater variability in the flux with a pronounced increase in the CO_2 emissions after rain events, which can be attributed to increased decomposition [26] or displacement of soil-stored CO_2 [30].

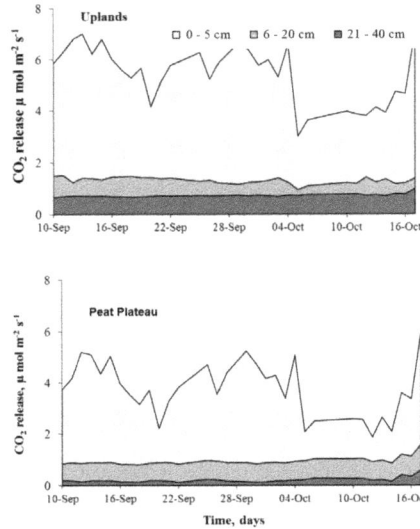

Figure 7. Comparative contribution of different soil layers to soil CO_2 efflux in uplands (top panel) and peat plateau (bottom panel) during 2008 at Fort Simpson (NWT) Canada.

3.5. Effect of Light on Surface Assimilation Rate

The surface assimilation rate was calculated as the difference between measured hourly NCE and ER, and was correlated to the simultaneously measured PAR values. Surface assimilation was significantly correlated with PAR at different landforms along the climatic gradient (Table 5). Correlation obeyed logarithmic relationship and was most significant in the areas of unimpeded solar radiation. In collapsed scar areas, seasonal assimilation was consistently higher (Table 6) than in either uplands or peat plateaus by as much as three times due to the abundance of sphagnum. In peat plateaus, assimilation rate was highly variable due to the frequent drying out of the lichens, which dominated the ground vegetation. With less shading in the peat plateau as compared to uplands (forest areas), significantly higher assimilation was observed (Table 6). During dry days, the surface assimilation was not measurable, and when surface was moist, assimilation by lichens was comparatively low (less than 0.05 g·CO_2·m^{-2}·h^{-1}) but increased logarithmically with illumination.

Table 5. Parameter values to estimate the effect of light on surface assimilation (g·CO_2·m^{-2}·h^{-1}) using Equation: $F'(CO_2) = a' \times \ln(Q)^{-b'}$ and coefficient of determination for landforms across different ecoregions (sites).

Site	Location	A'	b'	R^2
	Upland	0.41	0.15	0.57
Mid boreal	Peat Plateau	0.45	0.17	0.76
	Collapse Scar	1.27	0.26	0.81
	Upland	0.21	0.07	0.71
High boreal	Peat Plateau	0.20	1.14	0.55
	Collapse Scar	0.17	0.53	0.67
	Upland	0.18	0.93	0.69
Low Subarctic	Peat Plateau	0.11	0.61	0.93
	Collapse Scar	0.15	0.75	0.74
	Upland	0.16	0.64	0.66
High Subarctic	Peat Plateau	0.02	0.03	0.71
	Collapse Scar	0.38	0.12	0.87

Table 6. Cumulative seasonal (1 June to 30 September 2009) net assimilation rate at different landform positions (in $g \cdot C \cdot m^{-2} \cdot season^{-1}$).

Site	Upland	Peat Plateau	Collapse Scar
Low boreal	82.6 ± 4.01 Aa	139.2 ± 12.4 Ba	211.3 ± 9.68 Ca
High boreal	62.0 ± 6.58 Ab	92.0 ± 8.92 Bb	198.0 ± 5.24 Ca
Low Subarctic	56.7 ± 3.74 Ac	63.3 ± 7.69 Ac	127.4 ± 3.92 Cb
High Subarctic	38.12 ± 4.62 Ad	54.6 ± 8.16 Bc	86.5 ± 6.15 Cc

Capital letter represent the $p < 0.05$ between land form and lower case letter shows the $p < 0.05$ between sites.

3.6. Annual ER and NCE

A comparison of seasonal NCE among different sites (Figure 8) showed that net carbon exchange at the surface was a significantly larger source at the upland sites (38 to 576 $g \cdot C \cdot m^{-2} \cdot year^{-1}$) as compared to peat plateau (8 to 159 $g \cdot C \cdot m^{-2} \cdot year^{-1}$) while collapse scars ($-85$ to -23 $g \cdot C \cdot m^{-2} \cdot year^{-1}$) were C sinks at all locations along the transect. In general, annual NCE decreased with mean annual temperature, which decreased with increase in latitude, with high sub-arctic site showing a net assimilation integrated over upland, peat plateau, and collapse scar areas. The upland and peat plateau sites were, on average, net CO_2 sources to the atmosphere, while the collapse scar sites were, on average, net CO_2 sinks. The NCE measurements pertain only to the ground surface and short vegetation and do not include CO_2 uptake by trees and large shrubs. ER varied along the latitudinal gradient, for upland sites from 6.0 to 538 $g \cdot C \cdot m^{-2} \cdot year^{-1}$, for peat plateau from 2.0 $g \cdot C \cdot m^{-2} \cdot year^{-1}$ to $187g \cdot C \cdot m^{-2} \cdot year^{-1}$, followed by collapse scars with 60 to 214 $g \cdot C \cdot m^{-2} \cdot year^{-1}$ (Figure 9). Low soil temperature along with high water table at both upland and peat plateau sites in low and high subarctic regions resulted in lower ER. Since the water table was at the surface at all the collapse scar sites, ER was negligible. In mid and high-boreal zone, ER at the upland sites was two- to three-fold higher as compared to peat plateau sites with and without collapse scar.

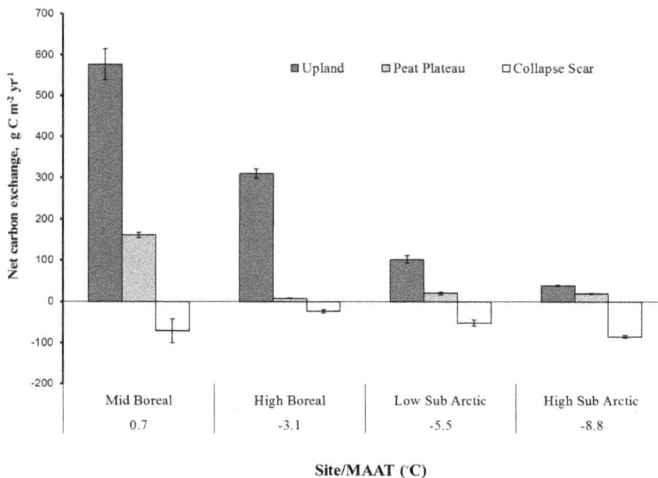

Figure 8. Seasonal net carbon exchange (NCE) for different sites and landform positions. MATT (°C) is mean annual air temperature.

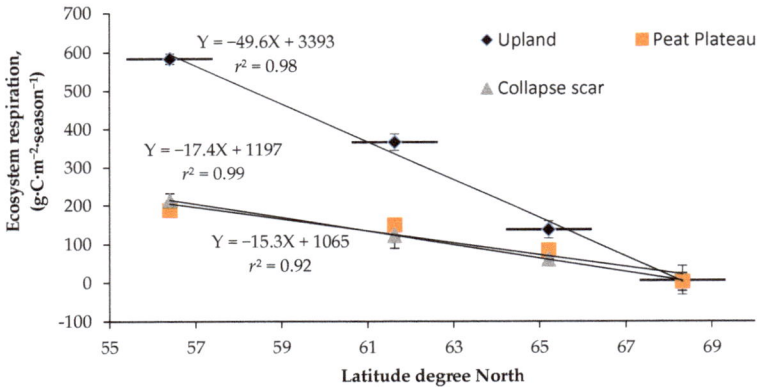

Figure 9. Cumulative seasonal surface fluxes for different landforms at sites along the latitudinal gradient.

4. Discussion

CO_2 exchange from the surface of permafrost-affected areas is determined by a combination of several factors, including soil microclimate [31–34], landform features [2,35–37], water table depth [38], extent of permafrost [2,39,40], and vegetation type [41,42]. There are two important contributors to the CO_2 flux: soil respiration and assimilation, and respiration of the ground cover vegetation. They vary differently in response to physiology and ecoclimate. NCE, assimilation rate and ER by surface vegetation were significantly affected by the local landform positions (i.e., uplands, peat plateau, and collapse scars), and different sites along latitude north–south gradient.

4.1. Spatial Variation in Net Carbon Exchange (NCE)

Net ecosystem exchange in a subarctic forest was lower than that of boreal forest due to extremely cold soils and high water table in the north. Bisbee [41] reported NCE to be 562 $g \cdot C \cdot m^{-2} \cdot year^{-1}$ for southern sites while Wang et al. [43] reported it to be 226–413 $g \cdot C \cdot m^{-2} \cdot year^{-1}$ for northern sites in the BOREAS study, which are similar to the NCE observed in the present study for mid- and high-boreal forest zones. In peat plateau, a presence of permafrost significantly decreased NCE from 160 $g \cdot C \cdot m^{-2} \cdot year^{-1}$ to 20–50 $g \cdot C \cdot m^{-2} \cdot year^{-1}$ among areas with sporadic and continuous permafrost. Similar NCE values were observed for a nutrient rich fen and sub-arctic palsa in Scandinavian peatlands [43,44]. However, it is not always possible to compare our seasonal NCE estimates with other studies, because of differences in climate, duration of the study period and variations in snow-cover period. Many studies have observed that spring conditions greatly influence seasonal and annual CO_2 exchange, where earlier snow melts and/or warmer spring temperatures almost universally enhance accumulated CO_2 uptake in high-boreal and subarctic regions [40,45,46]. Collapsed areas usually behaved as CO_2 sinks due to high productivity of bryophytes and restricted respiration in flooded soils. The length of the thawed period was significantly shorter in the northern sites, limiting both decomposition and growth periods, but this effect was apparently compensated by the increase in bryophyte assimilation due to longer daylight conditions in subarctic regions. It is likely that the combination of these two factors was responsible for similar values of NCE in collapsed areas at different sites as total assimilation varying between 20 and 70 $g \cdot C \cdot m^{-2} \cdot year^{-1}$. Our NCE estimates for collapse scar sites were in the range of values reported for low Arctic tundra sites [47–49] and also observed large inter-annual variability in NCE

4.2. Assimilation Rates across Different Landforms

Another important factor that determines ecosystem carbon exchange is photosynthetic activity (i.e., CO_2 assimilation rate) of other ground cover vegetation [50]. Assimilation of CO_2 by the bryophytes is affected by light intensity, moisture, temperature, and species composition. All these factors varied with ecozone and landform. Upland forests of the southern sites had greater shading than northern uplands and therefore had a limited assimilation rate of bryophyte layer, though the total length of the growing season was longer and the canopy provided some protection from drying. While assimilation rate of the ground cover vegetation was inhibited by shading, relatively good drainage of the upper soil layers, stable soil moisture levels, and abundant input of organic debris from tree crowns provided favorable conditions for high microbial respiration. Bryophyte layer of peat plateaus had a large proportion of lichens that are less effective in CO_2 assimilation [51] but are prone to drying, especially in non-treed northern sites. *Sphagnum* present at peat plateau sites may have been moisture-limited because *Sphagnum* moss photosynthesis is highly sensitive to moisture [52,53]. Tuittila et al. [54] measured the optimum water table level for photosynthesis by *S. angustifolium* and observed decreased photosynthesis rate at lower water table levels in bogs. Most effective bryophyte assimilation took place in the collapsed areas due to unimpeded illumination, earliest warming in the spring, unlimited supply of soil moisture, and relatively effective photosynthesis of mosses [55].

4.3. Spatial Variation in Ecosystem Respiration (ER)

Variation in soil temperature at the 5-cm depth explained most of the variability in ER. Q_{10} values for ER observed in the study (Tables 3 and 4) were within the range of values reported for boreal forest soils [56–58]. Many other studies on boreal forest stands [58,59] have shown that Ts (soil temperature) exerts a major influence on ER. Although, ER in uplands was primarily driven by soil temperature, high water table present in the collapse scar sites inhibited respiration [60]. Therefore the highest ER was observed in the southern-most location in the relatively dry upland soils. In the northern sites, it was limited by short frost-free periods, and frequently by water saturation in peat plateau and collapse scar sites [61]. Spatial variability in ER showed a strong relationship with latitudinal gradient which is a good proxy for the combined influence of radiation, length of growing season, temperature, precipitation, and vegetation cover (Figure 8). The different relationships for upland and peat plateau showed that the vegetation and climate were the most critical factors in regulating the spatial variability in ER with latitude [62,63]. As temperature, length of growing season and vegetation productivity decrease from south to north, ER decreases with latitude.

4.4. Quantitative Relationships between NCE and ER and Climate Variables

Soil temperature and moisture are important parameters in regulating surface CO_2 fluxes in boreal ecosystems [60,64–66]. Various empirical models have been used to quantify the relationship between ER and temperature, e.g., simple linear by Bisbee, [41] and Wang et al. [43] exponential by Lavigne et al., [64] and Rayment and Jarvis, [65]. We found that NCE and ER were highly correlated to soil temperature with an exponential relationship for all the sites. Increasing temperature can increase surface CO_2 fluxes through (a) activating dormant microbes and increasing microbial species richness, thus mineralization of C; (b) increasing root respiration by increasing photosynthates translocation from the aboveground part of the plant to roots; (c) accelerating gas exchange between ecosystems and the atmosphere [60]. However, the temperature sensitivity of NCE and ER is not constant, but tends to be modified by seasonal changes in soil moisture, root biomass, litter inputs, microbial populations, and other seasonally fluctuating conditions and processes [67]. Although temperature represented a significant control on the variability of the surface CO_2 fluxes in this study, other studies have found that water table depth or timing of the snowmelt play a more important role in predicting the CO_2 uptake capacity in many peatlands. For example, in a minerogenic peatland in Sweden, Peichl et al. [58] found that the growing season NCE was strongly correlated with pre-growing season

T_{air}, (air temperature) and suggested that less root damage would occur during warmer winters and access to water by the vegetation would be easier. Similarly, in a mild maritime climate ombrotrophic bog, winter temperature would be a main control on the inter-annual variability in NEE [61].

Another important driver of ER is the nature and source of the decomposing material. Landform is closely associated with vegetation type. While upland locations at all sites had black spruce stands, collapsed areas had only ground cover vegetation, and peat plateaus had trees in boreal ecozones but no trees in subarctic sites. Accordingly, these locations varied in the amount of plant debris input and its quality. Startsev and Lieffers [68] showed that decomposing litter generally trapped in the bryophyte layer contributed to the total CO_2 production, and provided bryophytes with a source of easily available nutrients and carbohydrates. In our study, the total amount of plant debris input was greater in the southern treed areas where trees were bigger and more productive, which was also an important factor in southern treed peat plateaus. In collapsed scar areas and northern subarctic treeless peat plateaus, the only source of decomposing organic matter was the annual production of ground vegetation and small shrubs, which provided considerably less decomposable material.

The type of ground vegetation could be another major variable to determine ER as upland sites were dominated by feathermoss in mid and high-boreal ecozones while peatland was dominated by the lichens. We found that Q_{10} values were highest at the peat plateau sites dominated by lichens. Bergeron et al. [62] also observed that Q_{10} was highest under lichen followed by sphagnum and lowest for feathermoss dominated understory vegetation. ER of bryophyte layer was also affected by temperature. The observed decrease in Q_{10} values from 5.4 and 5.5 to 2.2 and 1.2, upon thawing of peat plateau in low subarctic and high subarctic region respectively, were close to the range reported by previous studies [69,70].

Permafrost is another factor inhibiting respiration rates. ER was most active in the upper soil horizons (Figure 6). It appears that at upland and peat plateau sites, the depth of active layer was sufficient to support microbiological activity throughout the growing season, and therefore presence of permafrost had only a limited effect on ER. On the other hand, at collapse scar sites, accompanying factors such as length of time over which soil was frozen, the mean soil temperature throughout the year, and the presence of perched water table in the soil profile had a major impact on ER [38]. In collapse scar sites, lower ER could be attributed to slow decomposition rates of deep organic matter, low O_2 availability in saturated soils, low temperature, and substrate recalcitrance [36,37,71]. Although winter surface CO_2 fluxes were not measured in this study, Wang et al. [43] reported that in high-boreal forest, winter SR accounted for 5%–19% of the annual ER, which agrees well with the value reported in similar boreal forests [72] while Kim et al. [39] observed that wintertime soil CO_2 efflux contributed 24% of the annual CO_2 efflux in Alaska.

4.5. Implicatiuons of Changing Climate

Existing climatic records [73] indicate that climate change over the last five decades has increased mean annual air temperature at the study sites by 0.5–2 °C. Climatic change in is likely to affect NCE in several direct and indirect ways. Longer growing seasons, elevated CO2, and increased nutrients released from decomposing organic carbon may all stimulate plant growth [74] to increase C assimilation. Temperature-driven increases in soil respiration would likely be pronounced in the drained mid-boreal uplands where Q_{10} values are the largest. Higher respiration rates of drained upland soils combined with large input of organic debris from the tree crowns and limited assimilation of shaded bryophytes determined that uplands were consistent sources of CO_2, especially in the southern sites [43]. Low productivity of bryophytes in peat plateaus, in association with limited soil respiration due to greater water-saturated layer and permafrost, resulted in peat plateaus being near-neutral or minor sources of CO_2 with little difference across climatic gradient. Degradation of peat plateau to collapse scars will result in a dramatic change in hydrology and surface vegetation at the local level, and may result in a net increase in NCE, which would offset increase in ER depending on the extent of permafrost thawing. Accelerated collapsing of the peat plateau would subject soil to waterlogging, thereby decreasing aerobic respiration. Furthermore, with receding of permafrost, the connectivity of collapsing features would drain the landscape [75,76] resulting in increased active layer depth, thereby making the previously frozen organic matter available for decomposition. However, an increase in the rate of surface assimilation by bryophytes is to be expected in the expanding collapsed areas. With rising water table, the *Sphagnum* spp. found on hummocks tends to increase productivity [77]. However, lower water table and increase in mean annual air temperature would result in significant increase in surface CO_2 emissions, while only minor increase in surface assimilation rates can be expected [61]. As to whether collapse scar sites in boreal and subarctic regions are an annual source or sink for CO_2, we cannot know without estimating wintertime fluxes. Even small wintertime effluxes would offset growing season CO_2 sinks. There is clearly a pressing need for future research in this area.

5. Conclusions

This study elucidated the changes in the surface CO_2 exchange along the longitudinal gradient as influenced by the local landform and soil microclimate variations. Both ER and NCE decreased with latitude with the highest values observed at the upland and the lowest at peat plateau locations while collapse scar sites were C sinks. The strong relationship between ER and latitude can be used to estimate regional ER and the relationship can be refined with more data, especially wintertime measurements. Both NCE and ER were primarily controlled by soil temperature in the upper horizons along the climatic gradient. Environmental conditions and vegetation type determine the C source-sink relationship. With continuously waterlogged conditions at the collapsed scar sites, bryophyte NCE was highest, and was negligible at upland and peat plateau locations. On an average, upland forest sites and peat plateaus were respectively major and minor sources of CO_2, collapsed scar sites were sinks of CO_2.

Acknowledgments: This project was made possible with the generous support of the Government of Canada Program for International Polar Year, and Natural Resources Canada, Canadian Forest Service. Many thanks are also due to Steve Gooderham, Mike Gravel, Tom Lakusta and Paul Rivard, from the Forest Management Division, Government of the Northwest Territories. Without the valuable contributions of field and laboratory staff, this project would not have been possible and we would like to acknowledge the hard work of: Patrick Hurdle, Thierry Verem Sanders, Ruth Errington, and Catherine McNalty.

Author Contributions: J.S.B. conceived the idea and designed the experiments; N.S. performed the experiments; N.S. and J.S.B. analyzed the data; J.S.B., N.S. and R.S.J. wrote the paper.

Conflicts of Interest: The authors declare no conflict of interest. Funding agencies played no role in the design of the study; in the collection, analyses or interpretation of data; in the writing of the manuscript, or in the decision to publish the results.

References

1. Abbott, B.W.; Jones, J.B.J.; Schuur, E.; Chapin, F.; Bowden, W.B.; Bret-Harte, M.S.; Epstein, H.E.; Flannigan, M.D.; Harms, T.K.; Hollingsworth, T.N.; et al. Biomass offsets little or none of permafrost carbon release from soils, streams, and wildfire: An expert assessment. *Environ. Res. Lett.* **2016**, *11*, 034014. [CrossRef]
2. Schuur, E.G.; McGuire, A.D.; Schädel, C.; Grosse, G.; Harden, J.W.; Hayes, D.J.; Hugelius, G.; Koven, C.D.; Kuhry, P.; Lawrence, D.M.; et al. Climate change and the permafrost carbon feedback. *Nature* **2015**, *520*, 171–179. [CrossRef] [PubMed]
3. Schuur, E.A.; Bockheim, G.J.; Canadell, J.G.; Euskirchen, E.; Field, C.B.; Goryachkin, S.V.; Hagemann, S.; Kuhry, P.; Lafleur, P.M.; Lee, H.; et al. Vulnerability of Permafrost Carbon to Climate Change: Implications for the Global Carbon Cycle. *Bioscience* **2008**, *58*, 701–714. [CrossRef]
4. Hugelius, G.; Strauss, J.; Zubrzycki, S.; Harden, J.W.; Schuur, E.A.G.; Ping, C.-L.; Schirrmeister, L.; Grosse, G.; Michaelson, G.J.; Koven, C.D.; et al. Estimated stocks of circumpolar permafrost carbon with quantified uncertainty ranges and identified data gaps. *Biogeosciences* **2014**, *11*, 6573–6593. [CrossRef]

5. Tarnocai, C.; Canadell, J.G.; Schuur, E.A.G.; Kuhry, P.; Mazhitova, G.; Zimov, S.A. Soil organic carbon pools in the northern circumpolar permafrost region. *Glob. Biogeochem. Cycles* **2009**, *23*. [CrossRef]
6. Raich, J.W.; Schlesinger, W.H. The global carbon dioxide flux in soil respiration and its relationship to vegetation and climate. *Tellus B* **1992**, *44*, 81–99. [CrossRef]
7. Zhuang, Q.; McGuire, A.D.; O'Neill, K.P.; Harden, J.W.; Romanovsky, V.E.; Yarie, J. Modeling soil thermal and carbon dynamics of a fire chronosequence in interior Alaska. *J. Geophys. Res.* **2003**, *107*. [CrossRef]
8. Kurz, W.A.; Shaw, C.H.; Boisvenue, C.; Stinson, G.; Metsaranta, J.; Leckie, D.; Dyk, A.; Smyth, C.; Neilson, E.T. Carbon in Canada's boreal forest—A synthesis. *Environ. Rev.* **2014**, *21*, 260–292. [CrossRef]
9. Price, D.T.; Alfaro, R.I.; Brown, K.J.; Flannigan, M.D.; Fleming, R.A.; Hogg, E.H.; Girardin, M.P.; Lakusta, T.; Johnston, M.; McKenney, D.W.; et al. Anticipating the consequences of climate change for Canada's boreal forest ecosystems. *Environ. Rev.* **2013**, *21*, 322–365. [CrossRef]
10. Harden, J.W.; Koven, C.D.; Ping, C.L.; Hugelius, G.; McGuire, A.D.; Camill, P.; Jorgenson, T.; Kuhry, P.; Michaelson, G.J.; O'Donnell, J.A.; et al. Field information links permafrost carbon to physical vulnerabilities of thawing. *Geophys. Res. Lett.* **2012**, *39*, L15704. [CrossRef]
11. Thie, J. Distribution and Thawing of Permafrost in the Southern Part of the Discontinuous Permafrost Zone in Manitoba. *Arctic* **1974**, *27*, 189–200. [CrossRef]
12. Waddington, J.M.; Roulet, N.T. Carbon balance of a boreal patterned peatland. *Glob. Chang. Biol.* **2000**, *6*, 87–98. [CrossRef]
13. Mitsch, W.J.; Gosselink, J.G. *Wetlands*, 4th ed.; John Wiley and Sons: Hoboken, NJ, USA, 2007; p. 582.
14. Drewitt, G.B.; Black, T.A.; Nesic, Z.; Humphreys, E.R.; Jork, E.M.; Swanson, R.; Ethier, G.J.; Griffis, T.; Morgenstern, K. Measuring forest-floor CO_2 fluxes in a Douglas-fir forest. *Agric. For. Meteorol.* **2002**, *110*, 299–317. [CrossRef]
15. Green, R.N.; Trowbridge, R.L.; Klinka, K. Towards a taxonomic classification of humus forms. *For. Sci.* **1993**, *39*, 1–49.
16. Halsey, L.A.; Vitt, D.H.; Zoltai, S.C. Disequilibrium response of permafrost in boreal continental western Canada to climate change. *Clim. Chang* **1995**, *30*, 57–73. [CrossRef]
17. Moore, T.R.; Roulet, N.T.; Waddington, J.M. Uncertainty in Predicting the Effect of climatic Change on the Carbon Cycling of Canadian Peatlands. *Clim. Chang.* **1998**, *40*, 229–245. [CrossRef]
18. Kayranli, B.; Scholz, M.; Mustafa, A.; Hedmark, A. Carbon storage and fluxes within freshwater wetlands: A critical review. *Wetlands* **2010**, *30*, 111–124. [CrossRef]
19. Grosse, G.; Harden, J.; Turetsky, M.R.; McGuire, A.D.; Camill, P.; Tarnocai, C.; Frolking, S.; Schuur, E.A.G.; Jorgenson, T.; Marchenko, S.; et al. Vulnerability of high-latitude soil organic carbon in North America to disturbance. *J. Geophys. Res.* **2011**, *116*, G00K06. [CrossRef]
20. Liblik, L.; Moore, T.R.; Bubier, J.L.; Robinson, S.D. Methane emissions from wetlands in the discontinuous permafrost zone: Fort Simpson, NWT, Canada. *Glob. Biogeochem. Cycles* **1997**, *11*, 485–494. [CrossRef]
21. Turetsky, M.R.; Wieder, R.K.; Vitt, D.H. Boreal peatland C fluxes under varying permafrost regimes. *Soil Biol. Biochem.* **2002**, *34*, 907–912. [CrossRef]
22. Prater, J.L.; Chanton, J.P.; Whiting, G.J. Variation in methane production pathways associated with permafrost decomposition in collapse scar bogs of Alberta, Canada. *Glob. Biogeochem. Cycle* **2007**, *21*, GB4004. [CrossRef]
23. Flanagan, L.B.; Syed, K.H. Simulation of both photosynthesis and respiration in response to warmer and drier conditions in a boreal peatland ecosystem. *Glob. Chang. Biol.* **2011**, *17*, 2271–2872. [CrossRef]
24. Elberling, B.; Michelsen, A.; Schädel, C.; Schuur, E.A.G.; Christiansen, H.H.; Berg, L.; Tamstorf, M.P.; Sigsgaard, C. Long-term CO_2 production following permafrost thaw. *Nat. Clim. Chang.* **2013**, *3*, 890–894. [CrossRef]
25. Skinner, W.; Maxwell, B. Climatic Patterns, Trends and Scenarios in the Arctic. In Proceedings of the Sixth Biennial AES/DIAND Meeting on Northern Climate & Mid-Study Workshop of the Mackenzie Basin Impact Study, Yellowknife, NT, Canada, 10–14 April 1994; pp. 125–137.
26. Jassal, R.; Black, A.; Novak, M.; Morgenstern, K.; Nesic, Z.; Gaumont-Guay, D. Relationship between soil CO_2 concentrations and forest-floor CO_2 effluxes. *Agric. For. Meteorol.* **2005**, *130*, 176–192.
27. Millington, R.J.; Quirk, J.M. *Transport in Porous Media*; Elsevier: Amsterdam, The Netherlands, 1960; pp. 97–106.
28. Sallam, A.; Jury, W.A.; Letey, J. Measurement of gas diffusion coefficient under relatively low air-filled porosity. *Soil Sci. Am. J.* **1984**, *48*, 3–6. [CrossRef]

29. SAS Institute Inc. *The SAS System for Windows*, 9th ed.; SAS Institute Inc.: Cary, NC, USA, 2004.

30. Lee, X.; Wu, H.J.; Sigler, J.; Oishi, C.; Siccama, T. Rapid and transient response of soil respiration to rain. *Glob. Chang. Biol.* **2004**, *10*, 1017–1026. [CrossRef]

31. Bond-Lamberty, B.; Thomson, A. Temperature-associated increases in the global soil respiration record. *Nature* **2010**, *464*, 597–582. [CrossRef] [PubMed]

32. Bronson, D.R.; Gower, S.T.; Tanner, M.; Linder, M.; van Herk, I. Response of soil surface CO_2 flux in a boreal forest to ecosystem warming. *Glob. Chang. Biol.* **2008**, *14*, 856–867. [CrossRef]

33. Davidson, E.A.; Belk, E.; Boone, R.D. Soil water content and temperature as independent or confounded factors controlling soil respiration in a temperate mixed hardwood forest. *Glob. Chang. Biol.* **1998**, *4*, 217–227. [CrossRef]

34. Davidson, E.A.; Jassens, I.A. Temperature sensitivity of soil carbon decomposition and feedback to climate change. *Nature* **2006**, *440*, 165–173. [CrossRef] [PubMed]

35. Schuur, E.A.G.; Abbott, B.W.; Bowden, W.B.; Brovkin, V.; Camill, P.; Canadell, J.G.; Chanton, J.P.; Chapin, F.S., III; Christensen, T.R.; Ciais, P.; et al. Expert assessment of vulnerability of permafrost carbon to climate change. *Clim. Chang.* **2013**, *119*, 359–374. [CrossRef]

36. Clymo, R.S. The limits to peat bog growth. *Philos. Trans. Soc. R. Ser. B* **1984**, *3113*, 605–654. [CrossRef]

37. Wickland, K.P.; Striegl, R.G.; Neff, J.C.; Sachs, T. Effects of permafrost melting on CO_2 and CH_4 exchange of a poorly drained black spruce lowland. *J. Geophys. Res.* **2006**, *111*, G02011. [CrossRef]

38. Strachan, I.B.; Pelletier, L.; Bonneville, M.-C. Interannual variability in water table depth controls net ecosystem carbon dioxide exchange in a boreal peatland. *Biogeochemistry* **2016**, *127*, 99–111. [CrossRef]

39. Kim, Y.; Kim, S.D.; Enomoto, H.; Kushida, K.; Kondoh, M.; Uchida, M. Latitudinal distribution of soil CO_2 efflux and temperature along the Dalton Highway, Alaska. *Polar Sci.* **2013**, *7*, 162–173. [CrossRef]

40. Aurela, M.; Laurila, T.; Tuovinen, J.-P. The timing of snow melt controls the annual CO_2 balance in a subarctic fen. *Geophys. Res. Lett.* **2004**, *31*, LI6119. [CrossRef]

41. Bisbee, K.E.; Gower, S.T.; Norman, J.M.; Nordheim, E.V. Environmental controls on ground cover species composition and productivity in a boreal black spruce forest. *Oecologia* **2001**, *129*, 261–270. [CrossRef]

42. Yuan, W.P.; Luo, Y.Q.; Richardson, A.D.; Oren, R.; Luyssaert, S.; Janssens, I.A.; Ceulemans, R.; Zhou, X.H.; Grunwald, T.; Aubinet, M.; et al. Latitudinal patterns of magnitude and interannual variability in net ecosystem exchange regulated by biological and environmental variables. *Glob. Chang. Biol.* **2009**, *15*, 2905–2920. [CrossRef]

43. Wang, C.; Bond-Lamberty, B.; Gower, S.T. Carbon distribution of a well- and poorly-drained black spruce fire chronosequence. *Glob. Chang. Biol.* **2003**, *9*, 1–14. [CrossRef]

44. Aurela, M.; Lohila, A.; Tuovinen, J.-P.; Hatakka, J.; Riutta, T.; Laurila, T. Carbon dioxide exchange on a northern boreal fen. *Boreal Environ. Res.* **2009**, *14*, 699–710.

45. Griffis, T.J.; Rouse, W.R.; Waddington, J.M. Interannual variability of net ecosystem CO_2 exchange at a subarctic fen. *Glob. Biogeochem. Cycles* **2000**, *14*, 1109–1121. [CrossRef]

46. Lafleur, P.M.; Humphrey, E.R. Spring warming and carbon dioxide exchange over low Arctic tundra in central Canada. *Glob. Chang. Biol.* **2007**, *14*, 740–756. [CrossRef]

47. Kwon, H.-J.; Oechel, W.C.; Zulueta, R.C.; Hastings, S.J. Effects of climate variability on carbon sequestration among adjacent wet sedge tundra and moist tussock tundra ecosystems. *J. Geophys. Res.* **2006**, *111*, G03014. [CrossRef]

48. Groendahl, L.; Friborg, T.; Soegaard, H. Temperature and snowmelt controls on interannual variability in carbon dioxide exchange in the high Arctic. *Theor. Appl. Climatol.* **2007**, *88*, 111–125. [CrossRef]

49. Humphreys, E.R.; Lafleur, P.M. Does earlier snowmelt lead to greater CO_2 sequestration in two low Arctic tundra ecosystems? *Geophys. Res. Lett.* **2011**, *38*, L09703. [CrossRef]

50. Tuba, Z.; Csintalan, Z.; Proctor, M.C.F. Photosynthetic Responses of a Moss, *Tortula ruralis*, ssp. *ruralis*, and the Lichens *Cladonia convoluta* and *C. furcata* to Water Deficit and Short Periods of Desiccation, and Their Ecophysiological Significance: A Baseline Study at Present-Day CO_2 Concentration. *New Phytol.* **1996**, *133*, 353–361. [CrossRef]

51. Lund, M.; Lafleur, P.M.; Roulet, N.T.; Lindroth, A.; Christensen, T.R.; Aurela, M.; Chojnicki, B.H.; Flanagan, L.B.; Humphreys, E.R.; Laurila, T.; et al. Variability in exchange of CO_2 across 12 northern peatland and tundra sites. *Glob. Chang. Biol.* **2010**, *16*, 2436–2448. [CrossRef]

52. McNeil, P.; Waddington, J.M. Moisture controls on *Sphagnum* growth and CO_2 exchange on a cut-over bog. *J. Appl. Ecol.* **2003**, *40*, 354–367. [CrossRef]

53. Tuittila, E.-S.; Vasander, H.; Luine, J. Sensitivity of C sequestration in reintroduced *Sphagnum* to water-level variation in a cutaway peatland. *Restor. Ecol.* **2004**, *12*, 483–493. [CrossRef]

54. Moore, T.R.; Bubier, J.L.; Bledzki, L. Litter decomposition in temperate peatland ecosystems: The effect of substrate and site. *Ecosystems* **2007**, *10*, 949–963. [CrossRef]

55. Euskirchen, E.S.; Edgar, C.W.; Turetsky, M.R.; Waldrop, M.P.; Harden, J.W. Differential response of carbon fluxes to climate in three peatland ecosystems that vary in the presence and stability of permafrost. *J. Geophys. Res. Biogeosci.* **2014**, *119*, 1576–1595. [CrossRef]

56. Gaumont-Guay, D.; Black, T.A.; Barr, A.G.; Jassal, R.S.; Nesic, Z. Biophysical controls on rhizospheric and heterotrophic components of soil respiration in a boreal black spruce stand. *Tree Physiol.* **2008**, *28*, 161–171. [CrossRef] [PubMed]

57. Gaumont-Guay, D.; Black, T.A.; Griffis, T.J.; Barr, A.G.; Jassal, R.S.; Nesic, Z. Interpreting the dependence of soil respiration on soil temperature and water content in a boreal aspen stand. *Agric. For. Meteorol.* **2006a**, *140*, 220–235. [CrossRef]

58. Bergeron, O.; Margolis, H.A.; Coursolle, C. Forest floor carbon exchange of a boreal black spruce forest in eastern North America. *Biogeosciences,* **2009**, *6*, 1849–1864. [CrossRef]

59. Tang, J.W.; Dennis, D.B.; Qi, Y.; Xu, L.K. Assessing soil CO_2 efflux using continuous measurement of CO_2 profiles in soils with small solid-state sensors. *Agric. For. Meteo* **2003**, *118*, 207–220. [CrossRef]

60. Gaumont-Guay, D.; Black, T.A.; Griffis, T.J.; Barr, A.G.; Jassal, R.A.; Nesic, Z. Influence of temperature and drought on seasonal and interannual variations of soil, bole and ecosystem respiration in a boreal aspen stand. *Agr. Forest Meteorol.* **2006**, *140*, 203–219. [CrossRef]

61. Helfter, C.; Campbell, C.; Dinsmore, K.J.; Drewer, J.; Coyle, M.; Anderson, M.; Skiba, U.; Nemitz, E.; Billett, M.F.; Sutton, M.A. Drivers of long-term variability in CO_2 net ecosystem exchange in a temperate peatland. *Biogeosciences* **2015**, *12*, 1799–1811. [CrossRef]

62. Raich, J.W.; Potter, C.S.; Bhagawati, D. Interannual variability in global soil respiration, 1980–94. *Glob. Chang. Biol.* **2002**, *8*, 800–812. [CrossRef]

63. Rodeghiero, M.; Cescatti, A. Main determinants of forest soil respiration along an elevation/temperature gradient in the Italian Alps. *Glob. Chang. Biol.* **2005**, *11*, 1024–1041. [CrossRef]

64. Lavigne, M.B.; Ryan, M.G.; Anderson, D.E.; Baldocchi, D.D.; Crill, P.M.; Fitzjarrald, D.R.; Goulden, M.L.; Gower, S.T.; Massheder, J.M.; McCaughey, J.H.; et al. Comparing nocturnal eddy covariance measurements to estimates of ecosystem respiration made by scaling chamber measurements at six coniferous boreal sites. *J. Geophys. Res.* **1997**, *102*, 28977–28985. [CrossRef]

65. Rayment, M.B.; Jarvis, P.G. Temporal and spatial variation of soil CO_2 efflux in a Canadian boreal forest. *Soil Biol. Biochem.* **2000**, *32*, 35–45. [CrossRef]

66. Xu, M.; Qi, Y. Soil-surface CO_2 efflux and its spatial and temporal variations in a young ponderosa pine plantation in northern California. *Glob. Chang. Biol.* **2001**, *7*, 667–677. [CrossRef]

67. Peichl, M.; Öquist, M.; Löfvenius, M.O.; Ilstedt, U.; Sagerfors, J.; Grelle, A.; Lindroth, A.; Nilsson, M.B. A 12-year record reveals pre-growing season temperature and water table level threshold effects on the net carbon dioxide exchange in a boreal fen. *Environ. Res. Lett.* **2014**, *9*, 055006. [CrossRef]

68. Startsev, N.A.; Lieffers, V.J. Emission of nitrogen gas, nitrous oxide, and carbon dioxide on rehydration of dry feathermosses. *Soil Sci. Soc. Am. J.* **2007**, *71*, 214–218. [CrossRef]

69. Bubier, J.L.; Crill, P.M.; Moore, T.R.; Savage, K.; Varner, R.K. Seasonal patterns and controls on net ecosystem CO_2 exchange in a boreal peatland complex. *Glob. Biogeochem. Cycles* **1998**, *12*, 703–714. [CrossRef]

70. Lu, X.; Fan, J.; Yan, Y.; Wang, X. Responses of Soil CO_2 Fluxes to Short-Term Experimental Warming in Alpine Steppe Ecosystem, Northern Tibet. *PLoS ONE* **2013**, *8*. [CrossRef] [PubMed]

71. Yavitt, J.B.; Lang, G.E.; Wieder, R.K. Control of carbon mineralization to CH_4 and CO_2 in anaerobic, *Sphagnum*- derived peat from Big Run Bog. West Virginia. *Biogeochemisty* **1987**, *4*, 141–157. [CrossRef]

72. Winston, G.C.; Sundquist, E.T.; Stephens, B.B.; Trumbore, S.E. Winter CO_2 fluxes in a boreal forest. *J. Geophys. Res.* **1997**, *102*, 795–804. [CrossRef]

73. Environment and Climate Change Canada. Adjusted and Homogenized Canadian Climate Data (AHCCD). Available online: http://www.ec.gc.ca/dccha-ahccd/ (accessed on 14 November 2016).

74. Sistla, S.A.; Moore, J.C.; Simpson, R.T.; Gough, L.; Shaver, G.R.; Schimel, J.P. Long-term warming restructures Arctic tundra without changing net soil carbon storage. *Nature* **2013**, *497*, 615–618. [CrossRef] [PubMed]

75. Quinton, W.L.; Hayashi, M.; Chasmer, L.E. Permafrost-thaw-induced land-cover change in the Canadian subarctic: Implications for water resources. *Hydrol. Process.* **2013**, *25*, 152–158. [CrossRef]

76. Baltzer, J.L.; Veness, T.; Chasmer, L.E.; Sniderhan, A.E.; Quinton, W.L. Forests on thawing permafrost: Fragmentation, edge effects, and net forest loss. *Glob. Chang. Biol.* **2014**, *20*, 824–834. [CrossRef] [PubMed]

77. Bridgham, S.D.; Pastor, J.; Dewey, B.; Weltzin, J.F.; Updegraff, K. Rapid carbon response of peatlands to climate change. *Ecology* **2008**, *89*, 3041–3048. [CrossRef]

forests

MDPI

Article

Nutrient Resorption and Phenolics Concentration Associated with Leaf Senescence of the Subtropical Mangrove *Aegiceras corniculatum*: Implications for Nutrient Conservation

Hui Chen [1,2,†], Benbo Xu [1,†], Shudong Wei [1,2,*], Lihua Zhang [3,4,*], Haichao Zhou [2] and Yiming Lin [2]

[1]   College of Life Science, Yangtze University, Jingzhou 434025, China; chenhui_17@126.com (H.C.);
     benboxu@yangtzeu.edu.cn (B.X.)
[2]   Key Laboratory of the Ministry of Education for Coastal and Wetland Ecosystems, Xiamen University,
     Xiamen 361005, China; seapass2004@163.com (H.Z.); linym@xmu.edu.cn (Y.L.)
[3]   Key Laboratory of Coastal Environmental Processes and Ecological Remediation,
     Yantai Institute of Coastal Zone Research (YIC), Chinese Academy of Sciences (CAS), Yantai 264003, China
[4]   Shandong Provincial Key Laboratory of Coastal Zone Environmental Processes, YICCAS,
     Yantai 264003, China
*   Correspondence: weishudong2005@126.com (S.W.); lhzhang@yic.ac.cn (L.Z.); Tel.: +86-716-806-6858 (S.W.)
†   These authors contributed equally to this work.

Academic Editors: Scott X. Chang and Xiangyang Sun
Received: 31 August 2016; Accepted: 17 November 2016; Published: 22 November 2016

Abstract: *Aegiceras corniculatum* (L.) Blanco, a mangrove shrub species in the Myrsine family, often grows at the seaward edge of the mangrove zone in China. In the present study, seasonal dynamics of nutrient resorption and phenolics concentration associated with leaf senescence of *A. corniculatum* were investigated in order to evaluate its possible nutrient conservation strategies in the subtropical Zhangjiang river estuary. It was found that the nitrogen (N) and phosphorus (P) concentrations in mature leaves showed similar seasonal changes with the highest concentrations in winter and the lowest in summer, and were significantly higher than those in senescent leaves. The N:P ratios of mature leaves through the year were found to be less than 14, indicating that the *A. corniculatum* forest was N-limited. The nitrogen resorption efficiency (NRE) was higher than phosphorus resorption efficiency (PRE), and N resorption was complete. In addition, *A. corniculatum* leaves contained high total phenolics (TPs) and total condensed tannin (TCT) levels (both above 20%). TPs concentrations in mature and senescent leaves were all inversely related to their N or P concentrations. TPs:N and TCT:N ratios in senescent leaves were significantly higher than those in mature leaves. The obtained results suggested that high NRE during leaf senescence and high TPs:N and TCT:N ratios in senescent leaves might be important nutrient conservation strategies for the mangrove shrub *A. corniculatum* forest growing in N-limited conditions.

Keywords: *Aegiceras corniculatum*; nutrient resorption; phenolics; leaf senescence

---

1. Introduction

Mangrove species that flourish in low-nutrient environments have very efficient mechanisms for retaining and recycling nutrients [1]. *Aegiceras corniculatum* (L.) Blanco is a cryptoviviparous mangrove species, often grows at the seaward edge of the mangrove zone in China. This species seems most characteristic of the seaward mangal fringe and occurs typically as an isolated low shrub [2]. Compared with other mangrove species, *A. corniculatum* shrubs have high annual litter fall production in subtropical China [3]. The resulting litter from the leaf fall must be decomposed and the nutrients contained in that litter must be remineralized to become available again for plant uptake. However, most of this leaf fall is washed away by tides [4]. Such a high leaf fall rate will inevitably bring about a great loss of nutrients. Therefore, mangrove soils are generally low in nutrient concentrations [5], which can lead to the slow growth of mangrove species. However, mangroves are characterized by their high primary productivity [6]. To sustain the high levels of productivity, a substantial amount of nutrients is required. How mangrove plants, including *A. corniculatum*, cope with these conditions has provoked little attention [7].

Many previous studies showed that the concentrations of nutrients in the mangrove leaves varied with season and species [8,9]. In particular, seasonal changes in foliar nutrients occur in response to resorption before senescence [6]. Nutrient transfer from senescing leaves is the ecophysiological process by which plants withdraw nutrients from these leaves, making those nutrients available for later investment in new structures [10]. According to the reports by Lin et al. [8], Wei et al. [9], and Lin and Sternberg [11], the process of retranslocation is closely associated with leaf senescence and conservation of nutrients, and thus it is considered to be an important mechanism enabling plants to maintain growth in nutrient-poor sites. Nutrients may be used more efficiently at nutrient-poor sites, and this efficient nutrient use could reduce plant dependence on soil nutrient supply [12]. Feller [13], Reef et al. [14], and McKee et al. [15] suggested that nitrogen (N) and phosphorus (P) availability are important factors responsible for mangrove growth. Plant N:P ratio is a useful variable for consideration in ecological research because it reflects the gradual and dynamic nature of nutrient limitation rather than the fixed characteristics, such as N-limited versus P-limited [16].

Phenolic compounds, including tannins, are a significant component of plant secondary metabolites. In mangrove species, tannins are an abundant component, constituting as high as 20% dry weight [17]. Due to tannins being complex and energetically costly molecules to synthesize, their widespread occurrence and abundance indicate that tannins play an important role in plant function and evolution [18]. High tannin concentrations in plants are often associated with infertile site conditions. It has been suggested that there is an evolutionary advantage to higher tannin production. By reducing decomposition rates and decreasing N leaching potential, tannins may provide a nutrient conservation mechanism. In comparison to foliar nutrients, the detailed changes of leaf tannins between seasons and during leaf senescence of the mangrove species has been less intensively studied [8,9,19,20]. According to the reports by Serrano [21] and Northup et al. [22], this variability may determine not only the susceptibility of plants to herbivore attack, but also important aspects of nutrient cycling in terrestrial and aquatic ecosystems.

Therefore, nutrient resorption and high tannin production may be important for *A. corniculatum* to conserve nutrients in coastal environments. The objective of the present study was to evaluate possible nutrient conservation strategies of the mangrove shrub species *A. corniculatum* under low nutrient conditions. The questions asked here were regarding whether (1) leaf N and P concentrations follow a similar seasonal pattern, and decrease during leaf senescence; (2) a strong nutrient limitation will show higher resorption of the limiting nutrients; and (3) the production of phenolics will increase during leaf senescence under nutrient limited conditions.

2. Materials and Methods

2.1. Study Area

The experiments were carried out in Zhangjiangkou National Mangrove Nature Reserve (23°55′ N, 117°24′ E), which is located in the gulf of Gulei, Yunxiao, Fujian Province, in southern China. This nature reserve occupies 2358 hectares, and consists mostly of mangroves and salt marshes along the coastline of the Zhangjiang estuary. This nature reserve is the northernmost national nature mangrove in China, and it was included into the Ramsar List in 2008. The climate of the region is characteristic of a southern subtropical maritime monsoon climate. Based on Yunxiao Meteorological Administration, during 2001–2008 the mean air temperature was 21.15 ± 0.68 °C, 28.13 ± 0.28 °C, 24.07 ± 0.54 °C, and 15.34 ± 0.60 °C for spring, summer, autumn, and winter, respectively. The mean annual precipitation and evaporation were 187.21 mm and 1718.4 mm, respectively. The soil is clay, and the concentration of organic matter is 39.60 ± 1.90 mg/g. N and P concentrations in the soil are 2.20 ± 0.10 mg/g and 0.40 ± 0.00 mg/g, respectively [23]. The tide is irregular semidiurnal, the soil can be flooded for 1–2 h per day during spring tides, but the soil can be exposed for 3–7 days during the neap tide. The salinity of the soil is above 10‰, and the salinity of tidal water salinity ranges between 1‰ and 22‰ [24]. In the study site, *A. corniculatum* was the dominant mangrove species with a few *Kandelia candel*. The mean height of *A. corniculatum* plants was 2.8 m and the canopy density was 0.90.

2.2. Sample Collection

Leaf samples of *A. corniculatum* were collected from Zhangjiangkou National Mangrove Nature Reserve, Yunxiao, Fujian Province, China in 15 July 2009 (summer), 13 October 2009 (autumn), 15 December 2009 (winter), and 17 March 2010 (spring). Thirty trees with similar height and growth conditions were selected and labeled. The thirty trees were divided into five groups (six trees in one group) as five replications. The developmental stages of leaves were demarcated into two stages, green mature leaf and yellow senescent leaf. The green mature leaf was the third pair of developmentally matured leaves and did not show any sign of senescence. The yellow senescent leaf was collected by gently tapping the petiole. Only those senescent leaves that could be detached by this method were collected. Leaves damaged by insects and disease or mechanical factors were avoided. All samples were taken to the laboratory immediately after sampling, cleaned with distilled water, and then freeze dried using a desktop freeze-dryer at −56 °C for 72 h. The freeze-dried leaves were ground finely and stored at −20 °C prior to analysis.

2.3. Measurement of N and P Concentrations

The freeze-dried leaf samples were digested with sulfuric acid and hydrogen peroxide. The N concentration was determined based on the micro-Kjeldahl method described by Yoshida et al. [25], and the P concentration was evaluated according to the ascorbic acid-antimony reducing phosphate colorimetric method [26].

2.4. Determination of Total Phenolics (TPs), Extractable Condensed Tannins (ECT), Protein-Bound Condensed Tannins (PBCT), and Fiber-Bound Condensed Tannins (FBCT)

Established procedures demonstrated by Lin et al. [20] were used. TPs were measured with the Prussian blue method [27], and ECT, PBCT, and FBCT were determined by the butanol-HCl method [28] using purified tannins from *A. corniculatum* leaves as the standard. The concentration of total condensed tannins (TCT) was calculated by adding the respective quantities of ECT, PBCT, and FBCT [28].

2.5. Calculations

Resorption efficiency (RE) is the percentage of N or P recovered from the senescing leaves [29,30]. RE was calculated using the following equation: RE (%) = $(A_1 - A_2)/A_1 \times 100\%$. Where A_1 is N or P concentration in mature leaves, and A_2 is N or P concentration in senescent leaves.

Nutrient resorption proficiency referred to the minimum level to which a plant can reduce an element in senescing leaves. N and P resorption proficiency values were determined as absolute nutrient concentrations in senescent leaves [30]. Lower final nutrient concentrations correspond to higher proficiencies, which are usually expressed as percentages. This index's proxy seems to be more closely correlated than RE to the nutrient status of a population [31].

2.6. Statistical Analysis

All measurements were replicated five times. A one-way analysis of variance (ANOVA) was performed with season as the treatment factor. The Student-Newman-Keuls multiple comparison method was used to test significant differences between any two seasons. All analyses were performed by SPSS13.0 for Windows (SPSS Inc., Chicago, IL, USA).

3. Results

3.1. Seasonal Changes of N and P Concentrations, N:P Ratios, and Nutrient Resorption in Mature and Senescent Leaves of A. corniculatum

The N and P concentrations in mature leaves showed similar changes, with the highest concentrations in winter (17.94 ± 0.48 mg/g and 1.56 ± 0.05 mg/g) and the lowest in summer (9.27 ± 0.54 mg/g and 0.85 ± 0.04 mg/g), respectively. The N concentrations in senescent leaves ranged from 3.83 ± 0.15 mg/g to 4.77 ± 0.10 mg/g, with the lowest in spring and no significant differences among other three seasons. The P concentrations in senescent leaves were higher in summer (0.60 ± 0.04 mg/g) and autumn (0.59 ± 0.05 mg/g) than in winter (0.52 ± 0.03 mg/g) and spring (0.49 ± 0.03 mg/g) (Figure 1). Statistical results showed that the concentrations of N and P in mature leaves were significantly higher than those in senescent leaves in all seasons.

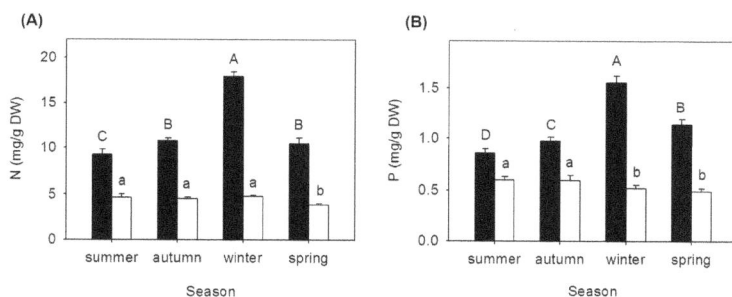

Figure 1. Seasonal changes in the (**A**) nitrogen (N) and (**B**) phosphorus (P) concentrations in leaves of *A. corniculatum*. Symbols are black bars for mature leaves and white bars for senescent leaves. Different capital letters indicate significant differences among seasons for N and P concentrations in mature leaves at $p < 0.05$. Different lowercase letters indicate significant differences among seasons for N and P concentrations in senescent leaves at $p < 0.05$.

The N:P ratios in mature leaves ranged from 9.23 ± 0.92 to 11.55 ± 0.55 with the lowest in spring. The nitrogen resorption efficiency (NRE) was significantly higher than phosphorus resorption efficiency (PRE) during leaf senescence in every season. The NRE and PRE followed the identical pattern, reaching the highest in winter (73.39% ± 1.06% and 66.51% ± 2.37%) and the lowest in summer (49.87% ± 4.39% and 30.01% ± 4.43%), respectively (Table 1).

The N concentrations in senescent leaves were all below 0.48%, and the P concentrations in senescent leaves were all above 0.05%, respectively (Table 1).

Table 1. Seasonal changes in the nitrogen and phosphorus ratios (N:P), nitrogen resorption efficiency (NRE), phosphorus resorption efficiency (PRE), nitrogen resorption proficiency (NRP), and phosphorus resorption proficiency (PRP) of *A. corniculatum* leaves.

Season	$N_m:P_m$	NRE (%)	PRE (%)	NRP (%)	PRP (%)
Summer	10.90 ± 1.13 a	49.87 ± 4.39 d	30.01 ± 4.43 d	0.46 ± 0.04 a	0.06 ± 0.00 a
Autumn	11.09 ± 0.41 a	58.18 ± 2.33 c	38.84 ± 5.94 c	0.45 ± 0.02 a	0.06 ± 0.00 a
Winter	11.55 ± 0.55 a	73.39 ± 1.06 a	66.51 ± 2.37 a	0.48 ± 0.01 a	0.05 ± 0.00 b
Spring	9.23 ± 0.92 b	63.36 ± 2.03 b	57.08 ± 3.48 b	0.38 ± 0.02 b	0.05 ± 0.00 b

N_m, P_m correspond to the N and P concentrations in mature leaves, respectively; Different letters in the same column indicate significant differences among seasons at $p < 0.05$.

There were significantly positive correlations between N and P concentrations in mature and senescent leaves. N concentrations in mature leaves and P concentrations in mature and senescent leaves were correlated with NRE and PRE, respectively. Meanwhile, the correlation between NRE and PRE was also highly significant (Table 2).

Table 2. Correlations between nitrogen (N) and phosphorus (P) concentrations, N and nitrogen resorption efficiency (NRE), P and phosphorus resorption efficiency (PRE), and NRE and PRE of *A. corniculatum* leaves.

Correlation	*F*	*R*	*p*
N_m-P_m	108.070	0.926	**<0.001**
N_s-P_s	5.765	0.493	**0.027**
N_m-NRE	55.731	0.869	**<0.001**
N_s-NRE	0.117	−0.080	0.737
P_m-PRE	91.214	0.914	**<0.001**
P_s-PRE	32.805	−0.804	**<0.001**
NRE-PRE	82.361	0.906	**<0.001**

N_m, P_m correspond to the N and P concentrations in mature leaves, respectively; N_s, P_s correspond to the N and P concentrations in senescent leaves, respectively. *p* values less than 0.05 are marked as bold numbers.

3.2. Seasonal Changes of Tannins Concentrations in Mature and Senescent Leaves of A. corniculatum

TPs concentrations in senescent leaves were higher than those in mature leaves. TPs concentrations reached the lowest in winter (210.19 ± 5.67 mg/g) for mature leaves. However, TPs concentrations in senescent leaves were the highest in spring (318.72 ± 12.66 mg/g) (Figure 2). TPs concentrations in mature and senescent leaves were all inversely correlated to N or P concentrations. A significant negative correlation between TPs concentrations in senescent leaves and N:P ratios in mature leaves was also observed (Table 3).

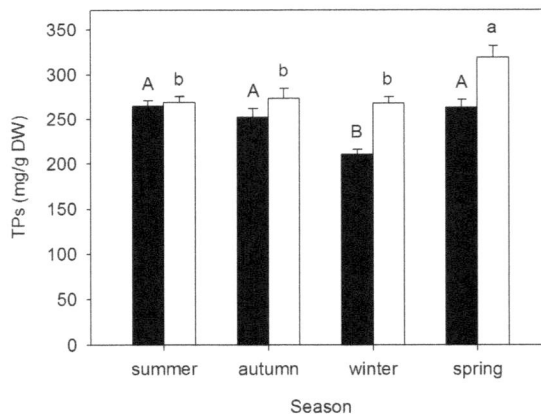

Figure 2. Seasonal changes in the total phenolics (TPs) concentrations in leaves of *A. corniculatum*. Symbols are black bars for mature leaves and white bars for senescent leaves. Different capital letters indicate significant differences among seasons for TPs concentrations in mature leaves at $p < 0.05$. Different lowercase letters indicate significant differences among seasons for TPs concentrations in senescent leaves at $p < 0.05$.

Table 3. Correlations between total phenolics (TPs) and nutrient concentrations (N and P), and N:P ratios of *A. corniculatum* leaves.

Correlation	*F*	*R*	*p*
TPs_m-N_m	122.338	−0.934	**<0.001**
TPs_s-N_s	34.399	−0.810	**<0.001**
TPs_m-P_m	56.098	−0.870	**<0.001**
TPs_s-P_s	11.631	−0.627	**0.003**
TPs_m-N_m:P_m	3.940	−0.424	0.063
TPs_s-N_m:P_m	12.143	−0.635	**0.003**

N_m, P_m correspond to the N and P concentrations in mature leaves, respectively; N_s, P_s correspond to the N and P concentrations in senescent leaves, respectively; TPs_m, TPs_s correspond to the TPs concentrations in mature and senescent leaves, respectively. *p* values less than 0.05 are marked as bold numbers.

ECT concentrations in mature leaves ranged from 190.77 ± 17.21 mg/g to 230.18 ± 33.10 mg/g, and remained relatively stable between the seasons. ECT concentrations in senescent leaves were significantly higher than those in mature leaves in winter and spring (Figure 3A).

PBCT in mature and senescent leaves were both higher in summer and winter than those in autumn and spring (Figure 3B). However, FBCT concentrations did not significantly change during leaf senescence, except in spring, and FBCT in mature leaves was the highest (1.56 ± 0.17 mg/g) in autumn (Figure 3C).

TCT concentrations had a similar changing trend to that of ECT concentrations during leaf senescence through the seasons (Figure 3D).

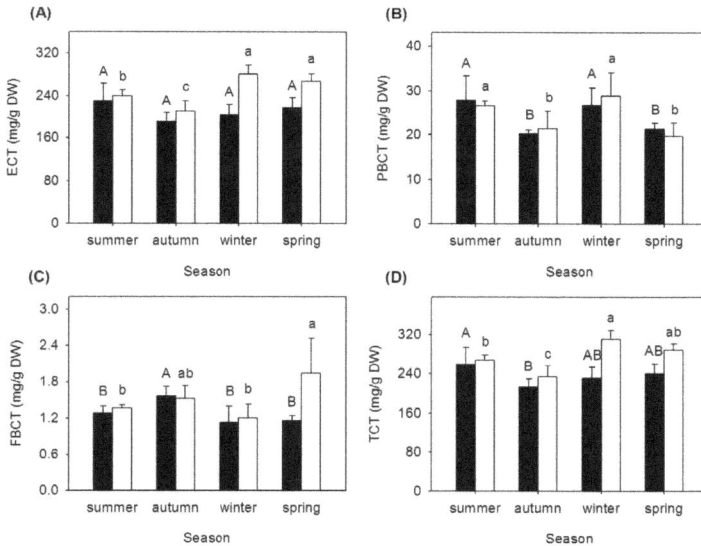

Figure 3. Seasonal changes in the concentrations of (**A**) extractable condensed tannins (ECT); (**B**) protein-bound condensed tannins (PBCT); (**C**) fiber-bound condensed tannins (FBCT); and (**D**) total condensed tannins (TCT) in leaves of *A. corniculatum*. Symbols are black bars for mature leaves and white bars for senescent leaves. Different capital letters indicate significant differences among seasons for ECT, PBCT, FBCT, and TCT concentrations in mature leaves at $p < 0.05$. Different lowercase letters indicate significant differences among seasons for ECT, PBCT, FBCT and TCT concentrations in senescent leaves at $p < 0.05$.

3.3. Seasonal Changes of TPs:N and TCT:N Ratios in Mature and Senescent Leaves of A. corniculatum

Seasonal changes of TPs:N and TCT:N ratios during leaf senescence are shown in Figure 4. TPs:N and TCT:N ratios in senescent leaves were significantly higher than those in mature leaves for all seasons. TPs:N and TCT:N ratios in mature leaves reached the highest in summer and the lowest in winter, while in senescent leaves they were higher in spring than in the other three seasons.

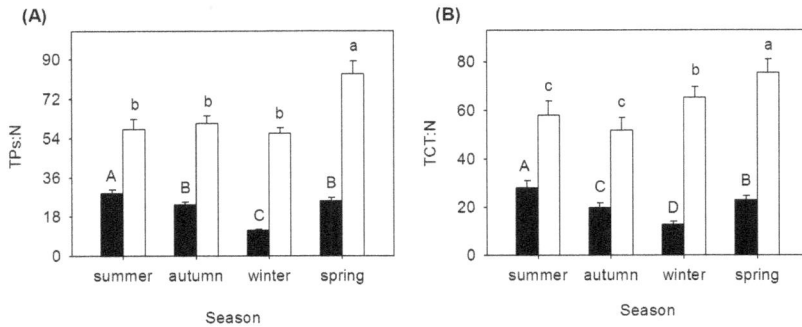

Figure 4. Seasonal changes of (**A**) total phenolics and nitrogen ratios (TPs:N) and (**B**) total condensed tannins and N ratios (TCT:N) in leaves of *A. corniculatum*. Symbols are black bars for mature leaves and white bars for senescent leaves. Different capital letters indicate significant differences among seasons for TPs:N and TCT:N ratios in mature leaves at $p < 0.05$. Different lowercase letters indicate significant differences among seasons for TPs:N and TCT:N ratios in senescent leaves at $p < 0.05$.

4. Discussion

N and P are the most common nutrients limiting plant growth [32]. The N and P concentrations in mature leaves of *A. corniculatum* followed a similar seasonal pattern, with the highest concentrations in winter and the lowest in summer (Figure 1). This was in agreement with the observations made in a study by Wang et al. [6], which showed higher N and P concentrations in cold seasons and lower concentrations in warm seasons for mature leaves. Similarly, Aerts et al. [33] also suggested that summer warming reduced N concentrations of mature and senescent leaves in *Rubus*. First, a portion of N and P were allocated to other plant parts (e.g., roots and fruits). Second, N and P concentrations were diluted by leaf mass accumulation during summer when *A. corniculatum* grew rapidly.

Nutrient resorption during leaf senescence serves to reduce the plant's dependence on current nutrient supply, and has been recognized as one of the most important strategies used by plants to conserve nutrients [8,9,34,35]. It has been estimated that approximately 50% of the nutrients (N and P) are resorbed during leaf senescence [29,36]. In the present study, the averaged NRE and PRE values were 61.20% ± 9.10% and 48.11% ± 15.31%, respectively (Table 1). These high N and P resorption values indicated that internal cycling of N and P can supply a significant fraction of the required nutrients for *A. corniculatum* growth. In addition, both NRE and PRE values were the highest in winter and the lowest in summer. This may indicate that *A. corniculatum* mainly resorbed N and P from senescent leaves in winter, but absorbed N and P from the soil in summer to adapt to low nutrient environments.

The N:P ratios of mature leaves are in response to natural variations in N and P supply, reflecting variation in N and P concentrations or both together [16]. A review of 40 fertilization studies revealed that an N:P ratio > 16 indicates P limitation on a community level, while an N:P ratio < 14 is indicative of N limitation [37]. As critical N:P ratios are successfully used to establish the nature of nutrient limitation in aquatic and agricultural ecosystems, the general concept may be applicable to a much wider range of ecosystems. N:P ratios in mature leaves of *A. corniculatum* in the Zhangjiang river estuary ranged from 9.23 to 11.55, and the lowest N:P ratio occurred in spring (Table 1). These results, both less than 14, suggested the presence of strong community level N limitation in this forest. Accordingly, N-limited *A. corniculatum* forest had significantly higher NRE than PRE. This result supported the hypothesis that a strong nutrient limitation will show higher resorption of the limiting nutrient.

Killingbeck [30] modified the concepts of nutrient resorption efficiency and proposed the use of "resorption proficiency", which is the level to which nutrient concentrations are reduced in senescent leaves. One of the important features of knowing the levels to which N and P can be reduced in senescent leaves is that these values offer an objective gauge by which resorption can be measured [30]. The potential or ultimate resorption proficiency, the lowest levels to which a nutrient is reduced in senescent leaves, is a reflection of biochemical limits to resorption and, unlike efficiency, is not subject to the temporal variations in nutrient concentrations present in green leaves and the timing of sampling [30,38]. For these reasons, resorption proficiency has been argued to be the more definitive measure of resorption success than efficiency [39]. By analyzing N and P levels in the senescent leaves of 89 species of deciduous and evergreen woody perennials, Killingbeck [30] proposed that in evergreen species, nutrient concentrations reduced below 0.7% N and 0.04% P or above 1.0% N and 0.05% P in senescent leaves are believed to represent complete or incomplete N and P resorption, respectively. In the present study, N concentrations in senescent leaves ranging from 0.38% to 0.48% were indicative of complete resorption of N; P concentrations in senescent leaves were all above 0.05%, reflecting that P resorption was incomplete (Table 1). Alongi et al. [5] argued that mangrove soils are generally low in nutrient concentrations, especially N. Our results suggested that there is greater need to resorb N from senescent leaves in N-limited mangrove forests. All the same, NRP and PRP values were lowest in spring when *A. corniculatum* grew rapidly. This is in agreement with the findings by Wei et al. [9], which suggested that lower N and P concentrations of senescent leaves in growing seasons in comparison to other seasons make it difficult to resorb additional nutrients.

N and P concentrations correlated with each other in mature and senescent leaves (Table 2). Chapin and Kedrowski [40] found a direct correlation between proportional nutrient retranslocation from the leaves during senescence and nutrient concentration in tree leaves. The significant positive correlation between N concentrations in mature leaves and NRE was consistent with the findings of Chapin and Kedrowski [40]. The significant negative correlation between P concentrations in senescent leaves and PRE was consistent with results reported in a previous study [8]. NRE had a significant positive correlation with PRE. This correlation suggested that N resorption may be controlled by biochemical processes similar to those that control P resorption, in accordance with the results of studies of wood species in tropical forests [41,42] and wetland graminoids [43]. However, other studies have found no significant correlations between the N and P resorption efficiencies of certain plants [40].

Traditionally, justification for the high metabolic cost associated with the production of tannins is attributed to improved herbivore defense [44]. In mangrove species, tannin is an abundant component as high as 20% dry weight which prevents damage from herbivory [17]. *A. corniculatum* leaves had high total phenolics and total condensed tannin levels (both above 20%), which increased during leaf senescence (Figures 2 and 3). The observed changes were in agreement with the findings reported for another mangrove species, *Rhizophora stylosa* [8]. As polyphenols are water-soluble and susceptible to leaching [45], leaching of polyphenols or tannins from green leaves by sporadic rain might be a cause for the net enrichment in senescent leaves [46]. However, Mafongoya et al. [47] indicated that much of the soluble carbon compounds, including polyphenols, are expected to be translocated from leaves during senescence. Constantinides and Fownes [48] found both an increase and a decrease in concentrations of polyphenols with the development and senescence of leaves, while Kuhajek et al. [49] observed no significant effects of leaf age on both condensed tannins and total phenolics. As for bound condensed tannins, only the FBCT concentrations in spring were significantly higher in senescent leaves than in mature leaves. These results were not exactly the same as the findings reported by Zhang et al. [34], which indicated that PBCT concentrations are significantly higher in senescent leaves than in mature leaves in all seasons.

The TPs concentrations in mature leaves were lowest in winter, and in senescent leaves concentrations were highest in spring (Figure 2). The TPs concentrations in mature and senescent leaves were all inversely related to N or P concentrations (Table 3). It is common to find a negative correlation between N and secondary compound concentrations, such as phenolics and tannins [50]. This pattern lends support to source-sink hypotheses, such as the carbon-nutrient balance (CNB) hypothesis and the growth-differentiation (GDB) hypothesis that predict increased C allocation to secondary C compounds under low nutrient conditions. Under nutrient limited circumstances, the diversion of C to secondary compounds may be essentially cost-free with respect to growth or reproduction [51]. Kandil et al. [52] suggested that the production and accumulation of the polyphenolics is favored in the high light, high temperature, high salinity, and low nutrient environment in which the mangroves occur, particularly when carbon used for growth or reproduction is precluded by lack of nutrient resources.

TPs:N and TCT:N ratios in senescent leaves were significantly higher than those in mature leaves (Figure 4). Condensed tannins are considered to have an important impact for N immobilization in soils, in particular during fall when fresh litter materials are introduced into soils [53]. Through both litter deposition and foliar leaching, soils in tannin rich plant communities receive appreciable tannin inputs. Because high tannin concentrations in plants are often associated with infertile site conditions [22], it has been suggested that there is an evolutionary advantage to higher tannin production. By reducing decomposition rates and decreasing N leaching potential, tannins may provide a nutrient conservation mechanism [8,54]. Therefore, parameters such as TPs:N and TCT:N ratios may be the best predictors of litter quality [55]. In the present study, higher TPs:N and TCT:N ratios in senescent leaves may slow down the rate of litter decomposition, thereby increasing the nutrient use efficiency and may be, in part, an adaptation to nutrient limited conditions.

In conclusion, N:P ratios of mature leaves are a good indicator for nutrient limitation in mangrove ecosystems. The results confirmed that resorption efficiency during leaf senescence depends on the type of nutrient limitation, and NRE was much higher than PRE in N-limited habitat. The resorption proficiency of N and P indicated that N resorption was complete and P resorption was incomplete. *A. corniculatum* leaves contained high total phenolics, extractable condensed tannin, and total condensed tannin levels. TPs:N and TCT:N ratios in senescent leaves were significantly higher than those in mature leaves. Our results indicated that high NRE during leaf senescence and high TPs:N and TCT:N ratios in senescent leaves might be important nutrient conservation strategies for the mangrove shrub *A. corniculatum* forest growing under the N-limited conditions. Furthermore, it has also been previously reported that mangroves are highly plastic in their responses to multiple abiotic and biotic factors such as destructive weather, herbivory, etc. As such, further work is required for understanding a more thorough temporal variation of nutrient resorption and phenolics concentration for *A. corniculatum* in long-term observations.

Acknowledgments: This work was supported by the National Natural Science Foundation of China (41201293), the WEL Visiting Fellowship Program of Xiamen University (WEL201504), and the Yangtze Youth Fund of Yangtze University (2015cqn60).

Author Contributions: Hui Chen, Benbo Xu, Shudong Wei, and Haichao Zhou designed and performed the experiments; Hui Chen, Shudong Wei and Lihua Zhang analyzed the data; Benbo Xu and Yiming Lin contributed reagents/materials/analysis tools; Hui Chen wrote the paper; Shudong Wei and Lihua Zhang revised the paper and modified the language.

Conflicts of Interest: The authors declare no conflict of interest.

References

1. Twilley, R.W.; Lugo, A.E.; Patterson-Zucca, C. Litter production and turnover in basin mangrove forests in southwest Florida. *Ecology* **1986**, *67*, 670–683. [CrossRef]
2. Tomlinson, P.B. *The Botany of Mangroves*; Cambridge University Press: New York, NY, USA, 1986.
3. Li, M. Nutrient dynamics of a Futian mangrove forest in Shenzhen, South China. *Estuar. Coast. Shelf. Sci.* **1997**, *45*, 463–472. [CrossRef]

4. Odum, W.E.; Heald, E.J. The detritus-based food web of an estuarine mangrove community. In *Estuarine Research*; Cronin, L.E., Ed.; Academic Press: New York, NY, USA, 1975; pp. 265–286.

5. Alongi, D.M.; Boto, K.G.; Robertson, A.I. Nitrogen and phosphorus cycles. In *Tropical Mangrove Ecosystems*; Robertson, A.I., Alongi, D.M., Eds.; American Geographysical Union: Washington, DC, USA, 1992; pp. 251–292.

6. Wang, W.Q.; Wang, M.; Lin, P. Seasonal changes in element contents in mangrove element retranslocation during leaf senescence. *Plant Soil* **2003**, *252*, 187–193. [CrossRef]

7. Cai, G. Salt content and regulation of mangroves. *Ecol. Sci.* **1990**, *2*, 123–127.

8. Lin, Y.M.; Liu, X.W.; Zhang, H.; Fan, H.Q.; Lin, G.H. Nutrient conservation strategies of a mangrove species *Rhizophora stylosa* under nutrient limitation. *Plant Soil* **2010**, *326*, 469–479. [CrossRef]

9. Wei, S.D.; Liu, X.W.; Zhang, L.H.; Chen, H.; Zhang, H.; Zhou, H.C.; Lin, Y.M. Seasonal changes of nutrient levels and nutrient resorption in *Avicennia marina* leaves in Yingluo Bay, China. *South. For.* **2015**, *77*, 237–242.

10. Millard, P.; Neilsen, G.H. The influence of nitrogen supply on the uptake and remobilization of stored N for the seasonal growth of apple trees. *Ann. Bot.* **1989**, *63*, 301–309.

11. Lin, Y.; Sternberg, L.D.L. Nitrogen and phosphorus dynamics and nutrient resorption of *Rhizophora mangle* leaves in south Florida, USA. *Bull. Mar. Sci.* **2007**, *80*, 159–169.

12. Pugnaire, F.I.; Chapin, F.S. Controls over nutrient resorption from leaves of evergreen Mediterranean species. *Ecology* **1993**, *74*, 124–129. [CrossRef]

13. Feller, I.C. Effects of nutrient enrichment on growth and herbivory of dwarf red mangrove (*Rhizophora mangle*). *Ecol. Monogr.* **1995**, *65*, 477–505. [CrossRef]

14. Reef, R.; Feller, I.C.; Lovelock, C.E. Nutrition of mangroves. *Tree Physiol.* **2010**, *30*, 1148–1160. [CrossRef] [PubMed]

15. McKee, K.L.; Feller, I.C.; Popp, M.; Wanek, W. Mangrove isotopic (δ^{15}N and δ^{13}C) fractionation across a nitrogen vs. phosphorus limitation gradient. *Ecology* **2002**, *83*, 1065–1075. [CrossRef]

16. Güsewell, S. N:P ratios in terrestrial plants: Variation and functional significance. *New Phytol.* **2004**, *164*, 243–266. [CrossRef]

17. Hernes, P.J.; Benner, R.; Cowie, G.L.; Goni, M.A.; Bergamaschi, B.A.; Hedges, J.I. Tannin diagenesis in mangrove leaves from a tropical estuary: A novel molecular approach. *Geochim. Cosmochim. Acta* **2001**, *65*, 3109–3122. [CrossRef]

18. Zucker, W.V. Tannins: Does structure determine function? An ecological perspective. *Am. Nat.* **1983**, *121*, 335–365. [CrossRef]

19. Lin, Y.M.; Liu, J.W.; Xiang, P.; Lin, P.; Ding, Z.H.; Sternberg, L.D.L. Tannins and nitrogen dynamics in mangrove leaves at different age and decay stages (Jiulong River Estuary, China). *Hydrobiologia* **2007**, *583*, 285–295. [CrossRef]

20. Lin, Y.M.; Liu, J.W.; Xiang, P.; Lin, P.; Ye, G.; Sternberg, L.D.L. Tannin dynamics of propagules and leaves of *Kandelia candel* and *Bruguiera gymnorrhiza* in the Jiulong River Estuary, Fujian, China. *Biogeochemistry* **2006**, *78*, 343–359. [CrossRef]

21. Serrano, L. Leaching from vegetation of soluble polyphenolic compounds, and their abundance in temporary ponds in the Doñana National Park (SW Spain). *Hydrobiologia* **1992**, *229*, 43–50. [CrossRef]

22. Northup, R.R.; Yu, Z.; Dahlgren, R.A.; Vogt, K.A. Polyphenol control of nitrogen release from pine litter. *Nature* **1995**, *377*, 227–229. [CrossRef]

23. Huang, G.M. Growth characteristics of *Spartina alterniflora* and its relative competitive ability with *Kandelia candle* in mangrove areas of the Zhangjiang estuary. Master's Thesis, Xiamen University, Xiamen, China, 2009.

24. Chen, H.; Lu, W.; Yan, G.; Yang, S.; Lin, G. Typhoons exert signifificant but differential impacts on net ecosystem carbon exchange of subtropical mangrove forests in China. *Biogeosciences* **2014**, *11*, 5323–5333. [CrossRef]

25. Yoshida, S.; Forno, D.A.; Cock, J.H.; Gomez, K.A. *Laboratory Manual for Physiological Studies of Rice*, 2nd ed.; The International Rice Research Institute: Manila, Philippines, 1972; pp. 7–9, 36–38.

26. Nanjing Institute of Soil Science. *Physical and Chemical Analysis of Soils*; Science and Technology Press: Shanghai, China, 1978; pp. 1–320.

27. Graham, H.D. Stabilization of the Prussian blue color in the determination of polyphenols. *J. Agric. Food Chem.* **1992**, *40*, 801–805. [CrossRef]

28. Terrill, T.H.; Rowan, A.M.; Douglas, G.B.; Barry, T.N. Determination of extractable and bound condensed tannin concentrations in forage plants, protein concentrate meals and cereal grains. *J. Sci. Food Agric.* **1992**, *58*, 321–329. [CrossRef]

29. Aerts, R. Nutrient resorption from senescing leaves of perennials: Are there general patterns? *J. Ecol.* **1996**, *84*, 597–608. [CrossRef]

30. Killingbeck, K.T. Nutrients in senesced leaves: Keys to the search for potential resorption and resorption proficiency. *Ecology* **1996**, *77*, 1716–1727. [CrossRef]

31. Rejmánková, E. Nutrient resorption in wetland macrophytes: Comparison across several regions of different nutrient status. *New Phytol.* **2005**, *167*, 471–482. [CrossRef] [PubMed]

32. Lambers, H.; Chapin, F.S.; Pons, T.L. *Plant Physiological Ecology*; Springer: New York, NY, USA, 1998.

33. Aerts, R.J.; Barry, T.N.; McNabb, W.C. Polyphenols and agriculture: Beneficial effects of proanthocyanidins forages. *Agr. Ecosyst. Environ.* **1999**, *75*, 1–12. [CrossRef]

34. Zhang, L.H.; Ye, G.F.; Lin, Y.M.; Zhou, H.C.; Zeng, Q. Seasonal changes in tannin and nitrogen contents of *Casuarina equisetifolia* branchlets. *J. Zhejiang Uni. Sci. B* **2009**, *10*, 103–111. [CrossRef] [PubMed]

35. Ye, G.F.; Zhang, S.J.; Zhang, L.H.; Lin, Y.M.; Wei, S.D.; Liao, M.M.; Lin, G.H. Age-related changes in nutrient resorption patterns and tannin concentration of *Casuarina equisetifolia* plantations. *J. Trop. For. Sci.* **2012**, *24*, 546–556.

36. Aerts, R.; Chapin, F.S. The mineral nutrition of wild plants revisited: A re-evaluation of processes and patterns. *Adv. Ecol. Res.* **2000**, *30*, 1–67.

37. Koerselman, W.; Meuleman, A.F.M. The vegetation N:P ratio: A new tool to detect the nature of nutrient limitation. *J. Appl. Ecol.* **1996**, *33*, 1441–1450. [CrossRef]

38. Kobe, R.K.; Lepczyk, C.A.; Iyer, M. Resorption efficiency decreases with increasing green leaf nutrients in a global data set. *Ecology* **2005**, *86*, 2780–2792. [CrossRef]

39. Wright, I.J.; Westoby, M. Nutrient concentration, resorption and lifespan: Leaf traits of Australian sclerophyll species. *Funct. Ecol.* **2003**, *17*, 10–19. [CrossRef]

40. Chapin, F.S.; Kedrowski, R.A. Seasonal changes in nitrogen and phosphorus fractions and autumn retranslocation in evergreen and deciduous taiga trees. *Ecology* **1983**, *64*, 376–391. [CrossRef]

41. Lal, C.B.; Annapurna, C.; Raghubanshi, A.S.; Singh, J.S. Effect of leaf habit and soil type on nutrient resorption and conservation in woody species of a dry tropical environment. *Can. J. Bot.* **2001**, *79*, 1066–1075.

42. Cai, Z.; Bongers, F. Contrasting nitrogen and phosphorus resorption efficiencies in trees and lianas from a tropical montane rain forest in Xishuangbanna, south-west China. *J. Trop. Ecol.* **2007**, *23*, 115–118. [CrossRef]

43. Güsewell, S. Nutrient resorption of wetland graminoids is related to the type of nutrient limitation. *Funct. Ecol.* **2005**, *19*, 344–354. [CrossRef]

44. Feeny, P. Seasonal changes in oak leaf tannins and nutrients as a cause of spring feeding by winter moth caterpillars. *Ecology* **1970**, *51*, 565–581. [CrossRef]

45. Hättenschwiler, S.; Vitousek, P.M. The role of polyphenols in terrestrial ecosystem nutrient cycling. *Trends Ecol. Evol.* **2000**, *15*, 238–243. [CrossRef]

46. Teklay, T. Seasonal dynamics in the concentrations of macronutrients and organic constituents in green and senesced leaves of three agroforestry species in southern Ethiopia. *Plant Soil* **2004**, *267*, 297–307. [CrossRef]

47. Mafongoya, P.L.; Giller, K.E.; Palm, C.A. Decomposition and nitrogen release patterns of tree prunings and litter. *Agrofor. Syst.* **1998**, *38*, 77–97. [CrossRef]

48. Constantinides, M.; Fownes, J.H. Nitrogen mineralization from leaves and litter of tropical plants: Relationship to nitrogen, lignin and soluble polyphenol concentrations. *Soil Biol. Biochem.* **1994**, *26*, 49–55. [CrossRef]

49. Kuhajek, J.M.; Payton, I.J.; Monks, A. The impact of defoliation on the foliar chemistry of southern rata (*Metrosideros umbellata*). *N. Z. J. Ecol.* **2006**, *30*, 237–249.

50. Mansfield, J.L.; Curtis, P.S.; Zak, D.R.; Pregitzer, K.S. Genotypic variation for condensed tannin production in trembling aspen (*Populus tremuloides*, Salicaceae) under elevated CO_2 and in high-and low-fertility soil. *Am. J. Bot.* **1999**, *86*, 1154–1159. [CrossRef] [PubMed]

51. Bryant, J.P.; Chapin, F.S., III; Klein, D.R. Carbon/nutrient balance of boreal plants in relation to vertebrate herbivory. *Oikos* **1983**, *40*, 357–368. [CrossRef]

52. Kandil, F.E.; Grace, M.H.; Seigler, D.S.; Cheeseman, J.M. Polyphenolics in *Rhizophora mangle* L. leaves and their changes during leaf development and senescence. *Trees* **2004**, *18*, 518–528. [CrossRef]

53. Maie, N.; Behrens, A.; Knicker, H.; Kogel-Knabner, I. Changes in the structure and protein binding ability of condensed tannins during decomposition of fresh needles and leaves. *Soil Biol. Biochem.* **2003**, *35*, 577–589. [CrossRef]

54. Kraus, T.E.C.; Dahlgren, R.A.; Zasoski, R.J. Tannins in nutrient dynamics of forest ecosystems—A review. *Plant Soil* **2003**, *256*, 41–66. [CrossRef]

55. Kraus, T.E.C.; Zasoski, R.J.; Dahlgren, R.A. Fertility and pH effects on polyphenol and condensed tannin concentrations in foliage and roots. *Plant Soil* **2004**, *262*, 95–109. [CrossRef]

forests

MDPI

Article

The Effects of Fertilization on the Growth and Physiological Characteristics of *Ginkgo biloba* L.

Jing Guo, Yaqiong Wu, Bo Wang, Yan Lu, Fuliang Cao and Guibin Wang *

Nanjing Forestry University, Co-Innovation Center for Sustainable Forestry in Southern China,
159 Longpan Road, Nanjing 210037, China; guojingg0921@163.com (J.G.); ya_qiong@126.com (Y.W.);
wangboo984@163.com (B.W.); 18260093866@163.com (Y.L.); fuliangcaonjfu@163.com (F.C.)
* Correspondence: guibinwang99@163.com; Tel.: +86-25-8542-8004

Academic Editors: Scott X. Chang and Xiangyang Sun
Received: 2 October 2016; Accepted: 22 November 2016; Published: 24 November 2016

Abstract: *Ginkgo biloba* L. is one of the most extensively planted and productive commercial species in temperate areas around the world, but slow-growth is the most limiting factor for its utilization. Fertilization is one of the key technologies for high quality and high forest yield. To better understand the impacts of fertilization on *Ginkgo* productivity, the effects of fertilization treatments (single fertilizer and combined fertilizer) on growth, nutrient content in *Ginkgo* leaves, and photosynthesis characteristics were studied in a 10-year-old *Ginkgo* plantation over two years. The single factor experiments suggested that DBH (diameter at breast height), H (height), NSL (length of new shoots), and V (trunk volume) showed significant differences between the different levels of single nitrogen (N) or phosphate (P) fertilizer application. Orthogonal test results showed that the nine treatments all promoted the growth of *Ginkgo*, and the formula (N: 400 g·tree^{-1}, P: 200 g·tree^{-1}, potassium (K): 90 g·tree^{-1}) was the most effective. G_s (stomatal conductance) and P_n (net photosynthesis rate) showed significant differences between the different amounts of single N or P fertilizer application, while single K fertilizer only affected P_n. Combined N, P, and K fertilizer had significant promoting effects on C_i (intercellular CO_2 concentration), G_s and P_n. N and P contents in *Ginkgo* leaves showed significant differences between the different amounts of a single N fertilizer application. A single P fertilizer only improved foliar P contents in *Ginkgo* leaves. A single K fertilizer application improved N and K content in *Ginkgo* leaves. The effects of different N, P, and K fertilizer treatments on the nutrient content of *Ginkgo* leaves were different.

Keywords: *Ginkgo*; fertilization; growth; photosynthesis; nutrient content

1. Introduction

Nutrient limitations develop when a stand's potential nutrient use cannot be met by the soil nutrient supply [1]. Improving stand nutrient supply through fertilization is a viable silvicultural option [2]. As an intervention strategy, large-scale application of fertilizers to forests has been implemented to accelerate growth of existing stands and shorten rotation times to overcome future projected timber shortfalls [3]. It is well established that fertilization, in particular nitrogen (N) and phosphate (P) applications, on nutrient limited sites increases tree productivity by increasing photosynthesis in the short term [4,5] and leaf area over the long term [6]. N is a vital constituent in proteins, nucleic acids, chlorophylls, and many secondary metabolites of plants [7]; therefore, it plays an essential role in the enzymatic activities of photosynthetic processes [8]. P plays a central role in almost all aspects of plant metabolism and is one of the nutrients that most commonly limits growth [9]. Potassium (K) is one of important nutrient elements and plays a major role in growth, modifying abundant enzyme activations and controlling cell osmoregulation [10]. Gross primary production (GPP) was also greatly increased by K fertilization as a result of lower stomatal and mesophyll resistance to CO_2 diffusion and higher photosynthetic capacity in the leaves [11,12]. In addition, potassium can also alleviate the harmful effects of abiotic stress [13–15].

The advantage of using chemical fertilizers is that nutrients are soluble and immediately available to the plants; therefore, the effect is usually direct and fast. Several studies have reported an increase in growing efficiency and higher enzymatic activities following fertilization [16,17]. Many studies have found that growing efficiency is largely unaffected by nutrient additions [18,19]. Different plants do not have the same nutrient demand and specific fertility targets [20]. The fast-growing broad-leaved tree *Populus alba* × *P. glandulosa* is more sensitive to increasing N availability [21], while the slow-growing *P. popularis* is more sensitive to decreasing N availability [22]. Among conifers, the fast-growing spruce families show more plasticity in biomass allocation than do the slowly growing ones under different nitrogen supplies [23], and the fast growing species *Pinus radiata* allocates more to the aboveground biomass under N and P fertilization [24].

Ginkgo is commonly referred to as Gongsun Tree or Duck Foot Tree, is one of the most ancient gymnosperms in the world, and is native to the temperate forests of China [25]. *Ginkgo* is an eco-economic tree species which is valuable for food, health care, medicine, timber, landscape, ecological protection, and scientific research. The great environmental adaptability and beauty of *Ginkgo* have made the tree a favorite species for planting as an ornamental tree throughout the world [26]. Its widespread use has been facilitated by its tolerance of air pollution and soil compaction. *Ginkgo* has a long history in leaf and nut production. However, it is still in the initial stage for timber cultivation in China. *Ginkgo* wood has also been highly valued for making furniture and handicraft articles, and *Ginkgo* wood is also an ideal material for making musical instruments [27]. Nowadays, *Ginkgo* wood is only used for carving and chopping blocks. The compression of *Ginkgo* wood was found to be stiffer than that of the other species [28]. Burgert et al. (2004) consequently concluded that there has been an evolutionary trend towards much more flexible compression wood [29]. Gong et al. (2009) reported higher lignin content in *Ginkgo* than in conifers [30]. *Ginkgo* wood was once used for pillars and rafters in the palaces and temples in ancient China. *Ginkgo* also has an anti-dust function, purifying the environment. One important role of *Ginkgo* is planting in rows in corridors, squares, or on both sides of a street as shade trees, as well as planting in gardens and vestibule entrances as landscape trees. *Ginkgo* wood has value for many purposes from the help of science and technology. In recent years, with the development of the *Ginkgo* industry, *Ginkgo* cultivation area continues to expand; many researchers have developed a variety of optimum fertilization schemes for *Ginkgo* plantations (especially leaf-producing plantations, nuts plantations, and seedlings) in China. However, studies concentrated on timber plantations are comparatively insufficient.

This study was conducted to examine the relationship between fertilization and growth, photosynthesis characteristics, and foliage nutrient content of pure *Ginkgo* timber plantations. The main objectives were to assess the effect of fertilization on growth, nutrient content in *Ginkgo* foliage, and photosynthesis characteristics of new *Ginkgo* shoots in a 10-year-old *Ginkgo* plantation over two years. Information generated from this study is expected to be of great value for providing optimal fertilization measures for improving the yield of *Ginkgo* timber forests and realizing balanced nutrient management for its fast growth. In addition, the application of a suitable fertilizer ratio after foliage nutrient content analysis, photosynthesis, and tree growth characteristics are some of the most effective ways to reduce cost and fertilizer waste.

2. Materials and Methods

2.1. The Study Site

The experimental field was located in Yellow Sea Forest Park, the east-central region of the Jiangsu Province, Eastern China (32°51′ N, 120°51′ E, 5 m above sea level). The study area belongs to a transition zone between a subtropical zone and warm temperate zone, with seasonal pluvial heat and significant monsoon activity. The study area is characterized by a mean annual temperature 15.0 °C, an annual rainfall of 1061.2 mm, an annual sunshine duration of 2209 h, and an annual frost-free period of 220 days. For determining the site fertility, soil samples from the experimental field were collected for physical and chemical characteristics before actual experiments. The soil was characterized as coastal sandy saline-alkali soil with a pH of 8.42, a bulk density of 1.28 $g \cdot cm^{-3}$, a total nitrogen of 0.75 $mg \cdot g^{-1}$, a total phosphorus of 0.26 $mg \cdot g^{-1}$, and a total potassium of 5.25 $mg \cdot g^{-1}$.

2.2. Materials and Experimental Design

This study was carried out in 2014 and 2015. The test forest was pure *Ginkgo* forest planted in 2005, with a vine space and row space of 2 m × 8 m. The trees came from the same variety. The experiment with different fertilizer treatments was designed based on a single factor design (EXP.1) and an orthogonal design (EXP.2). Fertilizer was applied to a depth of approximately 10–15 cm (to prevent losses to rain or wind) in several spots in a 70 cm radius circle around each tree in early April 2014 and early April 2015.

Single factor designs (unit: $g \cdot tree^{-1}$): EXP.1 was designed as single-factor experiment. The dosage of four treatments of single N fertilizer was 600, 400, 200, and 100, respectively. The dosage of four treatments of single P fertilizer was 800, 600, 400, and 200, respectively. The dosage of four treatments of single K fertilizer was 200, 90, 40, and 15, respectively. The amounts of fertilizers were in descending order. The fertilizers used in the experiments were urea (N), superphosphate (P), and potassium sulfate (K). There were a total of 13 treatments, and one of them was CK_1 (control, without fertilizer application). Each treatment had 10 *Ginkgo* trees with three replications in the trial. There were a total of 39 plots and 390 *Ginkgo* trees (average diameter at breast height was 13.37 cm, average height was 8.04 m, and their growth was uniform.) in the single factor tests. Additionally, there were two guard rows around the test plots.

Orthogonal designs (unit: $g \cdot tree^{-1}$): Second, the experiment with different fertilizer treatments (EXP.2) was designed based on orthogonal designs with three levels of urea, superphosphate, or potassium sulfate, for a total of 10 treatments one of them was CK_2 (control), without fertilizer application, which differed from the CK_1 above. Treatments were replicated three times (Table 1). There were a total of 30 plots and 300 *Ginkgo* trees (average diameter at breast height was 11.62 cm, average height was 5.78 m, and their growth was uniform) in the orthogonal tests. Orthogonal tests plots were in close proximity to the single-factor test spots.

Table 1. Orthogonal trial fertilizer treatments.

Treatment No.	Fertilization Level ($g \cdot tree^{-1}$)		
	Urea	Superphosphate	Potassium Sulfate
1	100	200	15
2	100	400	40
3	100	600	90
4	200	200	40
5	200	400	90
6	200	600	15
7	400	200	90
8	400	400	15
9	400	600	40
CK_2	0	0	0

CK_2: control in EXP.2.

2.3. Tree Growth Indictors' Measurement

Tree diameter at breast height (DBH), height (H), and length of new shoots (NSL) were measured in November 2015. Tree height was measured using a hypsometer (SGQ-1, Harbin, China). DBH and NSL were measured using diameter tape. We selected 21 *Ginkgo* trees of different diameter classes (2, 4, 6, 8, 10, 14, and 20) around the experiment plots, cut them down, and measured their DBH and H. Then, according to the standard volume equation and formula (Equation (1)):

$$V = a_0 D^{a1} H^{a2} \tag{1}$$

we used the R programming language (version 3.0, The University of Auckland, Auckland, New Zealand) software to obtain the trunk volume determination formula (Equation (2)):

$$LnV = 2.0135 \times lnD + 0.5685 \times lnH - 9.2452 \; (R^2 = 0.9819) \tag{2}$$

V: trunk volume (m^3), D: DBH (cm), H: height (m). We used this formula to calculate the *Ginkgo* trunk volume (V) on the test plots.

2.4. Determination of Photosynthetic Indexes

The photosynthetic characteristics were measured with a CIRAS-2 photosynthetic instrument (Hansatech, Norfolk, VA, USA). We selected sunny days in early August 2015 (08:30–11:30), and set the parameters: leaf temperature was not controlled, relative humidity was 85%, cylinder supply CO_2 concentration was 380 $\mu mol \cdot mol^{-1}$, and the light intensity of an artificial light source was 1200 $\mu mol \cdot m^{-2} \cdot s^{-1}$ (exceeding the *Ginkgo* light saturation point). Five trees were randomly selected from each replicate of each treatment on the test plots. Then, we selected mature healthy leaves of the fully-extended branches in the high position of the canopy. The net photosynthetic rate (P_n), transpiration rate (T_r), stomatal conductance (G_s), and intercellular CO_2 concentration (C_i) were measured.

2.5. Determination of Nutrient Concentration in Leaves

The sampling time was in August 2015. We selected five trees randomly from each replicate of each treatment in the test plots. We mined the new shoots in the middle part of the canopy with pole tree pruners. We collected the healthy 8–12 pieces of functional leaves without diseases and pest attack, and immediately put the leaves into an ice box, respectively. After the leaves were brought to the laboratory, the leaves were washed with deionized water, rapid fixed for 15 min at 105 °C, then dried to counterweigh at 70 °C, and, finally, crushed and passed through a 100-mesh sieve, severally (a total of 69 samples). Then, we weighed each sample to 0.1000 g and used concentrated H_2SO_4-$HClO_4$ to digest them. The nitrogen and phosphorus were analyzed with an AA3 continuous flow analytical system (Bran + Luebbe, Hamburg, Germany); potassium was analyzed via flame atomic absorption spectrometry (WFX-210, Beijing, China).

2.6. Statistical Analysis

Data are reported as the mean of three replicates ± standard deviation (SD), and all tests were performed using the Statistical Product and Service Solution statistical software program (IBM Inc., Chicago, IL, USA). One-way analysis of variance (ANOVA) was conducted to compare the effect of fertilization treatments on growth, photosynthetic indexes, and the element contents in leaves. ANOVA was performed at the $p < 0.05$ level of significance. Duncan's multiple-range test was performed for each variable (note: different lowercases in the table showed significant differences at the 0.05 level. Treatment means with different lowercases among treatments in the figure are significantly different at the 0.05 level).

3. Results

3.1. Effects of Fertilizer on Growth Indicators

In Yellow Sea Forest Park, DBH, H, NSL, and V in EXP.1 showed significant differences between levels of single N or P fertilizer applications, suggesting that N fertilizer application in excess of 100 g·tree^{-1} or P fertilizer application in excess of 400 g·tree^{-1} may be suitable for tree growth (Table 2; $p < 0.05$). However, growth responses were small and none of the variables showed a significant growth response to K fertilizer (Table 2; $p > 0.05$). Maximum dosage tended to have the greatest influences on the growth indicators for single N, P, and K fertilizers (N = 600 g·tree^{-1}, P = 800 g·tree^{-1}, or K = 200 g·tree^{-1}). After two years of continuous fertilization, a single N fertilizer application rate at 600 g·tree^{-1} resulted in an increase in V greater than 17.7% compared with unfertilized plots (CK$_1$). Meanwhile, a single P fertilizer application rate at 800 g·tree^{-1} resulted in an increase in V greater than 9.7% compared with unfertilized plots (CK$_1$). Therefore, single N or P fertilizer treatments promoted *Ginkgo* growth indicators significantly, while the promotion effect of single K fertilizer treatment was not significant. Additionally, orthogonal tests in EXP.2 showed that combined N, P, and K fertilizer had significant promoting effects on growth indictors (Table 3; $p < 0.05$); the maximum effect on the growth of *Ginkgo* was achieved at a high dosage of N. After two years of continuous fertilization, Treatment 7 resulted in an increase in V greater than 27.7% compared with unfertilized plots (CK$_2$).

Table 2. Effects of two years in EXP.1 of N, P, and K fertilizer (F) on the diameter at breast height (DBH, cm), height (H, m), length of new shoots (NSL, cm), and trunk volume (V, m^3) of *G. biloba* in 2014 and 2015 in Yellow Sea Forest Park in China.

F	DBH	H	NSL	V
N1 600	15.34 ± 0.396 a	9.85 ± 0.383 a	26.59 ± 1.204 a	0.0146 ± 0.000929 a
N2 400	15.11 ± 0.123 a	9.27 ± 0.492 b	25.56 ± 1.284 ab	0.0135 ± 0.000755 ab
N3 200	14.96 ± 0.301 a	9.17 ± 0.131 b	24.64 ± 2.840 ab	0.0133 ± 0.000448 b
N4 100	14.92 ± 0.407 a	9.14 ± 0.068 b	21.93 ± 2.635 bc	0.0132 ± 0.000369 b
CK$_1$ 0	14.35 ± 0.095 b	8.92 ± 0.158 b	20.23 ± 2.594 c	0.0124 ± 0.000256 b
P1 800	14.84 ± 0.063 a	9.45 ± 0.257 a	25.98 ± 1.33 a	0.0136 ± 0.000398 a
P2 600	14.42 ± 0.255 ab	9.13 ± 0.221 ab	24.02 ± 1.86 ab	0.0127 ± 0.000408 b
P3 400	14.23 ± 0.333 b	9.05 ± 0.182 ab	24.37 ± 0.688 ab	0.0124 ± 0.000514 b
P4 200	14.21 ± 0.292 b	8.83 ± 0.270 b	22.69 ± 0.419 bc	0.0121 ± 0.000464 b
CK$_1$ 0	14.35 ± 0.095 b	8.92 ± 0.158 b	20.23 ± 2.594 c	0.0124 ± 0.000256 b
K1 200	14.91 ± 0.499 a	9.01 ± 0.288 a	24.01 ± 0.844 a	0.0124 ± 0.000965 a
K2 90	14.65 ± 0.428 a	8.93 ± 0.195 a	23.71 ± 1.29 a	0.0117 ± 0.000445 a
K3 40	14.66 ± 0.553 a	8.88 ± 0.326 a	23.35 ± 2.14 a	0.0113 ± 0.000699 a
K4 15	14.38 ± 0.422 a	8.63 ± 0.459 a	24.49 ± 0.967 a	0.0114 ± 0.000993 a
CK$_1$ 0	14.35 ± 0.095 a	8.92 ± 0.158 a	20.23 ± 2.59 a	0.0124 ± 0.000256 a

Note: different lowercase letters in the table reflect significant differences between treatments. CK$_1$: control in EXP.1.

Table 3. Effects of two years in EXP.2 of combined N, P, and K fertilizer (F) on the diameter at breast height (DBH, cm), height (H, m), length of new shoots (NSL, cm), and trunk volume (V, m^3) of *G. biloba* in 2014 and 2015 in Yellow Sea Forest Park in China.

F	DBH	H	NSL	V
1	13.66 ± 0.908 bcd	7.45 ± 0.022 bc	24.15 ± 0.832 bcd	0.00983 ± 0.000675 cd
2	13.78 ± 0.825 abcd	7.49 ± 0.065 bc	26.606 ± 3.250 abc	0.00997 ± 0.000679 cd
3	13.2 ± 0.172 cd	7.42 ± 0.168 bc	26.00 ± 2.32 abc	0.00947 ± 0.000327 de
4	13.71 ± 0.233 bcd	7.65 ± 0.123 ab	25.90 ± 1.921 abc	0.01013 ± 0.000207 cd
5	13.88 ± 0.173 abc	7.73 ± 0.293 ab	25.26 ± 1.705 abc	0.01036 ± 0.000261 bc
6	13.94 ± 0.370 abc	7.55 ± 0.046 b	22.83 ± 0.656 cd	0.01017 ± 0.000304 cd
7	14.61 ± 0.347 a	8.01 ± 0.239 a	26.77 ± 4.070 abc	0.01130 ± 0.000302 a
8	14.50 ± 0.423 ab	7.81 ± 0.173 ab	28.89 ± 2.014 a	0.01094 ± 0.000240 ab
9	14.47 ± 0.208 ab	7.81 ± 0.205 ab	27.49 ± 1.980 ab	0.01092 ± 0.000304 ab
CK$_2$	12.92 ± 0.231 d	7.09 ± 0.0346 c	20.23 ± 2.594 d	0.00885 ± 0.000161 e

Note: different lowercase letters in the table reflect significant differences between treatments.

3.2. Effects of Fertilizer on Photosynthesis Indicators

In Yellow Sea Forest Park, P_n and G_s showed significant differences between levels of single N or P fertilizer application ($p < 0.05$), while T_r and C_i were not affected by fertilizer application (Table 4; $p > 0.05$). However, photosynthesis indicators showed no significant response to single K fertilizer treatment except P_n (Table 4; $p > 0.05$). After two years of continuous fertilization, a single N fertilizer application rate at 600 g·tree^{-1} resulted in an increase in P_n greater than 49.5% compared with unfertilized plots (CK$_1$). Meanwhile, single P fertilizer application rate at 800 g·tree^{-1} resulted in an increase in P_n greater than 27.6% compared with unfertilized plots (CK$_1$). Therefore, single N or P fertilizer treatments promoted *Ginkgo* P_n significantly, while the promotion effect of single K fertilizer was also significant at 90 g·tree^{-1} (K2). Additionally, orthogonal tests showed that combined N, P, and K fertilizer had a significant promoting effect on P_n and G_s ($p < 0.05$); while fertilization was not so important on T_r and C_i (Table 5; $p > 0.05$) in EXP.2, after two years of continuous fertilization, Treatment 9 resulted in an increase in P_n greater than 36.9% compared with unfertilized plots (CK$_2$).

Table 4. Average effects of two years N, P, and K fertilizer (F) in EXP.1 on the photosynthesis characteristics, net photosynthetic rate (P_n, µmol·m^{-2}·s^{-1}), transpiration rate (T_r, µmol·m^{-2}·s^{-1}), stomatal conductance (G_s, mol·m^{-2}·s^{-1}), and intercellular CO_2 concentration (C_i, µmol·mol^{-1}) in 2014 and 2015 in Yellow Sea Forest Park in China.

F	C_i	T_r	G_s	P_n
N1 600	278.56 ± 8.57 a	2.71 ± 0.252 a	167.64 ± 3.32 a	12.30 ± 0.472 a
N2 400	277.33 ± 4.98 a	2.68 ± 0.150 a	165.78 ± 2.55 a	11.74 ± 0.597 ab
N3 200	269.78 ± 5.10 a	2.43 ± 0.173 a	157.78 ± 4.43 a	10.25 ± 1.42 b
N4 100	267.79 ± 16.65 a	2.51 ± 0.284 a	156.67 ± 7.86 a	10.57 ± 0.635 b
CK$_1$ 0	267.33 ± 21.73 a	2.53 ± 0.208 a	127.33 ± 15.5 b	8.23 ± 0.833 c
P1 800	289.89 ± 3.29 a	3.03 ± 0.153 a	156.56 ± 5.34 a	10.50 ± 1.19 a
P2 600	275.11 ± 8.26 a	2.80 ± 0.152 a	139.78 ± 4.17 b	10.22 ± 1.06 ab
P3 400	266.89 ± 5.23 a	2.62 ± 0. 135 a	129.89 ± 8.06 b	8.77 ± 1.72 abc
P4 200	275.33 ± 19.50 a	2.67 ± 0.333 a	133.88 ± 6.02 b	7.14 ± 0.635 c
CK$_1$ 0	267.33 ± 21.73 a	2.53 ± 0.208 a	127.33 ± 15.50 b	8.23 ± 0.833 bc
K1 200	256.89 ± 1.92 a	2.60 ± 0.101 a	148.22 ± 11.5 a	10.39 ± 0.27 a
K2 90	251.00 ± 4.58 a	2.54 ± 0.081 a	133.11 ± 10.98 a	10.44 ± 0.44 a
K3 40	265.78 ± 7.93 a	2.46 ± 0.102 a	123.67 ± 15.94 a	8.37 ± 0.62 b
K4 15	266.79 ± 9.25 a	2.27 ± 0.095 a	121.56 ± 16.60 a	7.61 ± 0.37 b
CK$_1$ 0	267.33 ± 21.73 a	2.53 ± 0.208 a	127.33 ± 15.50 a	8.23 ± 0.83 b

Note: different lowercase letters in the table reflect significant differences between treatments.

Table 5. Average effects of two years of N, P, and K fertilizer (F) in EXP.2 on the photosynthesis characteristics, net photosynthetic rate (P_n, µmol·m^{-2}·s^{-1}), transpiration rate (T_r, µmol·m^{-2}·s^{-1}), stomatal conductance (G_s, mol·m^{-2}·s^{-1}), and intercellular CO_2 concentration (C_i, µmol·mol^{-1}) in 2014 and 2015 in Yellow Sea Forest Park in China.

F	C_i	T_r	G_s	P_n
1	239.00 ± 10.40 b	2.56 ± 0.34 a	122.33 ± 13.39 c	9.00 ± 0.91 cd
2	246.78 ± 27.03 b	2.41 ± 0.20 a	127.33 ± 13.00 bc	9.22 ± 0.74 bcd
3	243.00 ± 25.32 b	2.41 ± 0.47 a	136.11 ± 7.31 abc	9.36 ± 1.02 bcd
4	254.00 ± 8.09 b	2.61 ± 0.24 a	133.56 ± 9.52 bc	9.48 ± 0.17 bcd
5	247.33 ± 5.20 b	2.59 ± 0.34 a	141.34 ± 3.51 abc	9.71 ± 1.51 abcd
6	272.11 ± 17.88 ab	2.74 ± 0.15 a	140.11 ± 4.40 abc	10.11 ± 0.23 abc
7	271.78 ± 16.68 ab	2.69 ± 0.25 a	142.67 ± 3.51 abc	10.68 ± 1.22 abc
8	272.55 ± 20.51 ab	2.69 ± 0.08 a	146.89 ± 12.98 ab	10.71 ± 0.71 ab
9	290.33 ± 16.18 a	2.79 ± 0.16 a	154.33 ± 12.20 a	11.27 ± 0.38 a
CK$_2$	265.33 ± 18.73 ab	2.50 ± 0.11 a	129.13 ± 12.50 bc	8.53 ± 0.58 d

Note: different lowercase letters in the table reflect significant differences between treatments.

3.3. Effects of Fertilizer on Leaf N, P, and K Contents

In Yellow Sea Forest Park, N and P contents in leaves showed significant differences between levels of single N fertilizer application ($p < 0.05$), while K content was not affected by N fertilizer application (Figure 1; $p > 0.05$). N and K contents showed no significant response to single P fertilizer ($p > 0.05$), while P content showed a significant difference (Figure 2; $p < 0.05$). N and K contents showed significant differences between the different amounts of single K fertilizer application ($p < 0.05$), while P content was not affected by K fertilizer application (Figure 3; $p > 0.05$). After two years of continuous fertilization, a single N fertilizer application of 600 g·tree^{-1} resulted in an increase in N content greater than 16.4% and an increase in P content greater than 19.9% compared with unfertilized plots (CK$_1$). Meanwhile, a single P fertilizer application of 800 g·tree^{-1} resulted in an increase in P content greater than 23.8% compared with unfertilized plots (CK$_1$). A single K fertilizer application of 200 g·tree^{-1} resulted in an increase in N content greater than 12.2% and an increase in K content greater than 21.8% compared with unfertilized plots (CK$_1$). Therefore, single N fertilizer treatment promoted *Ginkgo* N and P content significantly, single P fertilizer treatment only promoted *Ginkgo* P content significantly, and the promotion effect of single K fertilizer treatment on N and K content were significant. Additionally, orthogonal tests in EXP.2 showed that combined N, P, and K fertilizer treatment had a significant promoting effect on N, P, and K content (Figure 4; $p < 0.05$). After two years of continuous fertilization, Treatment 9 resulted in an increase in N content greater than 13.4% compared with unfertilized plots (CK$_2$). Treatment 6 resulted in an increase in P content greater than 21.5% compared with unfertilized plots (CK$_2$). Treatment 7 resulted in an increase in K content greater than 16.7% compared with unfertilized plots (CK$_2$).

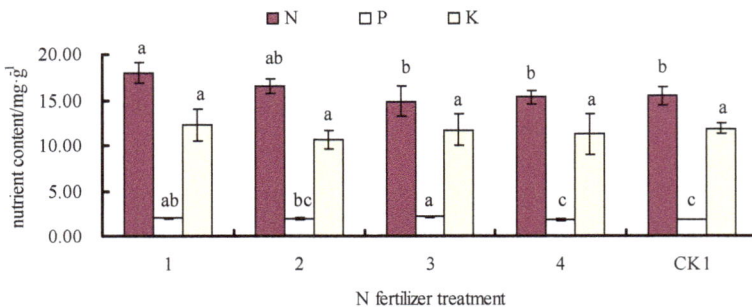

Figure 1. Effects of single N fertilizer application on *G. biloba* leaf N, P, and K contents in EXP.1.

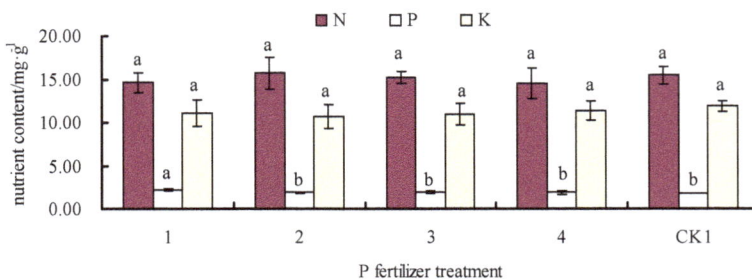

Figure 2. Effects of single P fertilizer application on *G. biloba* leaf N, P, and K contents in EXP.1.

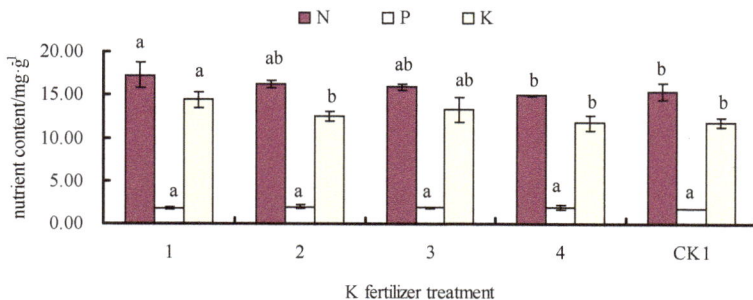

Figure 3. Effects of single K fertilizer application on *G. biloba* leaf N, P, and K contents in EXP.1.

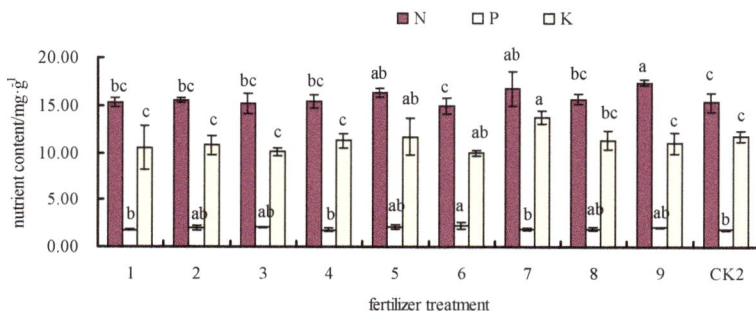

Figure 4. Effects of combined N, P, and K fertilizer application on *G. biloba* leaf N, P, and K contents in EXP.2.

4. Discussion

Nutrient deficiency is generally considered the major factor limiting growth, and fertilization is the main way to supplement the lack of nutrition. The fertilization effects have shown different results under different quantities of fertilizer, the number of years since fertilization treatment, and the nutrition status of the forest site [31]. Our study was carried out in 2014 and 2015. The single factor experiments suggested that DBH, H, NSL, and V showed significant differences between the different levels of single N or P fertilizer application. The orthogonal test results showed that the nine treatments all promoted the growth of *Ginkgo*, and Treatment 7 (N: 400 g·tree^{-1}, P: 200 g·tree^{-1}, K: 90 g·tree^{-1}) was the most effective one. The response to fertilization in *Ginkgo* plantations depended on the degree of mismatch between nutrient supply from the soil and the plant nutrient demand [32]. With an increment in the N fertilizer, there was also an increment in the growth, foliar nutrient status, and photosynthesis rate of the *Ginkgo* plantation. Phosphorus supply affected photosynthesis, leaf metabolites, and allocation to roots versus leaves and growth. In our study, the growth indicators improved with the P fertilizer dosage. However, excess P supply in the soil is a major environmental concern. The accumulation of P in the soil from applications of chemical fertilizer exceed the requirement of plant and can increase the risk of P movement to surface and groundwater [33]. Thus, we should be very circumspect in the application of P fertilizer. Many investigations have shown that improved N and P nutrition increased growth and yield in many tree species [34]. N and P deficiency cause negative effects on plants [35,36]. However, Gotore et al. (2014) proposed that single N application had negative effects on tree growth, while single P had positive effects on *Pinus patula* growth at Charter Forests. This may be because N is not a limiting factor to tree growth in the first few years after planting and may be readily available in the soil [19]. Additionally, Moilanen et al.

(2004) reported clearly that N fertilization is unnecessary on the nitrogen-rich sites, and additional N increased growth only at the most barren sites [37]. In our study, the promotion effect of a single K fertilizer on *Ginkgo* growth indicators was not significant, though it was better than the unfertilized plots. A previous study showed that fertilization with K was causing a decreased allocation to roots, which enables increased growth in height and leaf number [38]. Moreover, the viewpoint that an optimal potassium nutrition status can reduce the effects of abiotic stresses, such as drought, heat, high light intensity, or salinity has been well established [14,39,40]. The characteristic of our study site is coastal sandy saline–alkali soil; thus, K fertilizer is essential in Yellow Sea Forest Park. In general, N played a key role in the growth of *Ginkgo*, and P and K were important nutrient elements playing a major role in *Ginkgo* growth and development. According to the classification standard of soil nutrients in China, the analysis of soil elements in our experimental plots reveals that contents of N, P, and K are basically at low levels. Therefore, supplementary fertilization may be required on the experimental plots, particularly as the soil nutrient availability cannot meet the demand of the trees.

Photosynthesis is the basis of growth and yield formation, and the effects of nutrition on growth may be driven primarily by photosynthesis [41]. Fertilization can increase the photosynthetic efficiency, prolong the duration of photosynthesis, and enlarge the leaf area index (LAI). In the present study, G_s and P_n showed significant differences between the different levels of single N or P fertilizer applications, while single K fertilizer only affected P_n. Combined N, P, and K fertilizers had significant promoting effects on C_i, G_s, and P_n. N played a vital role in the enzymatic activities of photosynthetic processes, and N addition improved the *Ginkgo* photosynthesis rate and growth response. Among species there is often a strong positive correlation between maximum rates of photosynthesis and N [42]. Greater application of N fertilizer and greater allocation of N to metabolically active tissues, such as leaves and new stems, in the previous study promoted P_n by increasing LAI and chlorophyll concentration of leaves, which, in turn, supported better plant growth [43]. There was also a strong connection between P and the maximum rate of photosynthesis [44]; a partial explanation may lie in a positive relationship between the concentration of P and the amount of Rubisco. Inorganic phosphate (Pi) is a prerequisite for RuBP and also a competitive inhibitor of RUBP [45,46]. Additionally, there was a strong correlation between K concentrations and Rubisco [21]. Zorb et al. (2014) also reported that rates of photosynthesis were positively correlated with application rates of K [47]. K deficiency reduces photosynthesis by decreasing stomatal conductance [48]. However, the differences of stomatal conductance were not significant between different amounts of K fertilizer in our research. Possibly owing to the high mobility of K in both soils and plants, growth is expected to improve in response to added K, as shown in sandy soils [49]. Under this circumstance, deficiency of K may be so severe that fertilization is necessary to sustain tree growth until the end of rotation [50,51]. Transpiration and leaf water status are coordinated to minimize plant water loss and avoid hydraulic failure [52]. T_r of different levels of fertilization had no significant difference in our study. Probably because the climate characteristic of the site belongs to seasonal pluvial heat and significant monsoon, the rainfalls mainly occur in summer. Many studies have shown that growth of *Populus* was related to increasing water availability and fertilization on nutrient-limited soils [53,54]. Thus, sufficient rainfall and appropriate fertilization contributed to better growth in *Ginkgo* plantations in Yellow Sea Forest Park.

Nutrient concentrations of foliage have been accepted as adequate indicators of growth and soil fertility in forest stands [55,56]. Maintaining near-optimal foliar nutrient status of trees is especially critical to the maintenance of growth, the establishment of cold tolerance, and the evaluation of soil fertility at sites [56,57]. Our results suggested that N and P contents showed a significant influence of single N fertilizer application. Single P fertilizer only improved foliar P content, and high P fertilizer dosage had an inhibiting effect on K content. Single K fertilizer application improved N and K content in *Ginkgo* leaves. Effects of different combined N, P, and K fertilizer treatments on the nutrient content of *Ginkgo* leaves were different. In this study, we found that foliar N concentration was significantly correlated with N fertilizer application, which might result primarily from the supply status of soil nutrient elements. Moreover, foliar N concentration was significantly correlated with K fertilizer application, which might result from K-increased N use efficiency [58]. A previous study has showed that the maximum photosynthetic carbon assimilation rate positively correlated with foliar N concentrations [59]. A greater allocation of N to metabolically-active tissues, such as leaves and new stems in the fertilization treatment, increases the LAI and chlorophyll concentration of leaves, further supporting a better growth of *Ginkgo* plantations. In N-poor sites, particularly during cold growing seasons, the availability of N for trees may also be low and N deficiencies may be quite common [60,61]. Thus, it is an adaptation to store N in excess of current requirements for photosynthesis that can later be recycled and utilized during periods of lower N supply from the soil. P fertilization improved foliar P contents, and a high P fertilizer dosage had an inhibiting effect on K content. Our findings were consistent with the trend of a previous study of *Phoebe bournei* seedlings, where foliar N and P concentrations increased, but K concentration decreased as the phosphorus addition increased [62]. The results from Crous et al. (2008) showed a trend of increasing foliar K with an increase in the application of K when no phosphate fertilizer was applied. When P was applied with K on the plots with residual fertilizer, no such trend was observed [63]. Nakashgir (1992) reported nitrogen utilization of maize was accentuated when K application was supplemented [64]. We found that foliar K concentration was significantly correlated with K fertilizer application, which might owe to the high mobility of K in plants and soil. Additionally, light affects the internal redistribution of potassium, leading to higher concentrations in sun-exposed branches than in shaded ones [65,66]. The previous studies reported that potassium plays an important role in stomatal function by maintaining turgor pressure [67]. Growth is expected to improve in response to added K as shown in sandy soils [49]. According to the nutrient content contained in the *Ginkgo* leaves, the largest demand is N, followed by K, then P in this study. These results suggest that a new compound ratio of fertilizer is needed to optimize the growth of *Ginkgo* timber forest planted in coastal sandy saline–alkali soil because of different foliar N, P, and K concentrations under different fertilization treatments. According to the actual situation, we can suggest an optimal fertilization formula for improving the yield and quality of *Ginkgo* timber forests in Yellow Sea Forest Park.

5. Conclusions

The beneficial influences of single N, P, and K fertilization and combined N, P, and K fertilization on *Ginkgo* nutrition, photosynthesis, and growth were very obvious. This study confirmed the positive foliar nutrient content and the increased photosynthesis and growth responses with sufficient data, which indicated significantly better growth in fertilized stands compared to unfertilized stands. Single N, P, and K fertilizers and the combined N, P, and K fertilizer had positive effects on the growth of *Ginkgo*. Combined N, P, and K fertilizer and optional formulation are the better choices for achieving balanced nutrition of *Ginkgo* on our experimental plots. Sufficient nutrition supplements in barren places are significant to guarantee the growth of the forests. Efficient fertilizer regimes require an accurate knowledge of the nutrient status during the growth period to avoid any unforeseen deficiencies.

Acknowledgments: This study was supported by a Grant from the research program "Research on key technology of the cultivation of *Ginkgo* timber and medicinal forest" (2012BAD21B04) provided by National Science and technology support program, "Research and demonstration of sustainable management technology of *Ginkgo* timber forest in coastal beach" (BE2013443) provided by Science and technology support program of Jiangsu Province, the Priority Academy Program Development of Jiangsu Higher Education Institution (PAPD) and the Doctorate Fellowship Foundation of Nanjing Forestry University.

Author Contributions: Jing Guo and Yan Lu designed the study and conducted the field trial. Jing Guo, Yaqiong Wu, and Bo Wang were responsible for the samples collection and laboratory analysis. Jing Guo was responsible for the statistical analyses, with contributions from Fuliang Cao and Guibin Wang. Jing Guo wrote the paper.

Conflicts of Interest: The authors declare no conflict of interest.

References

1. Pathak, H.; Aggarwal, P.K.; Roetter, R.; Kalra, N.; Bandyopadhaya, S.K.; Prasad, S.; Keulch, H.V. Modelling the quantitative evaluation of soil nutrient supply, nutrient use efficiency, and fertilizer requirements of wheat in India. *Nutr. Cycl. Agroecosyst.* **2003**, *65*, 105–113. [CrossRef]
2. Fox, T.R.; Allen, H.L.; Albaugh, T.J.; Rubilar, R.; Carlson, C.A. Tree Nutrition and Forest Fertilization of Pine Plantations in the Southern United States. *South. J. Appl. For.* **2006**, *31*, 5–11.
3. Sophie, W.; Leigh, A.K.K.; Grayston, S.J. Effects of long-term fertilization of forest soils on potential nitrification and on the abundance and community structure of ammonia oxidizers and nitrite oxidizers. *FEMS Microbiol. Ecol.* **2011**, *79*, 142–154.
4. Gough, C.M.; Seiler, J.R.; Maier, C.A. Short-term effects of fertilization on loblolly pine (*Pinus taeda* L.) physiology. *Plant Cell Environ.* **2004**, *27*, 876–886. [CrossRef]
5. King, N.T.; Seiler, J.R.; Fox, T.R.; Johnsen, K.H. Post-fertilization physiology and growth performance of loblolly pine clones. *Tree Physiol.* **2008**, *28*, 703–711. [CrossRef] [PubMed]
6. Samuelson, L.J.; Butnor, J.; Maier, C.; Stokes, T.A.; Johnsen, K.; Kane, M. Growth and physiology of loblolly pine in response to long-term resource management: Defining growth potential in the southern United States. *Can. J. For. Res.* **2008**, *38*, 721–732. [CrossRef]
7. Luo, J.; Qin, J.J.; He, F.F.; Li, H.; Liu, T.X.; Polle, A.; Peng, C.H.; Luo, Z.B. Net fluxes of ammonium and nitrate in association with H$^+$ fluxes in fine roots of *Populus populari*. *Planta* **2013**, *237*, 19–31. [CrossRef] [PubMed]
8. Güsewell, S. N:P ratios in terrestrial plants: Variation and functional significance. *New Phytol.* **2004**, *164*, 243–266. [CrossRef]
9. Warren, C.R. How does P affect photosynthesis and metabolite profiles of *Eucalyptus globulus*? *Tree Physiol.* **2011**, *31*, 727–739. [CrossRef] [PubMed]
10. Coskun, D.; Britto, D.T.; Kronzucker, H.J. The physiology of channel-mediated K$^+$, acquisition in roots of higher plants. *Physiol. Plant.* **2014**, *151*, 305–312. [CrossRef] [PubMed]
11. Battie-Laclau, P.; Laclau, J.P.; Beri, C.; Mietton, L.; Muniz, M.R.A.; Arenque, B.C.; Piccolo, M.C.; Jordan-Meille, L.; Bouillet, J.P.; Nouvellon, Y. Photosynthetic and anatomical responses of *Eucalyptus grandis* leaves to potassium and sodium supply in a field experiment. *Plant Cell Environ.* **2014**, *37*, 70–81. [CrossRef] [PubMed]
12. Christina, M.; Maire, G.L.; Battie-Laclau, P.; Nouvellon, Y.; Bouillet, J.P.; Jourdan, C.; Goncalves, J.L.M.; Laclau, J.P. Measured and modeled interactive effects of potassium deficiency and water deficit on gross primary productivity and light-use efficiency in *Eucalyptus grandis* plantations. *Glob. Chang. Biol.* **2015**, *21*, 2022–2039. [CrossRef] [PubMed]
13. Mäser, P.; Gierth, M.; Schroeder, J.I. Molecular mechanisms of potassium and sodium uptake in plants. *Plant Soil* **2002**, *247*, 43–54. [CrossRef]
14. Cakmak, I. The role of potassium in alleviating detrimental effects of abiotic stresses in plants. *J. Plant Nutr. Soil Sci.* **2005**, *168*, 521–530. [CrossRef]
15. Li, X.Y.; Mu, C.S.; Lin, J.X.; Wang, Y.; Li, X.J. Effect of alkaline potassium and sodium salts on growth, photosynthesis, ions absorption and solutes synthesis of wheat seedlings. *Exp. Agric.* **2014**, *50*, 144–157. [CrossRef]
16. Albaugh, T.J.; Allen, H.L.; Dougherty, P.M.; Kress, L.W.; King, J.S. Leaf area and above- and belowground growth responses of loblolly pine to nutrient and water additions. *For. Sci.* **1998**, *44*, 317–328.
17. Balster, N.J.; Marshall, J.D. Eight-year responses of light interception, effective leaf area index, and stemwood production in fertilized stands of interior Douglas-fir (*Pseudotsuga menziesii* var. *glauca*). *Can. J. For. Res.* **2000**, *30*, 733–743. [CrossRef]
18. Samuelson, L.; Stokes, T.; Cooksey, T. Production efficiency of loblolly pine and sweetgum in response to four years of intensive management. *Tree Physiol.* **2001**, *21*, 369–376. [CrossRef] [PubMed]
19. Gotore, T.; Murepa, R.; Gapare, W.J. Effects of Nitrogen, Phosphorus and Potassium on the Early Growth of *Pinus patula* and *Eucalyptus grandis*. *J. Trop. For. Sci.* **2014**, *26*, 22–31.
20. Salifu, K.F.; Jacobs, D.F.; Birge, Z.K.D. Nursery nitrogen loading improves field performance of bareroot oak seedlings planted on abandoned mine lands. *Restor. Ecol.* **2009**, *17*, 339–349. [CrossRef]

21. Li, H.; Li, M.; Luo, J.; Cao, X.; Qu, L.; Gai, Y.; Jiang, X.; Liu, T.; Bai, H.; Janz, D. N fertilization has different effects on the growth, carbon and nitrogen physiology, and wood properties of slow-and fast-growing *Populus* species. *J. Exp. Bot.* **2012**, *63*, 6173–6185. [CrossRef] [PubMed]
22. Luo, J.; Li, H.; Liu, T.; Polle, A.; Peng, C.; Luo, Z. Nitrogen metabolism of two contrasting poplar species during acclimation to limiting nitrogen availability. *J. Exp. Bot.* **2013**, *64*, 4207–4224. [CrossRef] [PubMed]
23. Miller, B.D.; Hawkins, B.J. Ammonium and nitrate uptake, nitrogen productivity and biomass allocation in interior spruce families with contrasting growth rates and mineral nutrient preconditioning. *Tree Physiol.* **2007**, *27*, 901–909. [CrossRef] [PubMed]
24. Bown, H.E.; Watt, M.S.; Clinton, P.W.; Mason, E.G.; Whitehead, D. The influence of N and P supply and genotype on carbon flux and partitioning in potted *Pinus radiata* plants. *Tree Physiol.* **2009**, *29*, 857–868. [CrossRef] [PubMed]
25. Son, Y. Effects of Nitrogen Fertilization on Foliar Nutrient Dynamics in *Ginkgo* Seedlings. *J. Plant Nutr.* **2002**, *25*, 93–102. [CrossRef]
26. Son, Y.; Kim, H.W. Above-ground biomass and nutrient distribution in a 15-year-old ginkgo (*Ginkgo* biloba) plantation in Central Korea. *Bioresour. Technol.* **1998**, *63*, 173–177. [CrossRef]
27. Yuan, J.; Li, Q.; Xiao, G.L.; Zhu, S.H. Discussion on the utilization and development of ginkgo wood. *China For. Sci. Technol.* **2008**, *16*, 6–8.
28. Andersson, S.; Wang, Y.; Ponni, R.; Hanninen, T.; Mononen, M.; Ren, H.; Serimaa, R.; Saranpää, P. Cellulose structure and lignin distribution in normal and compression wood of the Maidenhair tree (*Ginkgo biloba* L.). *J. Integr. Plant Biol.* **2015**, *57*, 388–395. [CrossRef] [PubMed]
29. Burgert, I.; Fruhmann, K.; Keckes, J.; Fratzl, P.; Stanzl-Tschegg, S. Structure-function relationships of four compression wood types: Micromechanical properties at the tissue and fibre level. *Trees* **2004**, *18*, 480–485. [CrossRef]
30. Gong, Q.L.; Hu, A.H.; Xing, S.Y.; Wang, F. Research on systematic evolution of *Ginkgo biloba* based on chemical composition of wood. *Spectrosc. Spectr. Anal.* **2009**, *29*, 1512–1516.
31. Moilanen, M.; Hökkä, H. The growth response of scots pine to PK fertilization depends on the nutrient status of the stand on drained peatlands. *Suo* **2009**, *60*, 111–119.
32. Smethurst, P.; Baillie, C.; Cherry, M.; Holz, G. Fertilizer effects on LAI and growth of four *Eucalyptus nitens* plantations. *For. Ecol. Manag.* **2003**, *176*, 531–542. [CrossRef]
33. Grant, C.; Bittman, S.; Montreal, M.; Plenchette, C.; Morel, C. Soil and fertilizer phosphorus: Effects on plant P supply and mycorrhizal development. *Can. J. Plant Sci.* **2005**, *85*, 3–14. [CrossRef]
34. Bussi, C.; Smith, M.A.L. Effects of nitrogen and potassium fertilization on the growth, yield and pitburn of apricot (cv. Bergeron). *J. Hortic. Sci. Biotechnol.* **1998**, *73*, 387–392. [CrossRef]
35. Boyce, R.L.; Larson, J.R.; Sanford, R.L. Phosphorus and nitrogen limitations to photosynthesis in Rocky Mountain bristlecone pine (*Pinus aristata*) in Colorado. *Tree Physiol.* **2006**, *26*, 1477–1486. [CrossRef] [PubMed]
36. Turnbull, T.L.; Warren, C.R.; Adams, M.A. Novel mannose-sequestration technique reveals variation in subcellular orthophosphate pools do not explain the effects of phosphorus nutrition on photosynthesis in *Eucalyptus globulus* seedlings. *New Phytol.* **2007**, *176*, 849–861. [CrossRef] [PubMed]
37. Moilanen, M.; Silfverberg, K.; Hökkä, H.; Issakainen, J. Comparing effects of wood ash and commercial PK fertiliser on the nutrient status and stand growth of Scots pine on drained mires. *Balt. For.* **2004**, *10*, 2–10.
38. Santiago, L.S.; Wright, S.J.; Harms, K.E.; Yavitt, J.B.; Korine, C.; Garcia, M.N.; Turner, B.L. Tropical tree seedling growth responses to nitrogen, phosphorus and potassium addition. *J. Ecol.* **2012**, *100*, 309–316. [CrossRef]
39. Oosterhuis, D.M.; Loka, D.A.; Raper, T.B. Potassium and stress alleviation: Physiological functions and management of cotton. *J. Plant Nutr. Soil Sci.* **2013**, *76*, 331–343. [CrossRef]
40. Römheld, V.; Kirkby, E.A. Research on potassium in agriculture: Needs and prospects. *Plant Soil* **2010**, *335*, 155–180. [CrossRef]
41. Wang, X.G.; Zhao, X.H.; Jiang, C.J.; Li, C.H.; Cong, S.; Wu, D.; Chen, Y.Q.; Yu, H.Q.; Wang, C.Y. Effects of potassium deficiency on photosynthesis and photo-protection mechanisms in soybean (*Glycine max* (L.) Merr.). *J. Integr. Agric.* **2015**, *14*, 856–863. [CrossRef]
42. Warren, C.R.; Adams, M.A.; Chen, Z.L. Is photosynthesis related to concentrations of nitrogen and Rubisco in leaves of Australian native plants? *Funct. Plant Biol.* **2000**, *27*, 407–416. [CrossRef]

43. Pokharel, P.; Chang, S.X. Exponential fertilization promotes seedling growth by increasing nitrogen retranslocation in trembling aspen planted for oil sands reclamation. *For. Ecol. Manag.* **2016**, *372*, 35–43. [CrossRef]

44. Warren, C.R.; Adams, M.A. Phosphorus affects growth and partitioning of nitrogen to Rubisco in *Pinus pinaster*. *Tree Physiol.* **2002**, *22*, 11–19. [CrossRef] [PubMed]

45. Kumagai, E.; Araki, T.; Kubota, F. Effects of nitrogen supply restriction on gas exchange and photosystem 2 function in flag leaves of a traditional low-yield cultivar and a recently improved high-yield cultivar of rice (*Oryza sativa* L.). *Photosynthetica* **2007**, *45*, 489–495. [CrossRef]

46. Zafar, M.; Abbasi, M.K.; Khaliq, A. Effect of combining organic materials with inorganic phosphorus sources on growth, yield, energy content and phosphorus uptake in maize at Rawalakot Azad Jammu and Kashmir, Pakistan. *Arch. Appl. Sci. Res.* **2011**, *3*, 199–212.

47. Zorb, C.; Senbayram, M.; Peiter, E. Potassium in agriculture: Status and perspectives. *J. Plant Physiol.* **2014**, *171*, 656–669. [CrossRef] [PubMed]

48. Terry, N.; Ulrich, A. Effects of phosphorus deficiency on the photosynthesis and respiration of leaves of sugar beet. *Plant Physiol.* **1973**, *51*, 43–47. [CrossRef] [PubMed]

49. Warren, C.R.; McGrath, J.F.; Adams, M.A. Differential effects of N, P and K on photosynthesis and partitioning of N in *Pinus pinaster* needles. *Ann. For. Sci.* **2005**, *62*, 1–8. [CrossRef]

50. Moilanen, M.; Silfverberg, K.; Hokkanen, T.J. Effects of wood-ash on the tree growth, vegetation and substrate quality of a drained mire: A case study. *For. Ecol. Manag.* **2002**, *171*, 321–338. [CrossRef]

51. Silfverberg, K.; Issakainen, J.; Moilanen, M. Growth and nutrition of scots pine on drained and fertilized purple moor grass fens in central Finland. *Balt. For.* **2011**, *17*, 91–101.

52. Whitehead, D.; Beadle, C.L. Physiological regulation of productivity and water use in *Eucalyptus*: A review. *For. Ecol. Manag.* **2004**, *193*, 113–140. [CrossRef]

53. Dickmann, D.I.; Nguyen, P.V.; Pregitzer, K.S. Effects of irrigation and coppicing on above-ground growth, physiology, and fine-root dynamics of two field-grown hybrid poplar clones. *For. Ecol. Manag.* **1996**, *80*, 163–174. [CrossRef]

54. Thornton, F.C.; Bock, B.R.; Behel, A.D.; Houston, A.; Tyler, D.D. Utilization of waste materials to promote hardwood tree growth. *South. J. Appl. For.* **2000**, *24*, 230–237.

55. Barrongafford, G.A.; Will, R.E.; Burkes, E.C.; Shiver, B.; Teskey, R.O. Nutrient concentrations and contents, and their relation to stem growth, of intensively managed *Pinus taeda* and *Pinus elliottii* stands of different planting densities. *For. Sci.* **2003**, *49*, 291–300.

56. Tausz, M.; Trummer, W.; Wonisch, A.; Goessler, W.; Grill, D.; Jiménez, M.S.; Morales, D. A survey of foliar mineral nutrient concentrations of *Pinus canariensis* at field plots in Tenerife. *For. Ecol. Manag.* **2004**, *189*, 49–55. [CrossRef]

57. Jennifer, A.B.; Boerner, R.E.J. Nitrogen fertilization effects on foliar nutrient dynamics and autumnal resorption in maidenhair tree (*Gingko biloba* L.). *J. Plant Nutr.* **1994**, *17*, 433–443.

58. Pettigrew, W.T.; Meredith, W.R. Dry matter production, nutrient uptake, and growth of cotton as affected by potassium fertilization. *J. Plant Nutr.* **1997**, *20*, 531–548. [CrossRef]

59. Wright, I.J.; Reich, P.B.; Westoby, M.; Ackerly, D.D.; Baruch, Z.; Bongers, F.; Cavender-Bares, J.; Chapin, T.; Cornelissen, J.H.C.; Diemer, M. The worldwide leaf economics spectrum. *Nature* **2004**, *428*, 821–827. [CrossRef] [PubMed]

60. Sundström, E.; Magnusson, T.; Hånell, B. Nutrient conditions in drained peatlands along a north-south climatic gradient in Sweden. *For. Ecol. Manag.* **2000**, *126*, 149–161. [CrossRef]

61. Pietiläinen, P.; Kaunisto, S. The effect of peat nitrogen concentration and fertilization on the foliar nitrogen concentration of Scots pine (*Pinus sylvestris* L.) in three temperature sum regions. *Suo* **2003**, *54*, 1–13.

62. Wang, D.G.; Yin, G.T.; Yang, J.C.; Li, R.S.; Zou, W.T.; Jia, R.F. Effects of phosphorus fertilization on growth and foliar nutrient (N, P, K) of *Phoebe bournei* seedlings. *J. Nanjing For. Univ. (Nat. Sci. Ed.)* **2014**, *38*, 40–44.

63. Crous, J.W.; Morris, A.R.; Scholes, M.C. Growth and foliar nutrient response to recent applications of phosphorus (P) and potassium (K) and to residual P and K fertiliser applied to the previous rotation of *Pinus patula*, at Usutu, Swaziland. *For. Ecol. Manag.* **2008**, *256*, 712–721. [CrossRef]

64. Nakashgir, G.H. Influence of potassium on nitrogen utilization by maize under dryland conditions as affected by water storage. *Adv. Plant Sci.* **1992**, *5*, 134–142.

65. Nardini, A.; Grego, F.; Trifilò, P.; Salleo, S. Changes of xylem sap ionic content and stem hydraulics in response to irradiance in *Laurus nobilis. Tree Physiol.* **2010**, *30*, 136–140. [CrossRef] [PubMed]

66. Sellin, A. Experimental evidence supporting the concept of light-mediated modulation of stem hydraulic conductance. *Tree Physiol.* **2010**, *30*, 1528–1535. [CrossRef] [PubMed]

67. Pervez, H.; Ashraf, M.; Makhdum, M.I. Influence of potassium nutrition on gas exchange characteristics and water relations in cotton (*Gossypium hirsutum* L.). *Photosynthetica* **2004**, *42*, 251–255. [CrossRef]

MDPI

Article

Estimation of Nutrient Exports Resulting from Thinning and Intensive Biomass Extraction in Medium-Aged Spruce and Pine Stands in Saxony, Northeast Germany

Christine Knust *, Karoline Schua and Karl-Heinz Feger

TU Dresden, Institute of Soil Science and Site Ecology, 01737 Tharandt, Germany;
karoline.schua@tu-dresden.de (K.S.); karl-heinz.feger@tu-dresden.de (K.-H.F.)
* Correspondence: christine.knust@tu-dresden.de; Tel.: +49-351-463-31370

Academic Editors: Scott X. Chang and Xiangyang Sun
Received: 31 August 2016; Accepted: 19 November 2016; Published: 30 November 2016

Abstract: A growing interest in using forest biomass for bioenergy generation may stimulate intensive harvesting scenarios in Germany. We calculated and compared nutrient exports of conventional stem only (SO), whole tree without needles (WT excl. needles), and whole tree (WT) harvesting in two medium aged Norway spruce (*Picea abies* L. KARST.) and Scots pine (*Pinus sylvestris* L.) stands differing in productivity, and related them to soil nutrient pools and fluxes at the study sites. We established allometric biomass functions for each aboveground tree compartment and analyzed their nutrient contents. We analyzed soil nutrient stocks, estimated weathering rates, and obtained deposition and seepage data from nearby Level II stations. WT (excl. needles) and WT treatments cause nutrient losses 1.5 to 3.6 times higher than SO, while the biomass gain is only 1.18 to 1.25 in case of WT (excl. needles) and 1.28 to 1.30 in case of WT in the pine and spruce stand, respectively. Within the investigated 25-year period, WT harvesting would cause exports of N, K^+, Ca^{2+}, and Mg^{2+} of 6.6, 8.8, 5.4, and 0.8 kg·ha^{-1} in the pine stand and 13.9, 7.0, 10.6, and 1.8 kg·ha^{-1} in the spruce stand annually. The relative impact of WT and WT (excl. needles) on the nutrient balance is similar in the pine and spruce stands, despite differences in stand productivities, and thus the absolute amount of nutrients removed. In addition to the impact of intensive harvesting, both sites are characterized by high seepage losses of base cations, further impairing the nutrient budget. While intensive biomass extraction causes detrimental effects on many key soil ecological properties, our calculations may serve to implement measures to improve the nutrient balance in forested ecosystems.

Keywords: spruce; pine; thinning; aboveground biomass; energetic use; stand growth; nutrient contents; nutrient accumulation

---

1. Introduction

In order to mitigate fossil carbon dioxide emissions for energy generation, renewable energies such as bioenergy are being promoted as an alternative to fossil fuels. In Germany in 2013, renewable sources accounted for 12.3% of total energy consumption, of which biomass accounted for 7.6% [1]. Within the biomass sector, forest residues have been identified as a large underused source for potentially increasing the raw material supply. Already, about 11 million·m^3 of wood has been used directly for energy purposes, which accounts for about one fifth of the total annual harvest from forests. The potential for increasing forest biomass for energy (at a sustainable rate) is estimated to be between 12 and 19 million·m^3·year^{-1}, which could be achieved by increasing the utilization of forest residues and currently underused hardwood stands [2].

The process of extracting harvesting residues from forests, or whole tree harvesting, has a number of technological and ecological constraints. On the technological side, there is a high expenditure for the logistics due to the scattered location of the biomass and its low density, and thus transportability [3,4]. Advanced technologies and management practices could help to improve both feedstock quality and cost efficiency in the future [3,5]. The biomass market situation, and thus the actual price paid by biomass combustion facilities per unit of feedstock, will eventually determine economic feasibility of the utilization of forest residues for bioenergy.

On the ecological side, there is a threat of potentially high nutrient exports when extracting the nutrient-rich crown material, and thus a loss of productivity [6–10]. Also, if large parts of the harvesting residues are exported from the site, there will be fewer habitats for decomposer fauna and flora and less input material to refill the site-specific humus stock [11,12]. The nutrient issue could be attenuated by returning nutrients into the forests from sources such as wood ash from biomass combustion facilities [13–16]. Implementing wood ash recycling systems into practice would require thorough knowledge regarding the actual amounts of nutrients removed in management scenarios of various intensity [14]. Additionally needed is an estimation of the available soil nutrient pool, which plays a major role in assessing the impact of harvest-induced nutrient losses on site productivity, and thus on management sustainability in terms of maintaining nutrient reserves. Finally, site specific nutrient balances are needed to evaluate the impact of different harvesting intensities and to draw conclusions for adapted forest management.

Estimating nutrient fluxes associated with the extraction of aboveground biomass compartments under intensive management scenarios requires knowledge and estimation methods about the distribution of biomass and nutrients of all tree compartments. Yet yield tables and forest growth models typically used in forest management and planning in Germany concentrate on just the volume of the marketable round wood, which is typically the tree trunk up to a certain diameter (e.g., 7 cm over bark), depending on the current market situation. Information about the distribution of the biomass to above-ground tree compartments, such as the stem with a diameter smaller than 7 cm, branches, foliage, and dead branches is needed in order to estimate the profitability and the impact on the nutrient budget of going from classic stem-only (SO) harvesting to more intensive scenarios, such as whole tree without foliage (WT excl. needles) or whole tree (WT) harvesting [17,18]. Some studies on the impact of nutrient removals in intensive scenarios focus on the final felling and harvesting at the end of a stand's rotation time and its impact on the growth of the next stand generation [7,10]. The rationale for this approach is that the biomass and nutrient fluxes in final fellings, such as clear cuts or shelterwood cuttings, are especially large. Alternately, other studies have concentrated on the impact of intensive biomass exports in thinning operations of medium aged stands [10,19–22], which happen at a stage of stand development when the remaining trees exhibit high productivity, and thus require large amounts of nutrients for the buildup of aboveground biomass.

In our study, we aimed at quantifying the nutrient exports of thinning operations with variable intensities of biomass removal in two medium-aged coniferous stands representing typical site and stand conditions in the region of Saxony in northeast (NE) Germany and evaluating the impact of such treatments on the nutrient budget of the stands. We hypothesize that intensive harvesting scenarios (1) impose a strong negative effect on the nutrient balance of forest stands, even in the thinning stage; and (2) will lead to a more negative nutrient balance in the highly productive stands than in less productive stands, due to larger absolute amounts of biomass and thus nutrients removed. Therefore, we developed single-tree biomass equations to predict the dry mass of all aboveground tree compartments based on the tree's diameter at breast height (DBH) and analyzed nutrient contents of these compartments. Based on this knowledge, we estimated the biomass and nutrient extraction by thinning with three different management intensities: conventional SO, WT (excl. needles), and WT harvesting, which refers to aboveground biomass only. In all scenarios, stumps and roots are left in the stand. Also, we calculated the nutrients required to build up aboveground biomass within the examined time span. Finally, we set the amount of extracted and stored nutrients in relation to the site-specific nutrient stocks and fluxes in order to estimate the sustainability of intensive management scenarios.

2. Materials and Methods

2.1. Study Area and Sites

The study area was the Oberlausitz in Saxony, NE Germany. In the northern lowlands, the typically nutrient poor pleistocene soils are often covered with Scots pine (*Pinus sylvestris* L.) stands. Depending on the site potential which varies on a small scale, forests could be dominated by Scots pine, sessile oak (*Quercus petraea* (Mattuschka) Liebl.), pedunculated oak (*Q. robur* L.), silver birch (*Betula pendula* Roth), European beech (*Fagus sylvatica* L.), and various mixtures of these species. Of these species, pines are the most dominant in the study region, covering 69% of the forested area. In the hills in southern Oberlausitz, site conditions are generally more suitable for forestry, as soils are derived from weathered granodiorite bedrock overlain with a loess layer of variable thickness (often 10 to 40 cm). Many forest stands here are pure Norway spruce (*Picea abies* (L.) Karst.) plantations or spruce dominated, and spruce accounts for 13% of the forest cover, while under natural conditions most sites consist of beech dominated mixed forests. We selected two study stands to represent these contrasting stand types: a pine stand on a nutrient-poor Podzol-type soil with a low water holding capacity derived from pleistocene gravel and sand in the lowlands, and a spruce stand on a deep Cambisol-type soil in the southern hills (Table 1). Although the soil conditions at the spruce site were nominally more favorable than at the pine site, the base saturation of the mineral soil was substantially lower (Table 1), which was caused by atmospheric acid inputs from recent decades (notably from 1970 to 1990). Both stands were 38 years-old in 2013. At this age, thinning operations are carried out once or twice in ten years to increase stand stability and promote growth of the most favorable trees by reducing stand density.

Table 1. Site and stand characteristics of the studied Scots pine and Norway spruce stand in Saxony.

Ecosystem Component	Parameter	Unit	Scots Pine (Laußnitz)	Norway Spruce (Neusalza-Spremberg)
Site characteristics	Mean annual temperature	°C	9.4	8.1
	mean annual rainfall	mm	757	910
	altitude	m above sealevel	190	405
Soil characteristics	Soil type		Dystric Arenosol on pleistocenic sediment	Cambisol on weathered granodiorite overlain with loess
	Profile depth (root zone)	cm	35	60
	field capacity	mm·dm^{-1} ***	8.5	18.8
	Base saturation in the organic layer (mean ± SD)	%	30.4 ± 14.9	68.4 ± 43.5
	Base saturation in the mineral soil	%	8.2–3.4	4.5–1.7 ± 5.2–1.2
Stand characteristics in 2013	Age	Year	38	38
	Stand density	N·ha^{-1}	1850	725
	Stock density **		0.89	0.97
	Growing stock	m^3·ha^{-1}	199.0	400.5
	Average DBH *	cm	14.3	24.5
	Average height	m	13.9	23.8
	Mean annual increment	m^3·ha^{-1}·year^{-1}	10.2	16.2
	Basal area	m·ha^{-1}	29.9	34.1
	Average slenderness coefficient		0.95	0.97

* DBH: diameter at breast height; ** ratio of actual growing stock to potential maximum growing stock (according to yield table); *** mm or water per dm of soil depth.

2.2. Stand Biomass and Treatment

In 2013, inventories of the pine stand and the spruce stand were carried out on representative 0.2 ha subplots, determining the number of trees, the DBH, and the height of 10 trees representing the largest 10% of each stand. This data was used to model current stand properties and stand development for 25 years, using the individual tree based forest growth model BWinPro-S [23]. In this time span, forest management will focus on the process known as 'stand qualification', which is selective cutting to promote the growth of the healthiest trees. According to current management practices, we assumed moderate to heavy thinning from above in the spruce stand, and heavy thinning from below in the pine stand. The thinning intensity in terms of removed volume per hectare was adjusted in such a way to keep the degree of stocked area as well as the current increment constant, which resulted in the removal of about 60 m^3 every 5 years in the spruce stand and 45 m^3 every 10 years in the pine stand. This was done in order to maximize stand stability, and to achieve a high growth rate at the stand level as well as a reasonable DBH increment of the single trees.

Based on the inventories, nine to ten trees were selected across the range of DBH values in order to investigate the relative mass distribution as well as nutrient contents in the aboveground biomass compartments. The selected trees were felled and separated into the following compartments in the field: (1) stem wood (diameter > 7 cm) including bark; (2) tree top wood (diameter < 7 cm) including bark; (3) branches including bark and needles; (4) dead branches (Table 2). The fresh mass of these compartments was determined using a hanging scale mounted in the stand. Subsamples of each compartment were taken to the lab and dried at 60 °C until weight was constant. For the stem wood including bark compartment, we collected three 5 cm thick discs from the following tree heights: 1.3 m, 1/2 of the length of the stem wood, top of the stem wood. For the branches including needles, we collected one entire branch from each third whorl, starting at the youngest whorl. For the tree top wood we collected one 5 cm thick slice from the middle of the section, and for the dead branches we randomly collected three to five dead branches per tree. This resulted in relatively large amounts of sample material for the most important tree compartments in terms of mass and nutrient contents (i.e. stem wood, bark of the stem wood, branches and needles). In the lab, the subsamples were further divided into the following target compartments: stem wood, bark of the stem wood, tree top wood including bark, branches including bark, needles and dead branches (Table 2). During the drying and separating process, the water content of each compartment was determined as well as the wood/bark ratio and the branches/needles ratio. Thus, we were able to determine the dry mass of each compartment of each tree by simple ratio calculations. This data was used to fit allometric biomass models of the dry mass of each biomass compartment (M_{bc}) of the type $M_{bc} = a \times DBH \; (cm)^b$, where a and b are specific coefficients and the explaining variable is the DBH (cm), by using non-linear regressions of the original data. As our emphasis was to estimate nutrient stocks in each biomass component, we used the described independent biomass models for each component, instead of an approach focusing more on the additivity of the biomass equation.

Table 2. Partitioning of the aboveground biomass into tree compartments in the field and in the lab, number of samples taken per tree and compartment.

Compartment		Number of Samples
Field	**Lab**	**Samples/Tree**
Stem wood incl. bark	Stem wood > 7 cm	3
	Bark of stem wood	3
Tree top wood incl. bark	tree top wood incl. bark	1
Branches incl. bark and needles	Branches incl. bark	4–7, depending on tree size
	Needles	4–7, depending on tree size
Dead branches	Dead branches	3–5, depending on tree size

2.3. Nutrient Contents

Prior to chemical analysis, one composite sample per tree and compartment was created and ground to 0.25 mm (needles: 0.08 mm). For C and N analysis, aliquots of 10 mg of sample material were analyzed using a Vario EL III (Elementar Analysensysteme GmbH, Hanau, Germany). Nitrogen contents of low N biomass compartments (stem wood, bark, tree top wood, branches, dead branches) were determined with a Vario Max cube (Elementar Analysensysteme GmbH, Hanau,

Germany) using 300 mg aliquots of sample material. The contents of P, K, Ca, and Mg contents were analyzed after HNO_3 digestion using a CCD-ICP Spectrometer CIROS (Spectro Analytical Instruments, Kleve, Germany).

2.4. Soil Nutrient Stocks and Nutrient Balance

Soil profiles were dug and characterized in both stands, and each mineral soil horizon was sampled using metal rings, so that bulk density could be determined in addition to chemical analyses. We calculated the stock of exchangeable Ca^{2+}, K^+, and Mg^{2+} based on the cation exchange capacity (CEC_{eff}) after percolating 5 g subsamples with 1 M NH_4Cl and subsequent analysis in the CCD-ICP spectrometer CIROS (Spectro Analytical Instruments, Kleve, Germany). Total concentrations of N and P were determined after digestion in HNO_3, HF, and $HClO_4$, using the CCD-ICP spectrometer CIROS (Spectro Analytical Instruments, Kleve Germany). Based on horizon-wise nutrient contents and bulk densities, the exchangeable and total nutrient stocks per hectare were calculated for the root zone (35 cm at the pine site and 60 cm at the spruce site).

For the nutrient balance, data from the EU-monitoring Level II sites were used to estimate atmospheric deposition inputs and seepage outputs. For the pine stand, the Level II station "Laußnitz" was used, which is only 500 m away from the study site, while for the spruce stand data from the Level II station "Bautzen/Neukirch" was used, which is 15 km away from the study site, but is comparable in terms of site, soil, and stand conditions [24]. The weathering inputs were roughly estimated according to a method by [25]. First, a total weathering rate of 0.4–1.0 $kmol \cdot ha^{-1} \cdot year^{-1}$ was estimated according to literature values [26]. Then, the share of K^+, Ca^{2+}, and Mg^{2+} was determined in relation to their share in base saturation. Due to data scarcity, we could not set up nutrient budgets for P.

3. Results

3.1. Forest Development

Both the pine and the spruce stand are very productive. Thus, they exhibited a relatively high current increment which decreased slightly during the modeled period from 14.4 to 13.1 and 27.5 to 26.7 $m^3 \cdot ha \cdot year^{-1}$ at the pine and spruce site, respectively. The high productivity can be attributed both to proper forest treatments before the beginning of the study period, and to the favorable growing conditions. Figure 1 shows the development of the total aboveground biomass and the impact of the thinnings that were modeled to be carried out once in 10 years in the pine stand and once in 5 years in the spruce stand.

In both stands, the volume increment over the study period was slightly lower than the volume harvested in thinnings (132.5 vs. 141.1 $m^3 \cdot ha^{-1}$ in the pine stand and 277.1 vs. 318.9 $m^3 \cdot ha^{-1}$ in the spruce stand, Table 3). With regard to the further stand development after the study period, we set a slightly lower stocking density (Table 3). After the qualification phase focused in this paper, the pine stand will reach the dimensioning phase, in which no more thinnings are carried out and the trees are left to grow until they reach the target DBH (40 cm). Due to its high productivity, the spruce stand will more or less skip the dimensioning phase, because the largest trees will have already reached the target DBH (40 cm) after the qualification phase. Once they have reached it, forest management will focus on harvesting the mature trees and at the same time initiating stand regeneration.

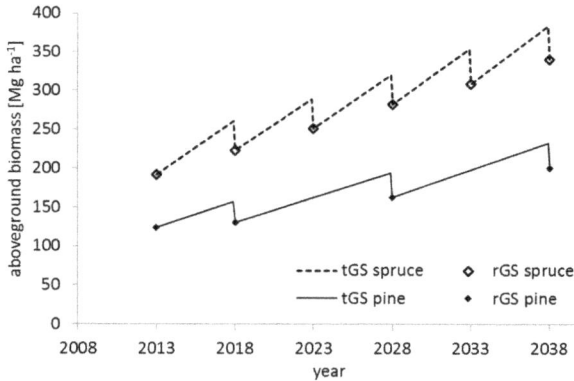

Figure 1. Development of total aboveground biomass (t·ha^{-1}) of the pine and spruce stand. tGS: total growing stock, rGS: remaining growing stock after thinning. Thinning of the pine stand in 2018, 2028, and 2038; thinning of the spruce stand is carried out in 2018, 2023, 2028, 2033, and 2038.

Table 3. Development of the remaining stand and sum of extracted volume in thinnings. Scots pine stand: three thinnings, Norway spruce stand: five thinnings during a 25-year period.

Date / Period	Current Increment	Stocking Density	Growing Stock	Extracted Volume
Pine	m^3·ha^{-1}·a^{-1}		m^3·ha^{-1}	m^3·ha^{-1}
2013	14.4	0.89	200.2	
2038	13.1	0.84	332.7	
Σ thinnings 2018–2038				141.1
Spruce				
2013	27.5	0.97	400	
2038	26.7	0.93	677.1	
Σ thinnings 2018–2038				318.9

3.2. Biomass and Nutrient Distribution in Tree Compartments

We achieved a satisfactory fit of the allometric biomass equations to our data for most of the biomass compartments. As an example, Figure 2 shows the biomass equation of the total aboveground biomass of the spruce stand. The equation parameters and the coefficients of determination are given in Table 4.

The coefficients of determination of the biomass functions were generally high, indicating a strong correlation of the DBH and the mass of the biomass compartments, especially in the case of stem wood, bark, branches, and needles, which are the most important aboveground biomass compartments with respect to their share of total aboveground biomass (Figure 3) and nutrient storage.

The development of the share of the biomass compartments in total aboveground biomass is shown in Figure 3, based on the biomass equations in Table 4. Obviously, the tree top wood incl. bark is relevant only for trees with a DBH < 20 cm, because a relatively large part of the stem is smaller than the defined 7 cm diameter. In larger trees, the stem wood with a diameter > 7 cm including bark accounts for 79% in both spruce and pine. The most important difference between the pine and spruce is the share of branches and needles, with pine trees having a larger share of branches at 14% and a smaller share of needles at only 4.4%, compared to 11.0% and 8.2% in spruce trees, respectively, due to their species-specific crown architecture and physiology. Figure 3 clearly shows the larger relative importance of low quality crown material in small trees, and thus in young stands. This probably makes whole tree harvesting for bioenergy purposes especially attractive in younger stands in order to increase the overall economic outcome of thinning operations.

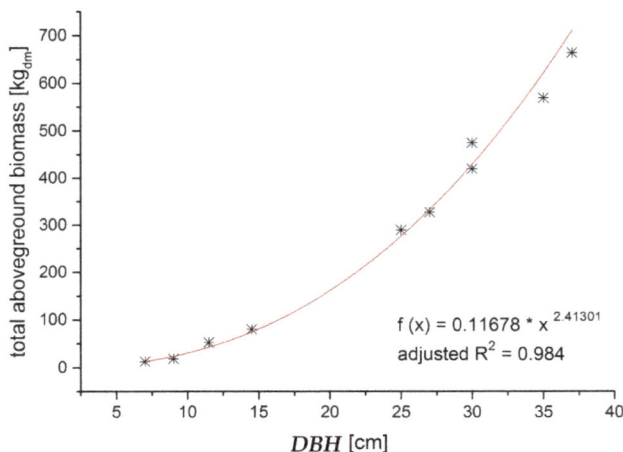

Figure 2. Relation of total aboveground biomass of the sampled spruce trees to DBH and biomass equation.

Table 4. Equation coefficients (a, b) and coefficients of determination (R^2) of the biomass equations $M_{bc} = a \times DBH \; (cm)^b$ for the aboveground biomass compartments of the pine and spruce stand.

Compartment	Pine Stand			Spruce Stand		
	a	*b*	R^2	*a*	*b*	R^2
total aboveground biomass	0.0786	2.539	0.98	0.11678	2.41301	0.98
stem wood	0.0075	3.2281	0.98	0.033	2.68382	0.98
bark of the stem wood	0.0052	2.5924	0.98	0.00926	2.3432	0.98
tree top wood incl. bark	174.34	−1.526	0.89	50.69505	−1.02053	0.78
branches	0.0008	3.3992	0.99	0.00542	2.66276	0.96
needles	0.0015	2.8154	0.94	0.01253	2.35013	0.94
dead branches	0.0002	3.3465	0.88	0.01722	1.91842	0.86

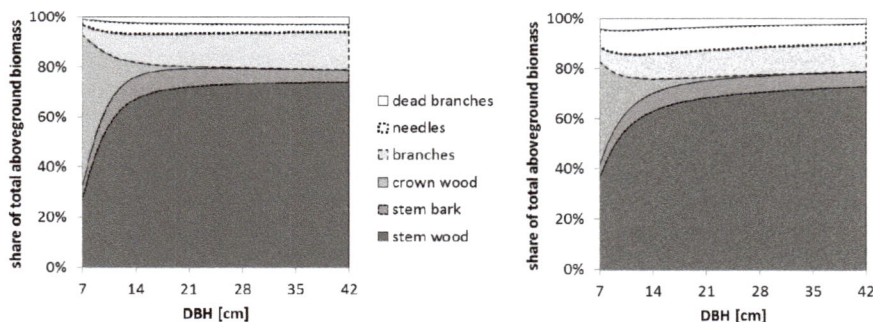

Figure 3. Share of aboveground biomass compartments of the examined Scots pine (**left**) and Norway spruce (**right**) stand at a DBH-range of 7 to 42 cm.

3.3. Nutrient Contents in Aboveground Biomass Compartments

Although nutrient contents for aboveground tree compartments were measured in various other studies, the results are often only partly comparable, because of differences in the defined compartments, in stand age and soil conditions [27]. Thus, with respect to the purpose of this study, we performed our own measurements of nutrient contents (Table 5).

Table 5. Nutrient contents (g·kg^{-1}) of aboveground biomass compartments of pine and spruce trees at the trial sites in East Saxony. Means and standard deviation, $n = 5$.

	Stem Wood		Stem Bark		Tree Top Wood		Branches		Needles		Dead Branches	
					Scots Pine							
N	0.72	(0.3)	2.76	(0.5)	1.98	(0.6)	4.24	(1.1)	16.4	(2.6)	2.24	(0.7)
P	0.04	(0.4)	0.26	(1.3)	0.17	(1)	0.40	(1.7)	1.41	(7.2)	0.09	(1.1)
K	0.36	(0.4)	1.25	(1.6)	0.91	(1.2)	1.78	(2.1)	6.06	(7.8)	0.26	(1.3)
Ca	0.72	(0.4)	6.86	(1.6)	1.34	(1)	3.11	(2.3)	4.03	(6.8)	2.06	(0)
Mg	0.17	(0.3)	0.46	(1.2)	0.35	(0.9)	0.50	(1.9)	0.74	(5.4)	0.21	(0)
					Norway Spruce							
N	0.80	(0.3)	3.78	(0.8)	2.07	(0.3)	4.49	(0.7)	13.2	(1.6)	1.86	(0.3)
P	0.06	(0)	0.54	(0.1)	0.31	(0.1)	0.60	(0.1)	1.36	(0.2)	0.10	(0)
K	0.46	(0.1)	2.70	(0.5)	1.23	(0.3)	2.73	(0.5)	5.10	(0.7)	0.28	(0.1)
Ca	0.90	(0.1)	7.55	(0.9)	1.27	(0.2)	3.16	(0.3)	4.16	(0.4)	2.98	(0.4)
Mg	0.14	(0)	0.98	(0.1)	0.34	(0.1)	0.63	(0.1)	0.90	(0.1)	0.29	(0.1)

The highest concentrations of all nutrients are generally found in the needles, with the exception of Ca, which shows the highest concentration in the bark. Magnesium concentrations differ between pine and spruce. In the pine trees, Mg concentrations are highest in the needles, followed by branches and bark; while in spruce trees, Mg is distributed similar to Ca with the highest concentrations in the bark, followed by needles and branches. The nutrient contents of the dead branches range somewhat in between those of branches and stem wood.

3.4. Biomass, Nutrient Extraction, Uptake, and Storage

Relative to the increment of the remaining stand, the harvested biomass in the thinnings of the 25-year study period are larger in the spruce stand compared to the pine stand (Table 6). While the biomass extraction in the WT scenario in the pine stand is lower than the biomass ingrowth (129 vs. 114 t·ha^{-1}), it exceeds the ingrowth in the spruce stand (150 vs. 201 t·ha^{-1}) due to a higher overall productivity of the spruce stand, and thus heavier and more frequent thinnings. When intensifying the biomass extraction from conventional stem only (SO) harvesting to whole tree harvesting (WT), the factor of increase is slightly higher in the pine stand compared to the spruce stand (factor of increase 1.30 in the pine stand compared to 1.28 in the spruce stand), due to the smaller dimension of the harvested pine trees, and thus a relatively larger proportion of crown material (see Figure 3). The difference in the factor of increase of WT (excl. needles), with a higher biomass extraction in case of pine compared to spruce (factor of increase 1.25 vs. 1.18), results from the higher share of branches in pine trees compared to spruce trees.

The relative loss of nutrients in intensive scenarios far exceeds the biomass gain, with factors of increase in whole tree harvesting between 1.7 (Ca, pine and spruce) and 3.6 (P, pine) to 3.4 (P, spruce), respectively. If needles are allowed to fall off before extracting crown material (WT excl. needles), the increase in nutrient exports is generally higher in pine trees than in spruce, which goes along with the relatively larger biomass extraction due to the higher share of branches in pine trees in the WT (excl. needles) scenario.

Table 6. Aboveground biomass increment and nutrient uptake/storage in the remaining stand.

Scots Pine Stand		Biomass	N	P	K	Ca	Mg
		t·ha^{-1}			kg·ha^{-1}		
Remaining stand Δ 2013–2038 (WT)		129	237	18	99	189	32
Σ thinnings	SO *	88	80	5	39	113	17
	WT (excl. needles) *	110	156	12	69	169	26
	WT *	114	230	18	96	187	30
Ratio Compared to SO							
thinnings	WT (excl. needles) *	1.25	2.0	2.4	1.8	1.5	1.5
	WT *	1.30	2.9	3.6	2.5	1.7	1.8
		Biomass	N	P	K	Ca	Mg
Norway Spruce Stand		t·ha^{-1}			kg·ha^{-1}		
remaining stand Δ 2013–2038 (WT) *		150	335	35	172	261	44
Σ thinnings	SO *	157	162	15	99	220	33
	WT (excl. needles) *	185	271	29	161	304	48
	WT *	201	486	51	245	372	63
Ratio Compared to SO							
thinnings	WT (excl. needles) *	1.18	1.7	1.9	1.6	1.4	1.5
	WT *	1.28	3.0	3.4	2.5	1.7	1.9

* SO: stem only, WT (exl. needles): whole tree without needles, WT: whole tree.

3.5. Evaluation of the Nutrient Budget as Influenced by Intensive Harvesting

For evaluation of the impact of the thinning scenarios of varying intensity on the nutrient regime of the study sites, annual rates of nutrient inputs and outputs were estimated and compared to the nutrient pools in the organic layer and the mineral soil in the rooting zone (Table 7). Both sites are characterized by relatively high atmospheric N inputs, which far exceed seepage loss and uptake by the trees. The result is a positive N balance on both stands regardless of the biomass extraction intensity, showing that the studied forest stands currently function as nitrogen sinks. The situation is fundamentally different for the base cations. Moderate atmospheric and weathering inputs are hardly sufficient to compensate for seepage loss and nutrient fixation in the biomass increment of the remaining stand. Even conventional SO harvesting would lead to negative nutrient balances and the more intensive biomass extraction scenarios would impair the nutrient balance even further. The only exception is K in the spruce stand, which would have a negative balance only under the WT scenario. Regardless of the intensity of biomass extraction, both sites are characterized by high seepage losses of base cations, in relation to the atmospheric and weathering inputs. In the case of Ca, soil available pools and seepage losses are both twice as high at the pine site compared to the spruce site, indicating that seepage losses of Ca are coupled to availability. On the other hand, this relation is much less pronounced in the case of Mg, and K does not show such a relation at all.

Table 7. Pools of total N and exchangeable K^+, Ca^{2+}, and Mg^{2+}, annual rates of nutrient inputs and outputs, as well as the nutrient balance under different biomass extraction scenarios (SO, WT excl. needles, WT) of the Scots pine stand and the Norway spruce stand.

		Scots Pine Stand				Norway Spruce Stand			
		N	K^+	Ca^{2+}	Mg^{2+}	N	K^+	Ca^{2+}	Mg^{2+}
Soil pools		kg·ha^{-1}				kg·ha^{-1}			
		2955	53	263	13	3599	172	127	34
		N	K	Ca	Mg	N	K	Ca	Mg

Table 7. *Cont.*

		Scots Pine Stand				Norway Spruce Stand			
Inputs		kg·ha^{-1}·year^{-1}				kg·ha^{-1}·year^{-1}			
	Deposition	19.3	2.1	3.3	0.7	36	4.8	4.9	1.3
	Weathering		4	9.8	0.5		7.6	7.5	1.2
Outputs		kg·ha^{-1}·year^{-1}				kg·ha^{-1}·year^{-1}			
	Seepage	1.5	3.1	15.5	3.2	1.9	2.9	7	3.9
	Uptake and storage in ingrowth	6.8	2.8	5.4	0.9	9.6	4.9	7.5	1.3
Thinning	SO	2.3	1.1	3.2	0.5	4.6	2.8	6.3	0.9
	WT excl. needles	4.5	2.0	4.8	0.8	7.7	4.6	8.7	1.4
	WT	6.6	2.8	5.4	0.8	13.9	7.0	10.6	1.8
Balance		kg·ha^{-1}·year^{-1}				kg·ha^{-1}·year^{-1}			
	SO	8.7	−0.9	−11.0	−3.4	19.9	1.8	−8.3	−3.6
	WT excl. needles	6.6	−1.8	−12.6	−3.7	16.8	0.0	−10.8	−4.0
	WT	4.4	−2.6	−13.2	−3.8	10.6	−2.4	−12.7	−4.5

4. Discussion

In this study we simulated relatively heavy thinnings in order to sustain stand stability to adapt the stands to climate change in the long term, and make use of the high forest productivity [28,29], which is accelerated by high atmospheric N inputs of about 19.3 and 36.0 kg·ha·year^{-1} at the pine and spruce site, respectively. The large difference in productivity between pine and spruce is both a species and a site effect, because the less demanding pine stands are usually grown on the least productive forest sites [19].

The biomass functions we established in order to model the mass of each aboveground compartment were easy to apply and fit our data very well. Only for the tree top wood of spruce and the dead branches, the R^2 was slightly lower than for the other compartments, probably because the tree top wood is a somewhat artificially defined compartment and the retention of dead branches on a tree depends on factors other than the tree's DBH, such as the actual stand density in the area surrounding the respective tree. Nevertheless, since these two compartments play a minor role for addressing the questions raised in this paper, and the R^2 was still sufficient, we retained the equations for use. The drawback of the biomass functions we established is that they are stand specific, and can therefore only be applied for the DBH range for which they were calibrated. As a result, we concentrated our study on just the thinning phase. In order to create more widely applicable biomass functions, more trees with a wider range of DBH, and from stands representing a wider range of productivity and stand density, would need to be measured. The biomass equations would then need to account for this greater variability by incorporating further explanatory variables such as height and stand density, in addition to the DBH [30].

One option proposed to increase the harvested biomass, and keep nutrient exports relatively low at the same time, is to leave tree tops in the forest for one year to allow the needles to fall off before collecting the tree tops (WT excl. needles scenario) [7,20]. The different pattern of crown biomass allocation between pine and spruce, with a greater share of branches in pine trees versus needles in spruce trees, shows that this method would lead to a greater biomass loss in spruce than in pine trees, when compared to the WT scenario. With respect to biomass feedstock quantities, this procedure would be more viable in pine than in spruce stands. If biomass feedstock quality is also considered, it would be preferable in any case to exclude the needles, as wood is a higher quality feedstock in terms of particle size distribution when chipping at the forest industries [3], as well as the behavior in the burning process.

The nutrient contents in the aboveground biomass compartments match with the findings of previous studies, as far as the compartments are comparable. Jacobsen et al. [31] performed nutrient analyses that are partly comparable, and found similar nutrient contents in the stem wood, bark, needles, and branches of pine and spruce trees. The nutrient contents reported by Weis and Göttlein [9] in the needles, bark, and wood of spruce trees are also within the range of our findings.

Due to the rough estimation method of weathering and the utilization of deposition and seepage values from similar and close-by sites (but not exactly the study sites), the nutrient balance should be interpreted with some caution. Nevertheless, it is still useful for evaluating the impact of biomass, and thus nutrient extraction, in the different thinning scenarios.

Our findings confirm our first hypothesis, that intensive harvesting scenarios impose a strong negative effect on the nutrient balance of forest stands even in the thinning stage. We demonstrated that intensive WT and WT (exc. needles) harvesting causes fairly high nutrient exports in the thinning phase of the trial sites. The biomass gain is far lower than the nutrient expenses, as the biomass gain is only 1.18 to 1.30 times compared to SO, but the ratio of nutrient losses is 1.4 to 3.6 compared to SO. The increase of nutrient losses from SO to WT harvesting are greater for N and P compared to the base cations, reflecting the over proportional abundance of N and P in the crown material, especially in the physiologically active needles. Yet the impact on the site-specific nutrient sustainability is far greater for the base cations than for N, because of excess atmospheric N inputs. We unfortunately have no data for the nutrient balance of P, but as P inputs into the forest ecosystem by weathering and deposition are usually low, a rather negative P balance similar to that of the base cations can be assumed. High seepage losses contribute strongly to the negative nutrient balance of the base cations. The seepage losses of base cations result from sulfur depositions during the 1970s to 1990s, which were greatly reduced by the end of the last century [32]. However, SO_4^{2-} from accumulated S pools still gets washed out of the soil profile, coupled with base cations to neutralize the charge [33]. This leads to negative nutrient balances for Ca^{2+} and Mg^{2+} already at the SO scenario, which decreases further in the WT (excl. needles) and the WT scenario, when K^+ also becomes negative.

According to our findings, we reject our second hypothesis, according to which the negative impact of intensive harvesting is greater in highly productive stands than in less productive stands, due to larger absolute amounts of removed biomass, and thus nutrients. In our study, the higher stand productivity of the spruce stand corresponded with a higher nutrient demand for uptake and storage, as well as higher nutrient losses due to biomass extraction, compared to the less productive pine stand. At the same time, these differences in productivity, and thus biomass-related nutrient fluxes, are balanced out by higher nutrient stocks and inputs through deposition (except for Ca), while seepage losses are equal to or even lower than the pine site. Thus, the relative impact of nutrient exports through intensive biomass extraction is almost equally severe in the spruce and pine stand, despite different absolute levels of biomass extraction.

Our findings should, however, be interpreted with some caution, as they are based on static assumptions regarding the development of forest growth, the nutrient concentrations in the biomass, and the weathering rate. In reality, increased nutrient scarcity due to intensive biomass extraction may cause reduced growth of the remaining stand, as has been shown in studies conducted in North American and Scandinavian forests [7,8,10], while other studies found no clear effect on tree growth [34,35]. A meta-analysis by Achat et al. [36] indicates an average growth reduction of 3%–7% after intensive biomass extraction. Declining pools of available nutrients may also cause reduced nutrient concentrations in the foliage of the remaining trees, which would also alter the nutrient balance. The nutrient balance is also known to be strongly influenced by the rate of mineral weathering [37], which is in practice very difficult to determine and could only be roughly estimated in our study. Furthermore, mineral weathering rates are not static, but mycorrhizal weathering may be accelerated if there are nutrient deficiencies [38]. Despite these uncertainties, our findings underline the possible nutrient balance risk of intensive biomass extraction from thinnings, which brings into question the nutrient balance sustainability of these intensive management practices.

Our study highlights that intensive harvesting scenarios like WT (excl. needles) and WT harvesting can have severe consequences for the nutrient balance of forest sites. Furthermore, intensive biomass removal is known to negatively affect the soil as the basis for biomass production. Reduced residue inputs reduce the organic carbon content [39] and mineralization rates and cause soil compaction and thus a reduced water holding capacity [36]. Based on the analysis of detrimental forest management practices (i.e., litter raking, fuelwood collection), the influence of biomass extraction on soil acidification was highlighted, because the uptake and storage of cations exceeds that of anions during tree growth and biomass buildup [40,41]. In order to mitigate the negative effects of intensive biomass extraction on the nutrient budget, the redistribution of wood ash is frequently discussed and has proven to be suitable to compensate for nutrient losses in forests, with the exception of N [13,42–45], and was previously found to be applicable in the form of bark-ash-pellets at the study sites [14].

5. Conclusions

Our study provides evidence that intensive biomass extraction may cause substantial losses of nutrients in pine and spruce stands, even in the thinning stage of stand development. This is amplified by the high productivity of both investigated stands, which is accelerated by high atmospheric N inputs. Despite the fact that the nutrient balance established in this study contains only a rough estimation of the weathering rate, it suggests that the budgets of base cations are already stressed due to high seepage losses caused by the still ongoing effects of very high S-driven acid deposition which occurred previously in the region. Therefore, the nutrient balance became negative even for SO harvesting in the case of Ca^{2+} and Mg^{2+}. Further intensification of biomass removal clearly impairs the nutrient budget, also for K^+, while the N balance remains positive even in the most intensive biomass removal scenario (WT). Even if a further development of the wood energy sector will increase the demand for forest biomass in the future, forest management should take the nutrient balance into account. Intelligent concepts to limit forest biomass extraction to a sustainable level need to be developed. At the same time, additional measures, such as wood ash recycling, may help to improve the nutrient balance of the base cations and P, without burdening the ecosystems with ever more N.

Acknowledgments: We acknowledge support by the Federal Ministry of Education and Research and the Open Access Publication Funds of the TU Dresden. The authors would like to thank Clemens Kurth for advice concerning the forest treatment and Henning Andreae, Frank Jacob, and Hendrik Horn for help with the biomass inventories and functions. We are very grateful for the valuable improvements of the manuscript provided by two anonymous reviewers.

Author Contributions: Christine Knust and Karl-Heinz Feger conceived and designed the experiments; Christine Knust and Karoline Schua performed the experiments; Christine Knust analyzed the data; Christine Knust wrote the paper.

Conflicts of Interest: The authors declare no conflicts of interest.

References

1. Fachagentur Nachwachsende Rohstoffe e.V. (FNR). *Basisdaten Bioenergie Deutschland 2014: Festbrennstoffe, Biokraftstoffe, Biogas*; FNR: Gülzow, Germany, 2014.
2. BMU; BMELV. *Nationaler Biomasseaktionsplan für Deutschland: Beitrag der Biomasse für Eine Nachhaltige Energieversorgung*; BMU, BMELV: Berlin, Germany, 2010.
3. Kizha, A.R.; Han, H.-S. Processing and sorting forest residues: Cost, productivity and managerial impacts. *Biomass Bioenergy* **2016**, *93*, 97–106. [CrossRef]
4. Wolfsmayr, U.J.; Rauch, P. The primary forest fuel supply chain: A literature review. *Biomass Bioenergy* **2014**, *60*, 203–221. [CrossRef]
5. Cremer, T. Bereitstellung von Holzhackschnitzeln Durch Die Forstwirtschaft: Produktivitätsmodelle als Entscheidungsgrundlage über Verfahren und Aushaltungsvarianten, Entwickelt auf der Basis Einer Metaanalyse. Ph.D. Thesis, University of Freiburg i.Br., Freiburg, Germany, 2008.
6. Vangansbeke, P.; de Schrijver, A.; de Frenne, P.; Verstraeten, A.; Gorissen, L.; Verheyen, K. Strong negative impacts of whole tree harvesting in pine stands on poor, sandy soils: A long-term nutrient budget modelling approach. *For. Ecol. Manag.* **2015**, *356*, 101–111. [CrossRef]

7. Egnell, G. Is the productivity decline in Norway spruce following whole-tree harvesting in the final felling in boreal Sweden permanent or temporary? *For. Ecol. Manag.* **2011**, *261*, 148–153. [CrossRef]

8. Walmsley, J.; Jones, D.; Reynolds, B.; Price, M.; Healey, J. Whole tree harvesting can reduce second rotation forest productivity. *For. Ecol. Manag.* **2009**, *257*, 1104–1111. [CrossRef]

9. Weis, W.; Göttlein, A. Nährstoffnachhaltige Biomassenutzung: Bei der Nutzung von Biomasse ist Vorsicht geboten: Nicht jeder Waldstandort verträgt den erhöhten Nährstoffentzug. *LWF Aktuell* **2012**, *90*, 44–47.

10. Helmisaari, H.-S.; Hanssen, K.H.; Jacobson, S.; Kukkola, M.; Luiro, J.; Saarsalmi, A.; Tamminen, P.; Tveite, B. Logging residue removal after thinning in Nordic boreal forests: Long-term impact on tree growth. *For. Ecol. Manag.* **2011**, *261*, 1919–1927. [CrossRef]

11. Clarke, N.; Gundersen, P.; Jönsson-Belyazid, U.; Kjønaas, O.J.; Persson, T.; Sigurdsson, B.D.; STUPAK, I.; Vesterdal, L. Influence of different tree-harvesting intensities on forest soil carbon stocks in boreal and northern temperate forest ecosystems. *For. Ecol. Manag.* **2015**, *351*, 9–19. [CrossRef]

12. Lassauce, A.; Lieutier, F.; Bouget, C. Woodfuel harvesting and biodiversity conservation in temperate forests: Effects of logging residue characteristics on saproxylic beetle assemblages. *Biol. Conserv.* **2012**, *147*, 204–212. [CrossRef]

13. Pitman, R.M. Wood ash use in forestry—A review of the environmental impacts. *Forestry* **2006**, *79*, 563–588. [CrossRef]

14. Knust, C.; Schua, K.; Göttlein, A.; Ettl, R.; Wolferstetter, T.; Feger, K.-H. Compensation of nutrient losses resulting from the intensified use of woody biomass using bark-ash-pellets. In *Bioenergy from Dendromass for the Sustainable Development of Rural Areas*; Butler Manning, D., Bemmann, A., Bredemeier, M., Lamersdorf, N., Ammer, C., Eds.; Wiley (Imprint), Inc.: Weinheim, Germany, 2015; pp. 269–286.

15. Demeyer, A.; Voundi Nkana, J.C.; Verloo, M. Characteristics of wood ash and influence on soil properties and nutrient uptake: An overview. *Bioresour. Technol.* **2001**, *77*, 287–295. [CrossRef]

16. Ingerslev, M.; Hansen, M.; Pedersen, L.B.; Skov, S. Effects of wood chip ash fertilization on soil chemistry in a Norway spruce plantation on a nutrient-poor soil. *For. Ecol. Manag.* **2014**, *334*, 10–17. [CrossRef]

17. Rademacher, P.; Schönfelde, E.; Meiwes, K.J. Elementgehalte in Baumkompartimenten von Fichte (*Picea abies* [L.] Karst), Kiefer (*Pinus sylvestris* [L.] Karst), Buche (*Fagus sylvatica* [L.] Karst), Eiche (*Quercus petraea und robur* [L.] Karst) und Douglasie (*Pseudotsuga menziesii* [L.] Karst). In *Möglichkeiten und Grenzen der Vollbaumnutzung: Ergebnisbericht (FKZ: 22015407)*; Nordwestdeutsche Forstliche Versuchsanstalt, Universität Göttingen: Göttingen, Germany, 2011; pp. 125–154.

18. Merino, A.; Balboa, M.A.; Rodríguez Soalleiro, R.; González, J.Á. Nutrient exports under different harvesting regimes in fast-growing forest plantations in southern Europe. *For. Ecol. Manag.* **2005**, *207*, 325–339. [CrossRef]

19. Tveite, B.; Hanssen, K.H. Whole-tree thinnings in stands of Scots pine (Pinus sylvestris) and Norway spruce (Picea abies): Short- and long-term growth results. *For. Ecol. Manag.* **2013**, *298*, 52–61. [CrossRef]

20. Zetterberg, T.; Olsson, B.A.; Löfgren, S.; Hyvönen, R.; Brandtberg, P.-O. Long-term soil calcium depletion after conventional and whole-tree harvest. *For. Ecol. Manag.* **2016**, *369*, 102–115. [CrossRef]

21. Jacobson, S.; Kukkola, M.; Mälkönen, E.; Tveite, B. Impact of whole-tree harvesting and compensatory fertilization on growth of coniferous thinning stands. *For. Ecol. Manag.* **2000**, *129*, 41–51. [CrossRef]

22. Jacobson, S.; Kukkola, M.; Mälkönen, E.; Tveite, B.; Möller, G. Growth response of coniferous stands to whole-tree harvesting in early thinnings. *Scand. J. For. Res.* **1996**, *11*, 50–59. [CrossRef]

23. Döbbeler, H.; Alber, M.; Schmidt, M.; Nagel, J.; Schröder, J. *BWINPro—Programm zur Bestandesanalyse und Prognose: Handbuch zur Gemeinsamen Version von BWINPro und BWINPro-S*; Version 6.3; Nordwestetsche Forstliche Versuchsanstalt, Göttingen, Germany and TU Dresden, Lehrstuhl für Holzmeßkunde und Waldwachstum: Tharandt, Germany, 2011.

24. Andreae, H. (Staatsbetrieb Sachsenforst, Kompetenzzentrum Wald und Forstwirtschaft, Saxony, Germany). Personal communication, 2014.

25. Von Wilpert, K.; Bösch, B.; Bastian, P.; Zirlewagen, D.; Hepperle, F.; Holzmann, S.; Puhlmann, H.; Schäffer, J.; Kändler, G.; Sauter, U.H. Biomasse-Aufkommensprognose und Kreislaufkonzept für den Einsatz von Holzaschen in der Bodenschutzkalkung in Oberschwaben. *Ber. Freibg. Forstl. Forsch.* **2011**, *87*, 155.

26. Hildebrand, E.E.; Wilpert, K. von; Kohler, M. *Ionenspeicher- und Mobilisierungspotentiale im Skelett und Feinboden des Standortes Conventwald bei Unterschiedlichen Depositionsszenarien: Abschlussbericht, FKZ PEF196009 und PEF 196010, BWC 99 009 und BWC 99010*; Universität Freiburg und Forstliche Versuchs- und Forschungsanstalt: Freiburg, Germany, 2000.

27. Rademacher, P. Contents of nutrient elements in tree components of economical important species in relation to their residual utilisation. *Holz Werkst.* **2005**, *63*, 285–296. [CrossRef]

28. Eisenhauer, D.-R.; Sonnemann, S. Waldbaustrategien unter sich ändernden Umweltbedingungen: Leitbilder, Zielsysteme und Waldentwicklungstypen. *Waldökol. Landsch. Nat.* **2009**, *8*, 71–88.

29. Fischer, H.; Wagner, S. Silvicultural responses to predicted climate change scenarios. *J. For. Ecol. For. Manag.* **2009**, *2*, 12–23.

30. Rumpf, S.; Nagel, J.; Schmidt, M. Biomasseschätzfunktionen von Fichte (*Picea abies* [L.] Karst), Kiefer (*Pinus sylvestris* [L.] Karst), Buche (*Fagus sylvatica* [L.] Karst), Eiche (*Quercus robur und petraea* [L.] Karst) und Douglasie (*Pseudotsuga menziesii* [L.] Karst) für Nordwestdeutschland. In *Möglichkeiten und Grenzen der Vollbaumnutzung: Ergebnisbericht (FKZ: 22015407)*; Nordwestdeutsche Forstliche Versuchsanstalt, Universität Göttingen: Göttingen, Germany, 2011; pp. 25–123.

31. Jacobsen, C.; Rademacher, P.; Meesenburg, H.; Meiwes, K.J. Gehalte chemischer Elemente in Baumkompartimenten: Literaturstudie und Datensammlung. In *Berichte des Forschungszentrums Waldökosysteme, Reihe B*; Universität Göttingen: Göttingen, Germay, 2003.

32. Oulehle, F.; Hofmeister, J.; Cudlín, P.; Hruška, J. The effect of reduced atmospheric deposition on soil and soil solution chemistry at a site subjected to long-term acidification, Načetín, Czech Republik. *Sci. Total Environ.* **2006**, *370*, 532–544. [CrossRef] [PubMed]

33. Wunderlich, S.; Raben, G.; Andreae, H.; Feger, K.-H. Schwefel-Vorräte und Sulfat-Remobilisierungspotential in Böden der Level-II-Standorte Sachsens. *AFZ-Der. Wald* **2006**, *60*, 762–765.

34. Kaarakka, L.; Tamminen, P.; Saarsalmi, A.; Kukkola, M.; Helmisaari, H.-S.; Burton, A.J. Effects of repeated whole-tree harvesting on soil properties and tree growth in a Norway spruce (*Picea abies* (L.) Karst.) stand. *For. Ecol. Manag.* **2014**, *313*, 180–187. [CrossRef]

35. Roxby, G.E.; Howard, T.E. Whole-tree harvesting and site productivity: Twenty-nine northern hardwood sites in central New Hampshire and western Maine. *For. Ecol. Manag.* **2013**, *293*, 114–121. [CrossRef]

36. Achat, D.L.; Deleuze, C.; Landmann, G.; Pousse, N.; Ranger, J.; Augusto, L. Quantifying consequences of removing harvesting residues on forest soils and tree growth—A meta-analysis. *For. Ecol. Manag.* **2015**, *348*, 124–141. [CrossRef]

37. Klaminder, J.; Lucas, R.W.; Futter, M.N.; Bishop, K.H.; Köhler, S.J.; Egnell, G.; Laudon, H. Silicate mineral weathering rate estimates: Are they precise enough to be useful when predicting the recovery of nutrient pools after harvesting? *For. Ecol. Manag.* **2011**, *261*, 1–9. [CrossRef]

38. Vadeboncoeur, M.A.; Hamburg, S.P.; Yanai, R.D.; Blum, J.D. Rates of sustainable forest harvest depend on rotation length and weathering of soil minerals. *For. Ecol. Manag.* **2014**, *318*, 194–205. [CrossRef]

39. Kreutzer, K. Ecological problems of full tree utilization. *Forstwiss. Cent.* **1979**, *98*, 298–308. [CrossRef]

40. Glatzel, G. The impact of historic land use and modern forestry on nutrient relations of Central European forest ecosystems. *Fertil. Res.* **1991**, *27*, 1–8. [CrossRef]

41. Feger, K.H. Bedeutung von ökosysteminternen Umsätzen und Nutzungseingriffen für den Stoffhaushalt von Waldlandschaften (Habilitationsschrift)—*Freiburger Bodenkundliche. Abhandlungen* **1993**, *31*, 237.

42. Ingerslev, M.; Skov, S.; Sevel, L.; Pedersen, L.B. Element budgets of forest biomass combustion and ash fertilisation—A Danish case-study. *Biomass Bioenergy* **2011**, *35*, 2697–2704. [CrossRef]

43. Kikamägi, K.; Ots, K.; Kuznetsova, T. Effect of wood ash on the biomass production and nutrient status of young silver birch (Betula pendula Roth) trees on cutaway peatlands in Estonia. *Ecol. Eng.* **2013**, *58*, 17–25. [CrossRef]

44. Saarsalmi, A.; Smolander, A.; Kukkola, M.; Moilanen, M.; Saramäki, J. 30-year effects of wood ash and nitrogen fertilization on soil chemical properties, soil microbial processes and stand growth in a Scots pine stand. *For. Ecol. Manag.* **2012**, *278*, 63–70. [CrossRef]

45. Solla-Gullón, F.; Santalla, M.; Rodríguez-Soalleiro, R.J.; Merino, A. Nutritional status and growth of a young Pseudotsuga menziesii plantation in a temperate region after application of wood-bark ash. *For. Ecol. Manag.* **2006**, *237*, 312–321. [CrossRef]

![forests logo] *forests*

MDPI

Article

Evaluation of Whole Tree Growth Increment Derived from Tree-Ring Series for Use in Assessments of Changes in Forest Productivity across Various Spatial Scales

Juha M. Metsaranta and Jagtar S. Bhatti *

Natural Resources Canada, Canadian Forest Service, Northern Forestry Centre, 5320 122 Street, Edmonton, AB T6H 3S5, Canada; juha.metsaranta@canada.ca
* Correspondence: jagtar.bhatti@canada.ca; Tel.: +1-780-435-7241

Academic Editors: Scott X. Chang and Xiangyang Sun
Received: 22 September 2016; Accepted: 22 November 2016; Published: 1 December 2016

Abstract: The inherent predictability of inter-annual variation in forest productivity remains unknown. Available field-based data sources for understanding this variability differ in their spatial resolution, temporal resolution, and typical units of measure. Nearly all other tree and forest characteristics are in practice derived from measurements of diameter at breast height (DBH). Therefore, diameter increment reconstructed annually from tree-ring data can be used to estimate annual growth increments of wood volume, but the accuracy and precision of these estimates requires assessment. Annual growth estimates for $n = 170$ trees sampled for whole stem analysis from five tree species (jack pine, lodgepole pine, black spruce, white spruce, and trembling aspen) in Western Canada were compared against increments derived from breast height measurements only. Inter-annual variability of breast height and whole tree growth increments was highly correlated for most trees. Relative errors varied by species, diameter class, and the equation used to estimate volume (regional vs. national). A simple example of the possible effect of this error when propagated to the stand level is provided.

Keywords: tree rings; volume increment; height increment; stem analysis

---

1. Introduction

Changes in climate and other environmental conditions can cause annual forest growth to deviate from its long-term historic average, with significant impacts on large-scale estimates of carbon stock changes and forest dynamics. Both spatial and inter-annual variability of tree growth and forest productivity are fundamental for the assessment of future forest management options as well as the annual reporting of forest carbon balances, but the inherent predictability of this inter-annual variation remains unknown [1]. Available data for understanding inter-annual variation differ in their temporal resolution, spatial coverage, and typical units of measure [2,3]. For example, eddy covariance sites provide data on ecosystem carbon fluxes with high temporal resolution. These data can be used to assess site-specific factors affecting inter-annual variation in fluxes, but the measurements are limited to a few sites over a limited number of recent years. Larger amounts of data may be obtained from forest inventory plots, where data such as the diameter, height, vital status, and species of trees are periodically recorded. Allometric relationships can be used to convert measured tree characteristics into stock and stock change estimates for various tree- and stand-level metrics such as wood volume or biomass. This approach provides large-scale estimates on changes resulting from both growth and mortality, allowing the assessment of long-term trends over landscapes [4]. However, the temporal resolution of these data is limited, typically only 5 to 10 years, so they may not provide the annual scale data required for analyzing tree response to annual climate fluctuations. Recent studies have noted the potential utility of tree ring data in terrestrial carbon cycle research [5], as it can lengthen the available time series of annually resolved estimates of ecosystem production or growth increment [3,6–8]. In these applications, ring width measurements are instead used to generate various tree or forest level estimates of annual growth increment. This is possible because past diameter at breast height (DBH), the most basic measurement of trees from which nearly all other tree and forest characteristics are derived, can be reconstructed annually from ring-width measurements. Therefore, tree-ring data could potentially form the basis for assessing stand-level changes in forest productivity, as well as in the development and evaluation of data products and ecosystem models [3,9,10].

An issue remains in that relatively little effort has been put into evaluating the accuracy of estimates derived from breast height, relative to actual whole tree growth increments [11–13]. For example, stem growth at breast height can over- or underestimate whole-tree annual increments under warm and dry climate [13–15]. Presently, the best available method for obtaining annual growth increment data is full stem analysis [16–18], which involves felling and sectioning the main stem at regular intervals. Annual wood volume and height increments can be calculated from ring-widths measured and cross-dated on these sections. Stem analysis is labour-intensive, so large-scale application for obtaining annual growth increment data is likely cost-prohibitive. However, it can provide a useful estimate of true annual growth increment, against which alternatives derived from data at breast height only, which is more widely available, can be compared. In this paper, we evaluate annual growth increment estimates of wood volume (m^3) derived from past diameters reconstructed from breast height ring widths compared to the same estimates obtained from full stem analysis data, when both estimates are available for the same tree. We did this using data from five species common in the Western Canadian boreal forest: lodgepole pine (*Pinus contorta* Dougl. *var. latifolia* Engelm.) jack pine (*Pinus banksiana* Lamb.), black spruce (*Picea mariana* (Mill.) B.S.P.), white spruce (*Picea glauca* (Moench) Voss.), and trembling aspen (*Poplus tremuloides* Michx.). The purpose of this evaluation is to determine the potential error associated with measures of volume growth increment, derived from tree ring measurements at breast height only, in terms of the actual value of the growth increment, as well as its pattern of inter-annual variation. We also apply our estimates of error to an example stand where tree-ring data have previously been used to estimate inter-annual variation in growth [2,3], to estimate the potential stand level magnitude of this error.

2. Materials and Methods

2.1. Stem Analysis Data Collection and Measurement

Full stem analysis data were obtained from 173 trees sampled at various points in time from 1994 to 2010 in the Western Canadian boreal forest. Samples were collected in four jurisdictions: Alberta, Saskatchewan, Manitoba, and the Northwest Territories, and for five species, lodgepole pine (*n* = 30) jack pine (*n* = 15), black spruce (*n* = 68), white spruce (*n* = 28), and trembling aspen (*n* = 31). The characteristics of the sample trees (at time of sampling) are further detailed in Table 1. These stem analysis data were acquired from several projects. Some were collected during the 1994 field campaign of the Boreal Ecosystem-Atmosphere Study (BOREAS) campaign from locations near Prince Albert, Saskatchewan, and Thompson, Manitoba. Sites, stands, soils, and the detailed stem analysis data and procedures are described elsewhere [19,20].

Briefly, tree height and DBH were measured on each sample tree prior to felling. Disks were cut at heights of 0.3 m, 1.0 m, 1.3 m, 2.0 m, and at 1 m intervals to the top of the stem. Fresh diameters were measured in the field for each disk. In the lab, each disk was dried in a 70 °C oven for three days, after which diameters and bark thickness were re-measured. For each disk, two radii were selected, 180° apart and avoiding compression wood. These radii were then sampled, X-rayed, and analyzed for ring width using DendroScan [21]. Additional samples were collected in 2005 and 2006 as a part of a study validating tree biomass models in the Canadian province of Alberta. The field methods were similar to those for the BOREAS project, with the exception that disks were sampled at slightly different heights: 0.3 m, 1.3 m, 2.8 m, and then at intervals of 2.5 m to the top of the stem. Disks were dried as for the BOREAS data. However, rather than processing for X-ray analysis with DendroScan, the samples were sanded with progressively finer sandpaper, scanned at 1600 dpi, and ring widths were then measured using CDendro (Cybis Elektronik & Data, Stockholm, Sweden). An additional 21 stem analyses were conducted during the summer of 2010 as part of a study validating tree biomass models for the Liard Valley, NWT, Canada. The disks on these trees were collected at heights of 0.3 m, 1.3 m, 5.0 m, and every 4.0 m to the top of the stem, and processed in the lab in the same way as the 2005–2006 samples. For all samples, ring widths were cross-dated between all the disks in each tree to ensure correct ring dating, and corrected by proportion to the mean fresh outside bark diameter of each disk.

Table 1. Summary of characteristics of trees sampled for stem analysis in this study.

Species	DBH (cm)	Height (m)	Age (Years)	Year Sampled (Year (*n*))
White spruce (SW) (*Picea glauca*)	23.0 (6.2, 51.4)	18.7 (6.7, 33.7)	86 (15, 226)	1994 (5) 2005 (4) 2006 (11) 2010 (8)
Black spruce (SB) (*Picea mariana*)	11.8 (5.0, 24.6)	10.8 (6.4, 20.9)	89 (18, 222)	1994 (26) 2005 (19) 2006 (16) 2010 (7)
Trembling aspen (TA) (*Populus tremuloides*)	21.4 (5.3, 54.8)	17.7 (6.5, 35.6)	69 (19, 176)	1994 (9) 2005 (15) 2010 (7)
Jack pine (PJ) (*Pinus banksiana*)	13.1 (7.0, 18.6)	12.3 (8.5, 15.8)	64 (52, 76)	1994 (15)
Lodgepole pine (PL) (*Pinus contorta*)	15.3 (4.2, 27.0)	14.7 (5.8, 28.1)	77 (14, 153)	2005 (24) 2006 (6)

For DBH, height, and age, the value in the table represents the mean of all sampled trees, and the values in parentheses are the minimum and maximum values for that species. For year sampled, values in parentheses are the number of trees of that species sampled in that year.

2.2. Stem Analysis Growth Estimates

The ring-width measurements for each disk section for each tree were input into a spreadsheet program that calculated annual estimates of tree height (m) and stem volume (m^3). Historical tree height (m) was interpolated assuming equal annual height growth within a log section, and assuming that each disk cuts through the middle of a growth cycle, and was referred to as stem analysis height (HTs). Volume was calculated from ring width measurements on each of the disk sections and summed for the entire tree for each year, and referred to as stem analysis volume (Vs). The volume for each log section was calculated using Smalian's formula, $V = (L(Ab + At)/2)$, for the top section from the volume of a paraboloid, $V = (LAb)/2$, and for the stump section from the volume of a cylinder $V = LAb$, where V is the volume of the section (cm^3), L is the length of the section (cm), Ab is the cross-sectional area of the base of the section (cm^2), and At is the cross-sectional area of the top of the section (cm^2). Note that, while the stem analysis-based measurements were taken as the true standard against which alternatives derived from the breast height sample were compared, they are not themselves without uncertainty. For example, estimates derived from stem analysis will differ depending on the method used to interpolate height between cross sections [18], which standard model is used to determine log volume [22,23], the number of cross-cuts sampled, and the number of radii measured along each sample [12], and because trees are not perfectly round in cross section [24,25].

2.3. Model-Based Growth Estimates

The breast height (1.3 m) sample taken from each tree was used as a reconstructed estimate of the past breast height diameter (DBH) of each tree, from which we then calculated annual increments of wood volume for comparison to the estimates derived from stem analysis. We describe the methods used for each of these in the following sections, and an additional summary is provided in Table 2.

Table 2. Summary of the volume and height estimation methodologies used in the comparisons.

Metric	Comparison	Description	References for Equations and Parameters
	(1) National DBH	(1.1) Volume is estimated from DBH only, using a national equation	[26]
(A) Volume	(2) National DBH and HT	(2.1) Volume is estimated from DBH and uncorrected height (HTr), using a national equation.	[26]; heights as in B1.1 or B1.2
		(2.2) Volume is estimated from DBH and corrected height (HTc) using a national volume equation	
	(3) Regional DBH and HT	(3.1) Volume is estimated from DBH and uncorrected height (HTr), using a regional equation.	[27] for the equation formulation, Regional parameters from published sources for Saskatchewan [28], Manitoba [29], Alberta [30], and the Northwest Territories [30]. Heights estimated as in B1.1 or B1.2
		(3.2) Volume is estimated from DBH and corrected height (HTc) using a regional equation	
(B) Tree height	(1) HT DBH model	(1.1) Heights estimated from diameters reconstructed on the breast height sample are compared to heights interpolated between stem analysis sections.	(1.1 and 1.2) Provincial parameter sets for trees in Manitoba [29]. Alberta parameter sets for other provinces (NWT, SK, and AB) [31]
		(1.2) Heights estimated as above, but a correction factor is calculated from the difference between measured and predicted height at time of sampling, and applied to the rest of the height time series.	

2.3.1. Height Estimation

Height was estimated from DBH using regional height–diameter equations, which use the Chapman-Richards function:

$$H = 1.3 + a\left(1 - e^{-b \cdot D_{ob}}\right)^c \tag{1}$$

where H is the tree height (m), D_{ob} is the outside bark diameter (DBH, cm), and a, b, and c are the parameters obtained from published sources for Alberta ([31], Table S1) and Manitoba ([29], Table S1). Trees in Saskatchewan and the Northwest Territories also used the Alberta parameters, as regional parameters were not available for these jurisdictions. We calculated two height increment series from the annually reconstructed DBH, (1) a raw height series (HTr), which used the height estimate directly from Equation (1), as well as (2) a corrected height series (HTc), where a ratio between the final predicted height from the raw height series and the measured tree height was used as a correction factor and applied to the raw height series to scale past predicted heights so that the final height in the corrected series was equal to the measured height at the time the tree was sampled.

2.3.2. Volume Estimation

Estimates of volume increment were obtained from the reconstructed DBH, both alone and in combination with height estimated using Equation (1), both in raw (HTr) and corrected form (HTc). We first used a national scale taper equation that uses DBH only as a predictor (VNd) [26]:

$$d^2_{ijkm} = DBH^2_{ijk} \frac{\beta_0 DBH^{\beta_1+\delta_{i1}+\delta_{ij1}} - h_{ijkm}}{\beta_0 DBH^{\beta_1+\delta_{i1}+\delta_{ij1}} - 1.3} \left(\frac{h_{ijkm}}{1.3}\right)^{2-(\beta_2+\delta_{i2}+\delta_{ij2}+\delta_{ijk})} + \varepsilon_{ijkm} \tag{2}$$

where d is the diameter at different cross-section heights (h), indices i, j, k, and m respectively refer to province, plot, tree, and cross section, β_0, β_1, and β_2 are fixed-effect parameters, and δ_i, δ_{ij}, and δ_{ijk} are random effects associated with the province, plot, and tree, respectively [26]. We also tested two methods that use both DBH and H as predictors. The first was also a national level taper equation (VNdh) [26]:

$$d^2_{ijkm} = DBH^2_{ijk} \frac{H_{ijk} - h_{ijkm}}{H_{ijk} - 1.3} \left(\frac{h_{ijkm}}{1.3}\right)^{2-(\beta_2+\delta_{i2}+\delta_{ij2}+\delta_{ijk})} + \varepsilon_{ijkm} \tag{3}$$

where d is the diameter at different cross-section heights (h), indices i, j, k, and m respectively refer to province, plot, tree, and cross section, β_2 is a fixed-effect parameter, and δ_i, δ_{ij}, and δ_{ijk}, are random effects associated with the province, plot, and tree, respectively [26] (Table S2). The second was a different taper equation (VRdh) [27], for which regional parameter estimates were generally available [28–30]:

$$d_i = a_0 DBH^{a1} a_2^{D_{ob}} X_i^{b1z_i^2+b2\ln(z_i+0.001)+b3\sqrt{z_i}+b4e^{z_i}+b5(DBH/H)} \tag{4}$$

where d is the inside bark diameter at height i along the stem (cm), DBH is the diameter at breast height (cm) of the tree outside bark, H is the total height (m), and X_i is $(1 - \sqrt{h_i/H}) / (1 - \sqrt{p})$, where h_i is the cross-section height i, p is the relative height of the inflection point, typically assumed to be 0.25, and z_i is the relative height (h_i/H) [27]. Parameters a_0, a_1, a_2, b_1, b_2, b_3, b_4, and b_5 were obtained from published sources Saskatchewan [28], Manitoba [29], and Alberta [30] (Table S3). Regional parameters were not available for the Northwest Territories; therefore, for trees in that jurisdiction, the Alberta parameters were used. Estimates of volume for each of the methods tested were obtained by numerical integration, and the volume increment from the difference in volume between subsequent years.

2.4. Comparison of Estimates

For volume increment, we conducted five comparisons, representing combinations of different volume equations (VNd, VNdh, and VRdh) and height estimation options (HTr (VNdhr or VRDhr) or HTc (VNdhc or VRdhc)), which were in each case assessed against stem analysis volume (Vs), which was considered true. Details on these comparisons are provided in Table 2. Differences were assessed using relative error,

$$100(\hat{Y}_i - Y_i)/Y_i \tag{5}$$

where $\hat{Y}_i$ is the estimated growth metric (height or volume) determined from the reconstructed DBH only, and Y_i is the same metric determined from the stem analysis for the ith year. For analysis, we grouped relative errors by species and 5 cm DBH class, and used the distribution of errors in each class as an indicator of the magnitude and potential significance of differences relative to the estimates derived from stem analysis. When the value zero (indicating no difference between the breast height and whole tree-derived volume increment) was below the 2.5th percentile of all differences in a class, we considered the estimates to be biased high. When zero was between the 2.5th and 25th percentiles, we considered them borderline high, between the 25th and 75th percentiles unbiased, between the 75th and 97.5th percentiles borderline low, and greater than the 97.5th percentile, low.

2.5. Inter-Annual Variation

An additional aim was to determine if annual increment values derived from stem analysis and their alternative were statistically in phase. To do this, we compared a first-differenced time series of each comparison variable and assessed the value of the cross correlation at lag zero, as well as determined the fraction of cases in which the cross correlation peaked at lag zero, relative to shifting the comparison time series forwards or backwards by up to five years.

2.6. Stand-Level Example Application

Errors at the tree-level were also propagated to the stand level for an example fixed area plot in Saskatchewan, Canada, for which samples for dendrochronological analysis were collected at breast height from all live and dead (standing and fallen) trees present at the time of sampling, and that has previously been used in assessments of mortality, competition dynamics, and ecosystem production [2,3]. The observed distribution of error for the regional volume model for jack pine using DBH and HTc as input was propagated using $n = 1000$ Monte Carlo simulations, under a simple assumption of an independence of errors among trees and size classes.

3. Results

3.1. Comparison of Estimates

Relative errors varied by species, diameter class, the equation used to estimate volume (regional vs. national), and whether or not the equation used only DBH or both DBH and HT as predictors (Figure 1). Estimates derived from breast height only were only biased high or low in relatively few species, diameter class, and volume equation categories ($n = 13/175$, 7%). Most (52%) showed no bias, and the rest (41%) were borderline high or low (Figure 1). The average range of errors across classes was ±50%, with some small differences among species. In general, the widest distribution of errors for any species or diameter class was for small trees (<10 cm DBH), and in particular for the national model that used only DBH as a predictor. For models that used both DBH and HT, the use of HTc rather than HTr, or a regional rather than a national model had only a marginal influence on the distribution of relative errors in any species or diameter class. The number of annual increment observations available for comparison by species and diameter class is provided in Table S4.

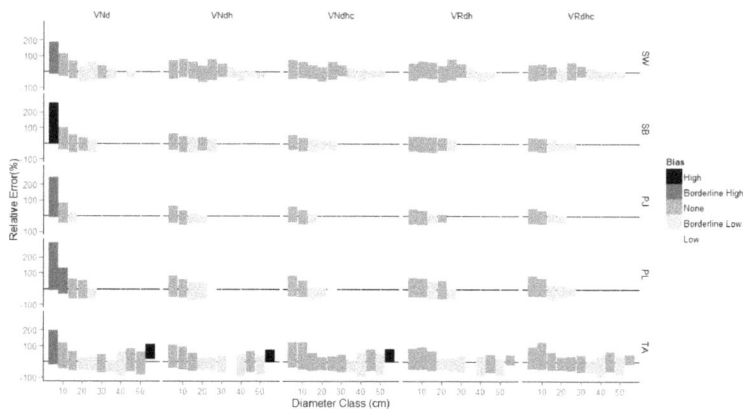

Figure 1. Relative error in estimation of volume increment from DBH reconstructed from tree-ring data, in comparison to volume increment derived from whole stem analysis for the same trees. Errors are plotted by species (rows) and 5 cm DBH class. The bars represent the range of the 2.5th and 97.5th percentiles of the relative errors in each class. Bars are plotted in different shades as a function of the percentile location of zero with the distribution of errors in a diameter class, and interpreted as described in the text. The columns represent different potential methods for estimating volume increment, national equations using DBH only (VNd), national equations using DBH and raw HT (VNdh), national equations using DBH and corrected height (VNdhc), regional equations using DBH and raw HT (VRdh), and regional equations using DBH and corrected height (VRdhc).

3.2. Inter-Annual Variation

For the assessment of the cross-correlation of inter annual variability between breast height and stem analysis-derived growth metrics, we show results for HTr only, as results for HTc are virtually identical. For volume increment, the cross correlation was maximized at lag zero for 94% of cases for estimates derived from DBH only using the national volume model (median 0.83 and 95% were between 0.20 and 0.99), and for 99% of samples for estimates derived from both DBH and HT for the national (median 0.92, 95% between 0.56 and 0.99), and regional models (median 0.93, 95% between 0.63 and 0.99).

3.3. Example Application

Results for the propagation of errors to a stand level example are provided in Figure 2. This particular example used a regional volume for jack pine in Saskatchewan and corrected estimate of height increment. Estimates of volume increment derived from breast height were within the propagated error range for ages less than 30 years. For ages greater than 30 years, the ranges fell below the point estimate derived from the breast height sample, indicating that it was biased to a high degree relative to whole tree increments. It is likely that many trees passed from a size class where the error distribution was unbiased (VRdhc for PJ, Figure 1), to one where the bias was borderline low. Relative to the median of the error simulations, the range of values typically spanned ±5%.

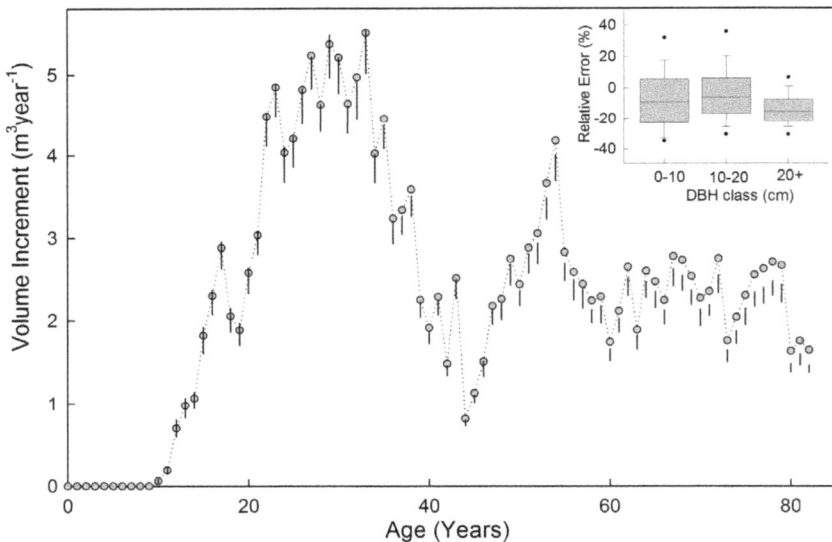

Figure 2. Time series of above ground volume increment ($m^3 \cdot ha^{-1} \cdot year^{-1}$) with stand age for a 900 m^2 fixed area jack pine plot in Saskatchewan, Canada, for which all live and dead trees present in the stand at age 82 were sampled at breast height for ring width measurement. The line plot represents volume increment derived from these measurements using a regional volume model using corrected height as input. Error bars represent the 2.5th and 97.5th percentiles of stand level estimates of volume increment, derived from diameter class-based relative errors (inset graph) derived from the observed difference in breast height and whole tree-derived growth increment for this species and volume estimation method.

4. Discussion

In this paper, we evaluated annual growth increment estimates of wood volume (m^3) derived from past DBH reconstructed from ring widths and compared these to estimates obtained from full stem analysis data, when both estimates are available for the same tree. This was done using data from $n = 170$ trees for five species common in the Western Canadian boreal forest: lodgepole pine, jack pine, black spruce, white spruce, and trembling aspen. The main results were that relative errors varied by species, diameter class, and the equation used to estimate volume and that the direction of inter-annual variability for whole tree and breast height estimates of volume increment were highly correlated for nearly all trees. In general, when estimates of volume increment are derived from models that use both DBH and HT as predictors, the total range of values spanned by the relative errors is narrower, relative to volume increment derived from a model using DBH only. The range of observed relative errors for national models was similar to that of regional models, at least for the species tested here. When tree-level errors were propagated to estimates of error for volume increment at the stand level, the overall range of errors was narrower, typically ±5% for our example jack pine stand, at least for our simple procedure that assumed the observed errors were independent. The overall range of errors is less than what is currently assumed to be the uncertainty of growth increments applied by Canada's National Forest Carbon Monitoring, Accounting, and Reporting system to estimate the carbon balance of Canada's managed forest [32], but it should be kept in mind that additional error would arise from model and measurement errors that were not accounted for, and at young ages from trees that have died and subsequently decayed such that they could not be detected when the stand was sampled [33]. In this analysis, breast height estimates were consistently high relative to whole tree estimates after age 30–40, but this is likely a result specific to this particular combination of tree species and volume estimation method.

Few other studies have attempted to compare the relative accuracy of different methods of measuring the annual incremental growth of trees. Similar to our analysis, standardized ring-width growth series at breast height and standardized whole-tree growth series of volume increment were highly correlated for high-elevation red spruce (*Picea rubens*) forests [11]. For European beech (*Fagus sylvatica*), the breast-height growth series were also highly correlated to whole-tree volume or biomass-increment series [13], but the breast-height series showed higher sensitivity to weather variation. An analysis of the relative accuracy of various full stem analysis sampling protocols for measuring volume increment found that it was possible to increase precision by increasing the density of sections sampled along the stem, and by measuring more radii at each sample [12]. Our results are essentially in line with these previous findings, in that we found that the direction of inter-annual variation in volume increment derived from breast height only was highly correlated with whole tree increment obtained by stem analysis. Our observation that using models with a height predictor generally improved the estimation of volume increment is interesting because this occurs even though total height estimates derived from stem analysis are calculated as the mean height increment between cross-cut segments. Interpolation between cross-cut segments means that stem analysis samples do not actually measure the inter-annual variability of height growth. There is no method presently available for retrospectively obtaining this information, except for some species that consistently grow a single annual whorl of branches. Ground-based measurements of height cannot resolve annual height increments for tall, slow growing trees [2,34], though height growth can be prospectively monitored with some accuracy by tree climbing [35].

Although it likely would have been of interest, we could not conduct a comparison for biomass increment because stem analysis does not provide a direct estimate of biomass that could be considered true. We could have compared estimates generated using reconstructed DBH only with estimates generated using DBH and HT, as alternative models that could have been applied do exist [36,37]. However, this analysis would have been a model inter-comparison exercise that compares the relative difference between estimates of biomass increment using DBH and HT relative to the use of DBH only, not an accuracy assessment directly. The estimation of biomass from tree level allometric equations remains challenging, and the current state of work in this area shows that changes in allometric equations can result in large variations in estimated biomass [38].

Some further challenges need to be considered when inferring above-ground forest increment from radial tree growth. Stand structure and density along with necessary biometric and metadata would be useful to upscale tree biomass estimates to the site level [39,40] and to efficiently select sample trees for estimating climate relationships in an unbiased manner [41]. An alternative, such as was used in our example application where tree level errors were propagated to the stand level, is to sample all live and dead trees in fixed area plots to reconstruct the growth of the whole stand [2,3], a method that has shown promise if the aim is to successfully detect climate and global change-related trends from tree-ring data [42], but remains challenged by the inability to sample trees that have died in the past and subsequently decayed [33]. In addition, replacing the relatively basic equations used in this study with more sophisticated methods that are sensitive to the effect of stand conditions on height–diameter relationships [43], tree volume [44,45], or biomass [46] should also improve the estimates of annual growth increment that could be obtained at both the tree and forest levels, for species where such models are available.

5. Conclusions

The ability to estimate annual growth increment over a long time period at a large spatial scale from field-based data would be a tremendous asset. Such data can be obtained by tree-ring analysis, which is increasingly being used in forestry applications to address questions around the effect of forestry practices, climate, insect outbreaks, global change, and other disturbances in commercially viable, closed-canopy forests as well as a wide range of unmanaged forest and woodland types. Traditionally, tree-ring data have been collected from old trees at climate-sensitive sites for reconstructing past climatic variation, and expressed in unitless ring-width indices. Therefore, observed causes and trends may not be applicable to more productive and managed forests. We have shown that translating these instead into tree- and forest-level estimates of annual growth increment in units that are of greater interest in forestry applications can provide reasonable estimates relative to alternative data on whole tree growth obtained by stem analysis.

Supplementary Materials: The following are available online at www.mdpi.com/1999-4907/7/12/303/s1, Table S1: Height–diameter model parameters [1,2], Table S2: National taper model parameters [3], Table S3: Regional taper model parameters [1,4–6], Table S4: Number of annual increment observations available for comparison by species and diameter class. Values in the diameter class column represent the largest diameter of tree in that class.

Acknowledgments: The authors would like to thank Shongming Huang and Yonghe Wang for comments on a previous version of the manuscript, and Thierry Varem-Sanders for extensive assistance with sample processing and data analysis. Funding for this study was provided in part by Canadian Forest Service and program of the Federal Panel on Energy Research and Development (PERD).

Author Contributions: Juha Metsaranta and Jagtar Bhatti both have contributed substantially to idea, data collection, data analysis and finally the manuscript writing.

Conflicts of Interest: The authors declare no conflict of interest.

References

1. Luo, Y.; Keenan, T.F.; Smith, M. Predictability of the terrestrial carbon cycle. *Glob. Chang. Biol.* **2015**, *21*, 1737–1751. [CrossRef] [PubMed]
2. Metsaranta, J.M.; Lieffers, V.J. Using dendrochronology to obtain annual data for modelling stand development: A supplement to permanent sample plots. *Forestry* **2009**, *82*, 163–173. [CrossRef]
3. Metsaranta, J.M.; Kurz, W.A. Inter-annual variability of ecosystem production in boreal jack pine forests (1975–2004) estimated from tree-ring data using CBM-CFS3. *Ecol. Mod.* **2012**, *224*, 111–123. [CrossRef]
4. Hember, R.A.; Kurz, W.A.; Metsaranta, J.M.; Black, T.A.; Coops, N.C.; Guy, R.D. Accelerated regrowth of temperate-maritime forests due to environmental change. *Glob. Chang. Biol.* **2012**, *18*, 2026–2040. [CrossRef]
5. Babst, F.; Alexander, M.R.; Szejner, P.; Bouriaud, O.; Klesse, S.; Roden, J.; Ciais, P.; Poulter, B.; Frank, D.; Moore, D.J.P.; et al. A tree-ring perspective on the terrestrial carbon cycle. *Oecologia* **2014**, *176*, 307–322. [CrossRef] [PubMed]

6. Graumlich, L.J.; Brubaker, L.B.; Grier, C.C. Long-term trends in forest net primary productivity: Cascade Mountains, Washington. *Ecology* **1989**, *70*, 405–410. [CrossRef]
7. Babst, F.; Bouriaud, O.; Alexander, R.; Trouet, V.; Frank, D. Toward consistent measurement of carbon accumulation: A multi-site assessment of biomass and basal area increment across Europe. *Dendrochronologia* **2014**, *32*, 153–161. [CrossRef]
8. Fang, O.; Wang, Y.; Shao, X. The effect of climate on the net primary production (NPP) of *Pinus koraiensis* in the Changbai Mountains over the past 50 years. *Trees* **2016**, *30*, 281–294. [CrossRef]
9. Bunn, A.G.; Hughes, M.K.; Kirdyanov, A.V.; Losleben, M.; Shishov, V.V.; Berner, L.T.; Oltchev, A.; Vaganov, E.A. Comparing forest measurements from tree rings and a space based index of vegetation activity in Siberia. *Environ. Res. Lett.* **2013**, *8*, 035034. [CrossRef]
10. Babst, F.; Bouriaud, O.; Papale, D.; Gielen, B.; Jansenns, I.A.; Nikinmaa, E.; Ibrom, A.; Wu, J.; Bernhofer, C.; Köstner, B.; et al. Above-ground woody carbon sequestration measured from tree rings is coherent with net ecosystem productivity at five eddy-covariance sites. *New Phytol.* **2014**, *201*, 1289–1303. [CrossRef] [PubMed]
11. LeBlanc, D.C. Relationship between breast-height and whole-stem growth indices for red spruce on Whiteface Mountain, New York. *Can. J. For. Res.* **1990**, *20*, 1399–1407. [CrossRef]
12. Newton, P.F. A stem analysis computational algorithm for estimating volume growth and its empirical evaluation under various sampling strategies. *Comput. Electron. Agric.* **2004**, *44*, 21–31. [CrossRef]
13. Bouriaud, O.; Breda, N.; Dupouey, J.-L.; Granier, A. Is ring width a reliable proxy for stem-biomass increment? A case study in European beech. *Can. J. For. Res.* **2005**, *35*, 2920–2933. [CrossRef]
14. Chhin, S.; Hogg, E.H.; Lieffers, V.J.; Huang, S. Growth–climate relationships vary with height along the stem in lodgepole pine. *Tree Phys.* **2010**, *30*, 335–345. [CrossRef] [PubMed]
15. Van der Maaten-Theunissen, M.; Bouriaud, O. Climate-growth relationships at different stem heights in silver fir and Norway spruce. *Can. J. For. Res.* **2012**, *42*, 958–969. [CrossRef]
16. Duff, G.H.; Nolan, N.J. Growth and morphogenesis in the Canadian forest species: I. The controls of cambial and apical activity in *Pinus resinosa*. *Can. J. Bot.* **1957**, *31*, 471–513. [CrossRef]
17. Shea, S.R.; Armson, K.A. Stem analysis of jack pine (*Pinus banksiana* Lamb.): Techniques and concepts. *Can. J. For. Res.* **1972**, *2*, 392–406. [CrossRef]
18. Dyer, M.E.; Bailey, R.L. A test of six methods for estimating true height from stem analysis data. *For. Sci.* **1987**, *33*, 3–13.
19. Halliwell, D.H.; Apps, M.J. *BOReal Ecosystem-Atmosphere Study (BOREAS) Biometry and Auxiliary Sites: Overstory and Understory Data*; Natural Resources Canada, Canadian Forest Service, Northern Forestry Centre: Edmonton, AB, Canada, 1997.
20. Varem-Sanders, T.M.L.; Cambpell, I.D. *BOReal Ecosystem-Atmosphere Study (BOREAS) Biometry and Auxiliary Sites: X-ray Densitometry of Tree Allometry Samples*; Natural Resources Canada, Canadian Forest Service, Northern Forestry Centre: Edmonton, AB, Canada, 1998.
21. Varem-Sanders, T.M.L.; Cambell, I.D. *DendroScan: A Tree-Ring Width and Density Measurement System*; UBC Press: Vancouver, BC, Canada, 1996.
22. Martin, A.J. Testing volume equation accuracy with water displacement techniques. *For. Sci.* **1984**, *30*, 41–50.
23. Figueiredo-Filho, A.; Schaaf, L.B. Comparison between predicted volumes estimated by taper equations and true volumes obtained by the water displacement technique (xylometer). *Can. J. For. Res.* **1999**, *29*, 451–461. [CrossRef]
24. Biging, G.S.; Wensel, L.C. The effect of eccentricity on the estimation of basal area and basal area increment of coniferous trees. *For. Sci.* **1988**, *34*, 621–633.
25. Bakker, J.D. A new, proportional method for reconstructing historical tree diameters. *Can. J. For. Res.* **2005**, *35*, 2515–2520. [CrossRef]
26. Ung, C.H.; Guo, X.J.; Fortin, M. Canadian national taper models. *For. Chron.* **2014**, *89*, 211–224. [CrossRef]
27. Kozak, A. A variable-exponent taper equation. *Can. J. For. Res.* **1988**, *18*, 1362–1368. [CrossRef]
28. Gál, J.; Bella, I.E. *New Stem Taper Functions for 12 Saskatchewan Timber Species*; Information Report NOR-X-338; Natural Resources Canada, Canadian Forest Service, Northern Forestry Centre: Edmonton, AB, Canada, 1994.
29. Klos, R. Ecologically Based Taper Equations for Major Tree Species in Manitoba. Master's Thesis, Lakehead University, Thunder Bay, ON, Canada, 2004.

30. Huang, S. *Ecologically Based Individual Tree Volume Estimation for Major Alberta Tree Species Report #1 Individual Tree Volume Estimation Procedures for Alberta: Methods of Formulation and Statistical Foundations*; Alberta Sustainable Resource Development, Public Lands and Forests Division: Edmonton, AB, Canada, 1994.

31. Huang, S. *Ecologically Based Individual Tree Volume Estimation for Major Alberta Tree Species Report #2 Ecologically Based Individual Tree Height-Diameter Models for Major Alberta Tree Species*; Alberta Sustainable Resource Development, Public Lands and Forests Division: Edmonton, AB, Canada, 1994.

32. Metsaranta, J.M.; Shaw, C.H.; Kurz, W.A.; Boisvenue, C.; Morken, S. Uncertainty of inventory-based estimates of the carbon dynamics of Canada's managed forest (1990–2014). *Can. J. For. Res.* **2016**. in review.

33. Metsaranta, J.M.; Lieffers, V.J.; Wein, R.W. Dendrochronological reconstruction of jack pine snag and downed log dynamics in Saskatchewan and Manitoba, Canada. *For. Ecol. Manag.* **2008**, *255*, 1262–1270. [CrossRef]

34. Hasenauer, H.; Monserud, R.A. Biased predictions for tree height increment models developed from smoothed 'data'. *Ecol. Mod.* **1997**, *98*, 13–22. [CrossRef]

35. Sumida, A.; Miyaura, T.; Torii, H. Relationships of tree height and diameter at breast height revisited: Analysis of stem growth using 20-year data of an even-aged *Chamaecyparis obtuse* stand. *Tree Phys.* **2013**, *33*, 106–118. [CrossRef] [PubMed]

36. Lambert, M.C.; Ung, C.H.; Raulier, F. Canadian national tree aboveground biomass equations. *Can. J. For. Res.* **2005**, *35*, 1996–2018. [CrossRef]

37. Ung, C.H.; Bernier, P.; Guo, X.J. Canadian national biomass equations: New parameter estimates that include British Columbia data. *Can. J. For. Res.* **2008**, *38*, 1123–1132. [CrossRef]

38. Weiskittel, A.R.; MacFarlane, D.W.; Radtke, P.J.; Affleck, D.L.R.; Temesgen, H.; Woodall, C.W.; Westfall, J.A.; Coulston, J.W. A call to improve methods for estimating tree biomass for regional and national assessments. *J. For.* **2015**, *113*, 414–424. [CrossRef]

39. Garcia, O. Sampling for tree-ring analysis. In Presented at Integrating Forest Information Over Space and Time, Canberra, Australia, 13–17 January 1992; pp. 110–128.

40. Osawa, A.; Abaimov, A.P.; Kajimoto, T. Feasibility of estimating total stem volume and aboveground biomass from measurement on the largest trees in even-aged pure stands. *Can. J. For. Res.* **2001**, *31*, 2042–2048. [CrossRef]

41. Mérian, P.; Bert, D.; Lebourgeois, F. An approach for quantifying and correcting sample size-related bias in population estimates of climate-tree growth relationships. *For. Sci.* **2013**, *59*, 444–452. [CrossRef]

42. Nehrbass-Ahles, C.; Babst, F.; Klesse, S.; Nöttzli, M.; Bouriaud, O.; Neukom, R.; Dobbertin, M.; Frank, D. The influence of sampling design on tree-ring based quantification of forest growth. *Glob. Chang. Biol.* **2014**, *20*, 2867–2885. [CrossRef] [PubMed]

43. Sharma, M.; Zhang, S.Y. Height-diameter models using stand characteristics for *Pinus banksiana* and *Picea mariana*. *Scand. J. For. Res.* **2004**, *19*, 442–451. [CrossRef]

44. Sharma, M.; Zhang, S.Y. Variable-exponent taper equations for jack pine, black spruce and balsam fir in eastern Canada. *For. Ecol. Manag.* **2004**, *198*, 39–53. [CrossRef]

45. Sharma, M.; Parton, J. Modelling stand density effects on taper for jack pine and black spruce plantations using dimensional analysis. *For. Sci.* **2009**, *55*, 268–282.

46. Almedag, I.S.; Stiell, W.M. Spacing and Age Effects on Biomass Production in Red Pine Plantations. *For. Chron.* **1982**, *58*, 220–224. [CrossRef]

![forests logo] *forests*

MDPI

Article

The Effect of Harvest on Forest Soil Carbon: A Meta-Analysis

Jason James * and Rob Harrison

School of Environmental and Forest Sciences, University of Washington, Box 352100, Seattle, WA 98195, USA; robh@uw.edu
* Correspondence: jajames@uw.edu; Tel.: +1-303-547-2792

Academic Editors: Scott X. Chang and Xiangyang Sun
Received: 28 September 2016; Accepted: 2 December 2016; Published: 7 December 2016

Abstract: Forest soils represent a substantial portion of the terrestrial carbon (C) pool, and changes to soil C cycling are globally significant not only for C sequestration but also for sustaining forest productivity and ecosystem services. To quantify the effect of harvesting on soil C, we used meta-analysis to examine a database of 945 responses to harvesting collected from 112 publications from around the world. Harvesting reduced soil C, on average, by 11.2% with 95% CI [14.1%, 8.5%]. There was substantial variation between responses in different soil depths, with greatest losses occurring in the O horizon (−30.2%). Much smaller but still significant losses (−3.3%) occurred in top soil C pools (0–15 cm depth). In very deep soil (60–100+ cm), a significant loss of 17.7% of soil C in was observed after harvest. However, only 21 of the 945 total responses examined this depth, indicating a substantial need for more research in this area. The response of soil C to harvesting varies substantially between soil orders, with greater losses in Spodosol and Ultisol orders and less substantial losses in Alfisols and Andisols. Soil C takes several decades to recover following harvest, with Spodosol and Ultisol C recovering only after at least 75 years. The publications in this analysis were highly skewed toward surface sampling, with a maximum sampling depth of 36 cm, on average. Sampling deep soil represents one of the best opportunities to reduce uncertainty in the understanding of the response of soil C to forest harvest.

Keywords: forest management; harvest; soil carbon; soil order; deep soil; meta-analysis

---

1. Introduction

Forest ecosystems contain 1240 Pg C [1,2], which represents as much as 80% of aboveground terrestrial C and 70% of all soil organic C [3–5]. The relative proportion of forest C found in soils varies among biomes, ranging from roughly 85% of the terrestrial C pool in boreal forests, to 60% in temperate forests, to 50% in tropical rainforests [1,6]. The net balance of soil C in forests relies upon large rates of detrital inputs (61.4 Pg C year^{-1}) and respiratory losses (60 Pg C year^{-1}), which together represent substantial yearly turnover in the soil C pool [7]. By altering the rates of detrital inputs and respiratory outputs in soils, the extent and intensity of forest harvest can have substantial impacts not only on ecosystem function but also on atmospheric chemistry and global climate [6,8,9].

C is one of the principal components of soil organic matter (SOM), a key component of soil that plays an important role in many biological, chemical, and physical properties [10–12]. SOM provides a crucial source of energy and nutrients for soil microbes, buffers soil pH, and helps to stabilize soil structure [12,13]. Along with nitrogen and phosphorus, SOM is considered a critical indicator for soil health and quality.

Thus, soil C is an essential component of forest C accounting, yet many models assume that only surface soil responds to forest management and that soil C returns to equilibrium within 20 years after harvest [14]. Recent national or global assessments of forest C lack any mention of mineral soil

C [15–17], implicitly assuming that soil C remains constant after forest harvest. Furthermore, carbon monitoring programs include soil C inconsistently. For example, the American Carbon Registry [18] and the Verified Carbon Standard [19] do not require or specify protocols for soil C measurements. The Intergovernmental Panel on Climate Change (IPCC) inventory standards [20] assume constant mineral soil C in Tier 1, with an option for inclusion of national soil C inventories only if preferred by a particular agency, and the U.S. Forest Service Inventory and Analysis Program [21] specifically limits soil sampling to 20 cm depth. The inclusion of soil in models of ecosystem C following harvest can have significant effects. For example, in a model of the forest C pool change following intensive bioenergy harvest, Zanchi et al. [22] show that the inclusion of soil increases the C payback period by approximately 25 years when substituting forest bioenergy for coal. Thus, the inclusion or exclusion of soil in ecosystem C models and ecological monitoring programs can have a major impact on forest policy when attempting to mitigate climate change through forest management [14].

Ambiguity about the effect of forest harvesting on soil C has persisted in the literature, likely exacerbated by the inherent spatial and temporal variability in soil measurements that can obscure the results of even the most well-designed studies [23]. By gathering the results from many studies that apply similar treatments, meta-analysis can overcome the high levels of spatial and temporal variability to provide cumulative answers that may not have been evident within individual sites [23,24]. Previous meta-analyses on the effect of harvesting on soil C have found either minimal effects on soil C pools [25] or substantial (30%) loss to O horizon pools with little change to mineral soil C [9]. Variation in soil C response has been shown to significantly differ among soil types and different harvesting strategies [9].

Studies of soil C change due to harvest have historically been strongly biased toward surface sampling [26]. Nave et al. [9] reported a mixed response to harvest in deeper soil (20–100 cm depth), ranging from a slight average decrease (−5%) in studies that reported C pools to a large average increase (+20%) in studies that reported only C concentration. Several recent reviews have highlighted the need for greater sampling of deep soil [26–28], especially as the shifting paradigm of SOM research has come to reject the assumption that deep soil C cannot not change on timescales relevant to anthropogenic C emissions [29–31]. Resolving the response of deep soil horizons to harvesting is important because these horizons occupy a much greater volume than surface O and A horizons. Even small changes in subsurface C can exacerbate or compensate for changes in surface soil C, and neither the magnitude nor direction of subsoil C change is clear from previous research.

The process of meta-analysis is necessarily cumulative, with each iteration updating previous analyses to further constrain the error in effect size estimates and to extend the scope of analysis. Thus, the objective of our meta-analysis is to update and extend the findings of Nave et al. [9] with respect to five major research questions:

(1) What is the overall effect of forest harvesting on soil C pools?
(2) How does the effect of forest harvest on soil C change with soil depth?
(3) To what extent does the effect of harvesting differ among soil orders?
(4) Do site pretreatment strategies or increasing harvesting intensity (i.e., whole tree harvest) moderate or accentuate harvesting impacts on soil C?
(5) How long does soil C take to recover from harvest across different soil types?

2. Materials and Methods

Meta-analysis is a cumulative activity which builds upon previous research and meta-analyses on similar research questions. Our meta-analysis builds upon the work of Nave et al. [9] and Johnson and Curtis [25] by updating their results with studies published between 2008 and 2016. The database published by Nave et al. [9] was independently recreated from each of 75 references. Metadata for each study was verified, and additional metadata such as the sampling depth of each response ratio was gathered. A total of 8 effect sizes differed in our dataset from the Nave et al. [9] database, all of which were either additional data for mineral soils or a split of one effect size into two based upon sampling depth.

To add to this database with studies published between 2008 and 2016, we searched the peer-reviewed literature for relevant studies using the online database Web of Science with combinations of the terms: forest, timber, harvest, logging, soil C, soil organic matter, and management. No climate criteria was used to screen studies. To be included in the meta-analysis, publications had to report both a control as well as harvested treatments. Both pretreatment soil C and unharvested reference stands were considered acceptable controls, and measurements of reference stands were considered the superior control. For forest chronosequence studies, soil C data from the oldest stand was used as the control. A minimum stand age of 30 years was considered acceptable for control stands, although most studies used controls of considerably greater age. Nave et al. [9] found that studies reporting only soil C concentration data yielded different conclusions about the direction of harvest effects than those studies reporting soil C pool data. Consequently, soil C pool data was used in our meta-analysis when both concentration and pool data were available.

We collected potentially useful predictor variables of soil C response from each publication, including soil order, geographic region, and time since harvest (Table 1). Binning of continuous predictor variables (such as precipitation) was carried out in the same intervals as Nave et al. [9] for ease of comparison. Each study was categorized by harvest, residue management, and site preparation strategies. Harvesting technique was categorized as sawlog when only the merchantable bole (stem) was removed from the site or whole tree harvest (WTH) when the tops, limbs, and foliage were removed in addition to the bole. To test the response of soil C at different depths, data from each study was separated into one of five groups: O horizon, top soil (0–15 cm), mid soil (15–30 cm), deep soil (30–60 cm), and very deep soil (60–100+ cm). A sixth group called whole soil was assigned to studies that aggregated mineral soil samples instead of reporting results at separate depths. Several studies aggregated soil data from 0–100 cm, which reduced the number of unique deep and very deep soil observations even though these depths were separately sampled.

Table 1. Factors gathered as potential predictor variables in this meta-analysis.

Factor	Levels	
Reporting units	Pool (Mg·ha^{-1}), concentration (% or mg·g^{-1})	
Soil Depth	O horizon	Forest Floor
	Top Soil	0–15 cm
	Mid Soil	15–30 cm
	Deep Soil	30–60 cm
	Very Deep	60–100+ cm
Overstory species	Hardwood, conifer/mixed	
Soil order	Alfisol, Andisol, Entisol, Inceptisol, Mollisol, Spodosol, Oxisol, Ultisol	
Geographic group	NE North America, NW North America, SE North America, SW North America, Europe, Asia, Pacific (Australia, New Zealand)	
Harvest type	Clearcut, thin	
Harvest intensity	Stem only, whole tree	
Residue management	Removed, spread	
Site preparation	Broadcast burn, tillage/scarification	
Soil texture	Fine (mostly silt or clay), coarse (mostly sand), organic	
Time since harvest	Continuous	
Mean Annual Temperature	0–5, 5–7.5, 7.6–10, 10.1–15, 15.1–20, >20 (°C)	
Mean Annual Precipitation	<500, 500−750, 751−1000, 1001−1400, 1401−1800, >1800 (mm)	

Our meta-analysis estimates the magnitude of change in soil C using the ln-transformed response ratio R, which is defined as

$$\ln (R) = \ln \left(\frac{\overline{X_T}}{\overline{X_C}} \right) \tag{1}$$

where $\overline{X_T}$ is the mean soil C value of treatment (harvested) observations, and $\overline{X_C}$ is the mean soil C value of control observations for a given set of experimental conditions at a specific site and depth. Multiple response ratios were recorded for each publication, with the number of response ratios (k) depending upon the number of experimental conditions imposed and the number of samples taken by depth. For example, a publication that reports the results of two thinning treatments and two clear-cut treatments at three depth increments (forest floor, top soil, and mid soil) versus a control would yield 12 response ratios. R is a unit-less measure of effect size, which allows comparison among studies that report data in different units [24]. By back transforming $\ln(R)$, $[(e^{(\ln(R))} - 1) \times 100]$, mean response ratios can be interpreted as the percentage change in soil C relative to the control. Estimates of the standard deviation and sample size for each $\overline{X^T}$ and $\overline{X^C}$ were not available in several publications. Consequently, an accurate estimate of total heterogeneity (Q_T) for the dataset was not possible. Subsequent partitioning of Q_T into within- and among-group heterogeneity (Q_W and Q_A, respectively) for random and mixed effect models (as is customary for meta-analyses) was not possible [24]. Instead, we used nonparametric resampling techniques (bootstrapping) to estimate confidence intervals around mean effect sizes in an unweighted meta-analysis [9]. Adams et al. [32] recommend bootstrapping confidence intervals for ecological meta-analyses, and show that confidence bounds based on this method are more conservative than standard meta-analyses. Bootstrapping was implemented using the bootES package [33] in R [34]. For all statistical tests in our analysis, $\alpha = 0.05$.

Although not exhaustive, the database we compiled from the literature search contained 945 soil C response ratios from 112 publications published between 1979 and 2016. Roughly half the dataset was comprised of response ratios analyzed by Nave et al. [9]. The full dataset is available as Supplementary Material, including maximum sampling depth and the number of response ratios from each paper (Appendix A).

3. Results

3.1. Overall Effect and Change with Depth

Across all studies, harvesting led to a significant average decrease in soil C of 11.2% relative to control (Figure 1). Whether the response to harvest was reported as pools or concentrations had a large impact on the estimated effect of harvest on soil C, with mean response for studies reporting C concentration units (%, mg·g^{-1}, etc.) 16.2% higher (with a 95% CI [20.9%, 11.8%]) than studies reporting C pool units (Mg·ha^{-1}, tons·ha^{-1}, etc.). Concentration responses are higher than pool responses at all soil depths, except for very deep and whole mineral soil, which did not have enough concentration response ratios to construct separate confidence intervals (Figure 1). Consequently, all subsequent analyses focused on the subset of data reporting soil C pools.

Several different soil layers show significant losses of C due to harvesting. Overall, O horizons lost 30.2% of their carbon as a result of harvesting. Losses from top soil were much smaller, although the estimated loss when reported in pool units was significant (-3.3%). In mid (15–30 cm) and deep soil (30–60 cm), the average loss of soil C was greater than topsoil, although the smaller number of response ratios for these depths resulted in more poorly constrained estimates. Studies only reporting C concentration observed a 14.5% increase in deep soil (30–60 cm), although the sample size was relatively small. The overall effect in very deep soil (60–100+ cm) was significant, with an average loss of 17.7%. Unfortunately, this region of the profile was not frequently sampled (21 response ratios out of 945 total), and consequently the 95% confidence interval is quite wide.

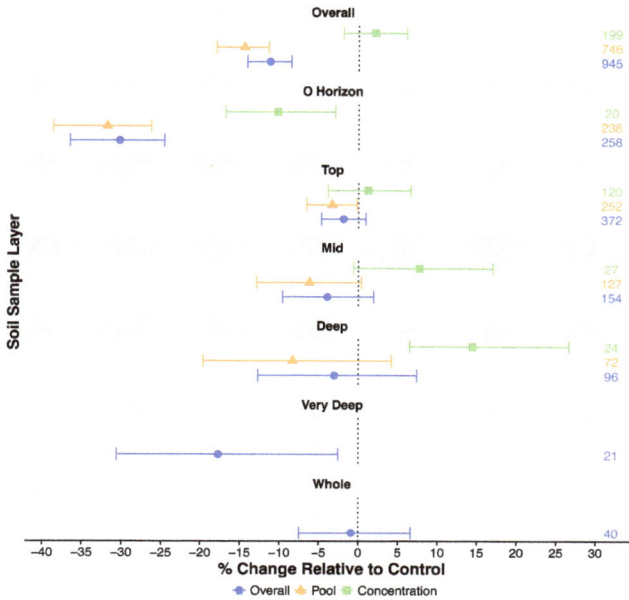

Figure 1. Response of soil C to forest harvesting, overall and faceted by soil depth. All points are back-transformed mean effect sizes ± 95% confidence intervals calculated by nonparametric bootstrap. The number of response ratios (k) that make up each mean effect is listed on the right. Mean effects with confidence intervals overlapping the dashed line (0%) show no significant change in soil C due to harvesting. Within each facet, mean effect sizes are shown for the overall effect as well as separately for studies reporting C pool units or concentration units.

3.2. Effect of Harvesting across Soil Orders

The effect of harvesting on soil C differs between soil orders (Figure 2). For the Alfisols and Inceptisols, there are significant losses in O horizon C pools (−12.0% and −45.4%, respectively), but no significant loss in the mineral soil. Mollisols lost an average of 17.7%, although neither O horizon nor mineral soil responses were significantly different from 0. In several cases, small samples sizes made separate testing of organic and mineral soil impossible within a single order (Andisols, Entisols, Oxisols). However, in each of these cases the overall effect was significant. Soil C increased by 24.5% on average in Andisols, but decreased by 18.8% in Entisols and 30.9% in Oxisols. The number of response ratios was more concentrated in the Alfisol, Inceptisol, Spodosol, and Ultisol orders, although a large number of publications did not report information on soil classification. The response to harvesting in Spodosols is substantial (−19.0% overall), with significant losses in both the O horizon (−36.4%) and moderately less in the mineral soil (−9.1%). Likewise, Ultisols lost significant soil C in response to harvesting (−24.7% overall), with the most substantial losses occurring in the O horizon (−66.0%) rather than in the mineral soil (−11.9%).

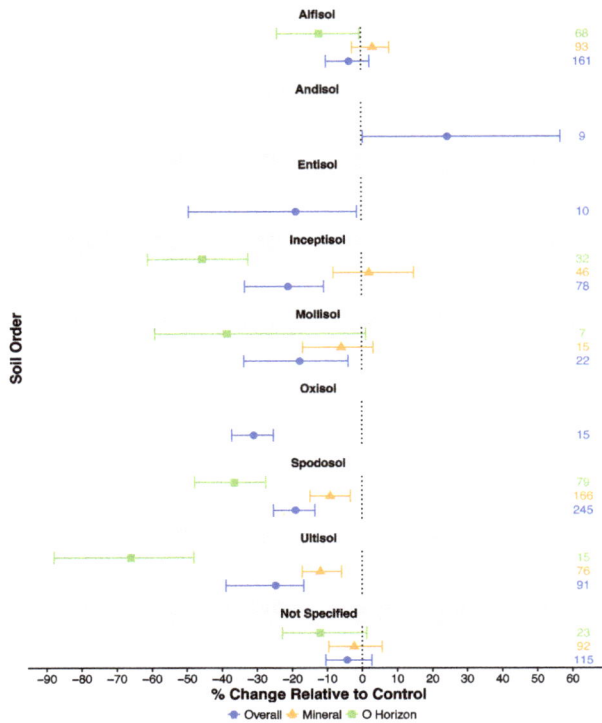

Figure 2. Response of soil C to harvesting in different soil orders. Mean effect sizes ± 95% confidence intervals calculated by nonparametric bootstrap are shown for all response ratios in each soil order (Overall) and broken out into mineral soil or O horizon. The number of response ratios (*k*) comprising each mean effect are listed on the right. Effect sizes were calculated only on response ratios reported in pool units (*k* = 746).

3.3. Differences in Response to Harvest between Forest Types

The response of soil C to harvest differs between hardwood and coniferous/mixed forest types (Figure 3). The decline in O horizon C pools is significantly greater in conifer/mixed forests (−38.1%) compared to hardwood forests (−25.4%). Differences between forest types were not significant for any mineral soil layer. However, the decline in soil C after harvest was significant for hardwood forests but not conifer/mixed forests in deep soil (30–60 cm) and in studies reporting whole mineral soil C pools. Also in these studies, the difference between hardwood and conifer/mixed forest response to harvest is marginally significant ($p < 0.1$). The number of observations are highly concentrated in O horizon and top soil, consequently limiting the precision of mean effect size estimates in deeper layers. No observations for hardwood forest were made in very deep soil (60–100+ cm).

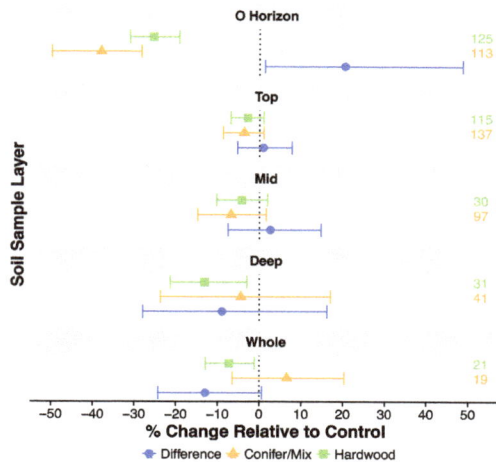

Figure 3. Response of soil C harvesting at different depths in soil, broken down by hardwood or conifer/mixed forest types. Mean effect sizes ± 95% confidence intervals calculated by nonparametric bootstrap are shown for hardwood and conifer/mixed forests. Blue circles show the mean difference between these forest types (Hardwood–Conifer/Mix) ± 95% confidence interval for the difference. Differences are calculated on the logarithmic effect size scale, and then back-transformed to % change, and thus do not necessarily add up on the % change scale. The number of response ratios (*k*) in each forest type at each depth is listed on the right. Data for very deep soil is not shown because there were no observations for this soil layer in hardwood forests.

3.4. Harvest Intensity, Residue Management and Site Pretreatment

Differences in forest management strategies can significantly impact the response of soil C to harvesting (Figure 4). While there was no significant overall difference observed between thinning and clear-cut harvesting, less C was lost from mineral soils under clear-cut harvesting compared to thinning (+9.3%). Likewise, harvest intensity significantly changed the response of mineral soil C, with soils undergoing whole tree harvesting losing 13.3% less C than bole-only harvesting. Possible mechanisms for these counter-intuitive results are considered in Discussion Section 4.5.

The practice of broadcast burning sites in preparation for planting after a harvest leads to significant additional losses of soil C, with burned soils losing 15.2% more C than soils with no pretreatment. This effect is especially severe in the O horizon (40.9% additional loss than if sites were not burned), and somewhat curtailed in the mineral soil (8.3% additional loss). The wide 95% CI for the estimate of differences in O horizon responses due to burning reflects disparities in burn severity and treatment implementation among different studies.

Spreading of residual materials across harvested sites (by chipping tops and limbs or other methods) resulted in significant additional loss of soil C (−10.9%), with these extra losses occurring mostly in the mineral soil (−17.5%). On the other hand, residue removal resulted in no significant additional losses to soil C.

Tillage is sometimes used to prepare soils for planting after harvest, either to create raised planting beds or to prepare the soil seed bed. This intensive style of site preparation did not result in significant losses in soil C, especially in the mineral soil. However, very large losses were reported in the O horizon (mean effect = −37.1%) with a very wide confidence interval due to a small number of observations. Additional study of the effect of tillage would help to reduce this error.

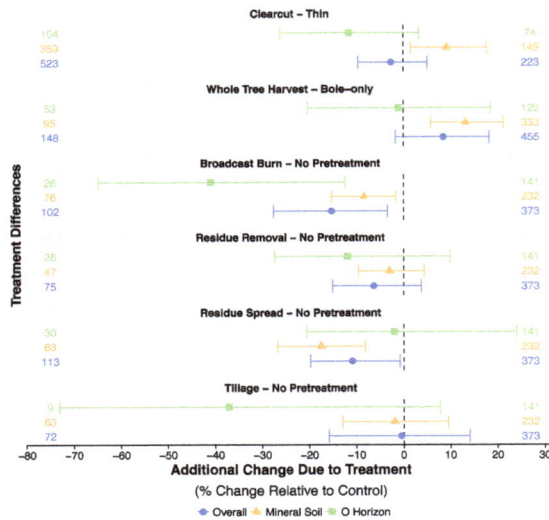

Figure 4. Differences in response of soil C to harvesting between treatment strategies. Differences are calculated by subtracting [more intensive treatment] − [less intensive treatment], such that positive differences represent reduced loss of C due to more intensive treatment, and negative differences represent increased losses of C due to more intensive treatment. Point estimates are back-transformed differences between mean effect sizes ± 95% confidence intervals calculated by nonparametric bootstrap. Mean effect differences with confidence intervals overlapping the dashed line (0%) show no significant difference between the harvesting, residual management, or pretreatment strategies. The number of response ratios (k) for the intensive treatment in each comparison appear on the left and for the less intensive treatment on the right.

3.5. Recovery of Soil C after Harvest

The recovery time for soil C following harvest differs among soil orders (Figure 5). Only 4 soil orders contained enough observations over time to model recovery times: Alfisols, Inceptisols, Spodosols, and Ultisols. We modeled time as a second degree polynomial (Time + Time^2) separately for O horizons and mineral soils for each soil order (Table 2).

Table 2. Linear regression coefficients and significance for second degree polynomial model of response of soil C to harvesting over time.

Coefficient	Estimate	SE	t-Value	p-Value
Intercept (Alfisol, mineral soil)	12.702	3.587	3.541	0.0004
O horizon	−21.475	3.766	−5.703	<0.0001
Inceptisol	−10.876	5.717	−1.902	0.0577
Spodosol	−14.717	4.320	−3.407	0.0007
Ultisol	−24.776	5.391	−4.596	<0.0001
Time	−67.834	41.56	−1.632	0.10325
Time^2	120.412	40.361	2.983	0.0030
Residual SE: 40.24 on 533 *DF*				
F-Statistic: 10.74 on 6 and 533 *df*, $p < 0.0001$			$R^2 = 0.108$	Adj. $R^2 = 0.098$

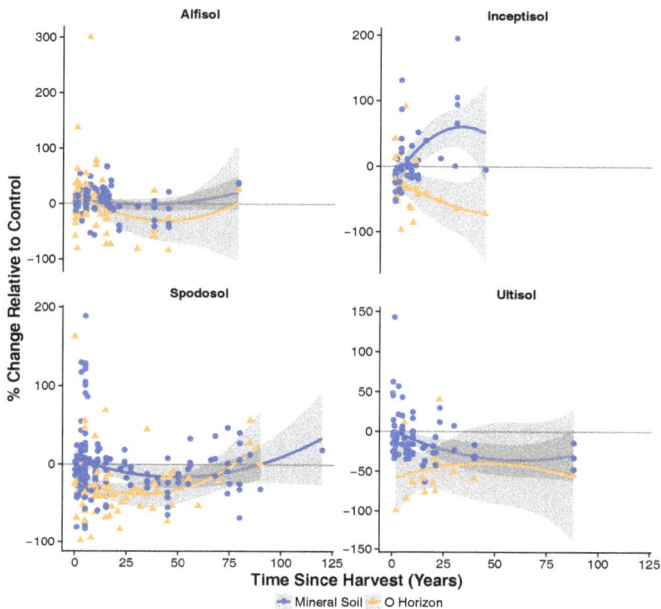

Figure 5. Temporal patterns in both O horizon (yellow triangle) and mineral soil (blue circle) C pools for Alfisol, Inceptisol, Spodosol, and Ultisol orders. Other soil orders are not shown due to an inadequate number of response ratios over time. Regression lines show trends with time using a second order polynomial. For the overall model, $F = 9.205$ on 7 and 532 degrees of freedom, Adj. $R^2 = 0.096$, and $p < 0.0001$ (Table 2).

4. Discussion

4.1. Overall Effect of Harvesting on Soil C

Our results reveal that across many publications in the literature there is a significant loss of soil C in response to harvest (−11.2% overall, −14.4% for studies reporting C pools). This estimate is slightly greater than that found by Nave et al. [9], who reported −8% change relative to control. The difference between these estimates derives from additional losses reported in mineral soil, since the effect of harvesting on O horizon C is identical between this study and Nave et al. [9] (−30%). Indeed, while no significant loss of soil C due to harvesting was reported in previous meta-analyses on the subject [9,25], this analysis reveals significant if small losses in various mineral soil layers. Our meta-analysis has roughly double the number of responses than previous meta-analyses on the subject, and consequently has greater statistical power. In particular, this has allowed us to break down the response of mineral soil C to harvest into more depth increments to better characterize how response is moderated or accentuated by depth.

4.2. Depth Distribution of Soil C Response to Harvest

The response of soil C to harvest differs among depths in the soil. O horizons show the most substantial declines (by percentage), although the O horizon is typically a smaller pool of C than mineral soil horizons. Consequently, smaller absolute declines in O horizon C pools can lead to larger response ratios. Forest type significantly alters the response of O horizons to harvesting, with hardwood forests undergoing less drastic losses than conifer and mixed forests (Figure 3). This result is in contrast with Nave et al. [9], who found that conifer O horizon soil C declines significantly less than hardwood forest floors. Coniferous forest litter is thought to be more chemically recalcitrant to decomposition because of higher C/N and lignin/N, as well as slower N mineralization rates [35,36]. The trend for more soil C loss from coniferous forest floors could be due to differences in the harvesting techniques utilized for each forest type. On the other hand, less change in soil C in coniferous forests in deeper mineral soils could suggest that some of the additional loss in O horizon C pools is the result of translocation of C into mineral soil rather than mineralization to CO_2. Whatever the case, the mechanism for this difference is not clear and warrants additional study.

In mineral soils, the relative response to harvest is typically less than the O horizon, but this small relative loss might correspond to a larger absolute loss of C in the mineral soil in many forests. The major exception to this pattern are Spodosols, which can contain larger proportions of total soil C in deep, acidic O horizons. Declines in top soil C pools were modest (-3.3%) but still significant (Figure 1). Mean effect size estimates become more negative with soil depth, although these estimates are not significant. The overall estimate of change in very deep soil (60–100+ cm) shows substantial and significant loss of C (-17.7%). This estimate, however, only covers a small number of observations (21) from Spodosol, Ultisol, Alfisol, and Inceptisol soil orders and completely excludes hardwood forests. The lack of observations in deeper soil horizons leads to very wide confidence intervals.

On average, the maximum depth of soil sampled by the publications in this meta-analysis was 35.9 cm (Figure 6). The average depth of sampling for each response ratio in the database is even more surface-skewed at 21.3 cm. Many of the observations down to 100 cm in the literature only report treatment differences for the whole mineral soil profile (0–100 cm), which eliminates any possibility of understanding the relative response of different horizons or depths. The scarcity of observations in deep soil is incongruous with the increasing loss of soil C with depth relative to control observed in this analysis. More important than the magnitude or significance of the harvest response in very deep soil is the conclusion that much greater attention should be paid to deep soil C pools in both individual forest manipulation experiments and broad-scale C inventory.

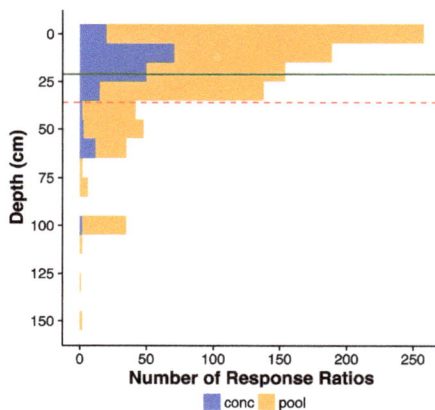

Figure 6. Number of response ratios plotted by the maximum depth of sampling for each observation. Response ratios calculated from concentration are in blue, and from pools in orange. The average depth of all response ratios is denoted by the solid green line (21.3 cm, *n* = 945). The average maximum sampling depth for all 112 publications in the meta-analysis is denoted by the dashed red line (35.9 cm).

While soil C in deep soil is much less concentrated than in O and A horizons, subsurface soil represents a much greater volume of soil than surface soil, especially in older/more well developed soil orders like Alfisols, Ultisols, and Oxisols. Some major regions for forestry contain substantial portions of total soil C in deep horizons. For example, 38% of total soil C was below 50 cm and 24.1% below 1 m in production forest soils in the Pacific Northwest [37]. The imprint of biological activity extends many meters into soil, even into the C horizon [38]. Globally, the average maximum rooting depth for trees is ~7 m [39], far outreaching even the deepest observations in this database. Harvesting disrupts the continued growth and turnover of roots extended deep into soil by mature trees, which in turn disturbs the steady state of C cycling in deep soil by changing environmental conditions (temperature, moisture) as well as the type and rate of C inputs. Furthermore, the flush of nitrate and dissolved organic matter that frequently follows harvest [40,41] could prime the breakdown of older, subsurface C by providing a spike in nutrient availability and labile energy sources [31,42,43]. Alteration of aboveground ecosystems can cause changes in subsurface soils. For example, Mobley et al. [44] observed that, over a period of several decades following afforestation of agricultural land, modest C gains in surface soil were more than offset by large losses in soil C below 30 cm. Neither the response of deep soil C to harvest nor the mechanisms for that change have been sufficiently resolved in the literature, and future work to address these questions are necessary.

4.3. Differences in Soil C Response to Harvest among Soil Orders

Substantial variation in response to harvest was observed among soil orders. Several soil orders had very few response ratios (Andisols, Entisols, Mollisols, and Oxisols), which greatly widens confidence intervals. Nonetheless, significant changes in soil C in response to harvest were observed for all four of these orders. Andisols were the only order to show a significant average increase in soil C in response to harvest. This likely stems from Andisols particular mineralogy, which is often characterized by short-range-order minerals like allophane and imogolite [45–47]. The capacity for these types of minerals to adsorb organic matter makes Andisol soil C especially resistant to loss after harvesting. Alfisols also appear to be resistant to loss of soil C after harvesting, with relatively small loss in O horizons (−12.0%) being the only significant effect. All other soil orders have significant overall losses in soil C, roughly −20% for Entisols, Inceptisols, Mollisols, Spodosols, and Ultisols. The uneven distribution of observations among soil orders (most response ratios in the database are from Alfisols, Inceptisols, Spodosols, and Ultisols) results in substantial differences in the size of confidence intervals among different orders. Unfortunately, many studies did not report soil taxonomic information, and thus 115 response ratios could not be assigned to a soil order. The lack of studies on Andisols is curious given the importance of these soils to forestry in several regions such as the Pacific Northwest, USA and New Zealand. Several studies on Andisol and other under-represented soil orders were excluded from this analysis because of a lack of appropriate controls.

4.4. Recovery of Soil C after Harvest

Recovery of soil C after harvesting can take several decades [9]. O horizon pools decline more severely than mineral soil pools, especially in the first several decades (Figure 5). In Spodosols, O horizons recovered from harvesting after 60–85 years, while mineral soil recovered over a longer period of 75–100+ years. While the response to harvest was less severe in mineral soils, the longer recovery period implies either lagged response time between forest floor and mineral soils or differences in the decay rate constants leading to longer-term changes in mineral soil C compared to the forest floor. In the case of Alfisols and Inceptisols, soil C in mineral pools increased or stayed the same after harvest, while O horizons declined. However, the observations of harvest effects on Alfisols, Inceptisols, and Ultisols were largely confined to within the first 50 years post-harvest. Consequently, an estimate for the recovery period of soil C pools in these soil orders cannot be assessed with much confidence. Continued observation of existing harvesting experiments in other soil orders must be made to better characterize changes in soil C over time. For Andisols, Entisols, Mollisols, and Oxisols, only a few time points have been documented, and much further study will be necessary to understand recovery of soil C after harvest.

The modeled recovery time has a fairly low adjusted R^2 (0.1) and thus a low predictive capacity. Substantial variation in the response to harvest exists within each soil order, reflecting differences in tree species, harvest intensity, and pretreatment strategies, among other factors. Moreover, soil orders are hardly homogeneous, and differences in the response of soil C among lower levels of classification within each order could be as important as order-level differences. Nonetheless, the substantial and significant differences between orders considered in the model suggest that both the resistance of soil C to change and the recovery period of soil C following harvest (resilience) consistently varies among soil types. Compared with 20-year recovery periods assumed by many models [14], our results indicate that soil C recovery takes place over at least triple that time frame for both O horizons and mineral soil in many cases.

While forests >30 years of age were considered acceptable controls for this analysis, the preponderance of data in this meta-analysis show decreases in soil C relative to control at time = 30 years. Consequently, studies that use mature second growth stands barely over this threshold for experimental controls likely underestimate the response of soil C due to harvesting treatments. Depending upon the site conditions and soil order, control stands of at least 50–75+ years since harvest would be recommended, with older stands being more accurate controls.

4.5. The Effect of Harvest Strategies on Soil C

Differences in harvesting and soil pretreatment strategies significantly impact the loss of soil C after harvest. Curiously, despite the greater relative losses of soil C in O horizons, significant differences between harvest intensities and pretreatment strategies were only found in the mineral soil with the exception of broadcast burning (Figure 4). The reduced loss of soil C from mineral soil observed in treatments with greater harvest intensity (+9.3% for clearcut, +13.3% for whole tree harvest) runs counter to the intended effect of these experimental treatments on soil C. One possibility is that increased harvest intensity reduces the quantity of dissolved organic matter and inorganic nutrients leached into the mineral soil, thus reducing the priming [42,48,49] of mineral soil C mineralization through less addition of energy-rich substrates and nutrients. Another possibility is that response of soil C to increased harvest intensity is soil-type specific, and thus an aggregate analysis such as this is subject to bias by unequal sampling of different soil orders. Whatever the case, this dataset cannot identify the specific mechanism(s) driving this difference, and further study is warranted.

Tilling of forest soils prior to planting should intuitively disrupt O horizons to a greater extent than less intensive practices. However, due to the very small number of observations of this practice in the dataset, the large mean treatment effect on soil C was could not be differentiated from 0. By mixing organic material into the surface mineral soil, tilling could increase top soil C in the short term and possibly prime additional breakdown of C over time. In regions where this practice is used, additional research could help to reveal the mechanisms driving change in the soil C of O horizons and mineral following tillage.

Broadcast burning led consistently to additional loss of soil C in both O horizons and mineral soil. The large additional reduction in O horizon C (-40.9%) is expected given that such a treatment is intended to reduce slash on site to facilitate planting. The loss of carbon after harvest extends into deep soil, especially following slash burning (Figure 7). Although there are few observations in very deep soil (60–100+ cm), burning appears to especially exacerbate C losses in this layer. This result is despite the direct effects of fire (such as soil heating and nutrient volatilization) being highly attenuated with depth [50,51]. Levels of mineralized nitrogen (NH_4^+ and NO_3^-) and soluble sugars spike within the first year following fire, leading to increased microbial biomass N and N leaching loss [52]. Thus, the flush of nutrients and organic matter into deeper mineral soil following post-harvest broadcast burning has the potential to impact soil C dynamics throughout the soil profile. The number of observations in deep and very deep layers is small, and consequently additional research is necessary to better differentiate between harvesting and fire effects in deep soil horizons.

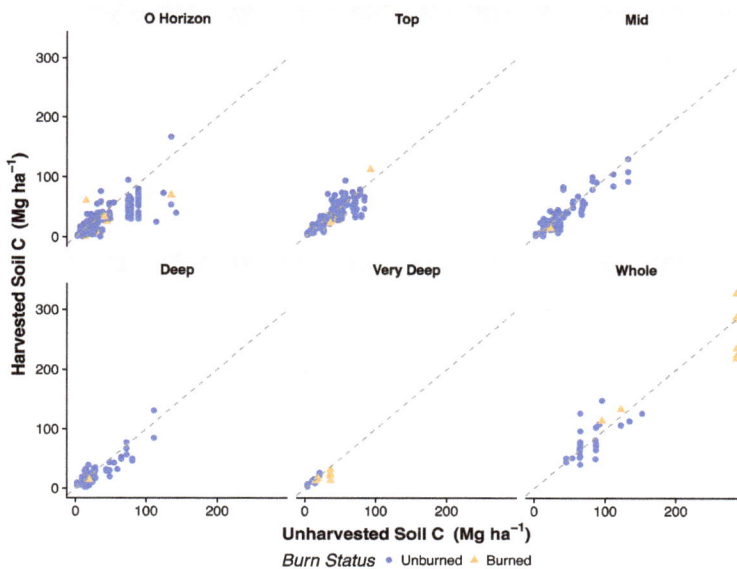

Figure 7. Absolute change in soil C due to harvest for each soil depth in this analysis (O horizon, top, mid, deep, very deep, and whole mineral soil). Different points show burned (yellow triangle) and unburned (blue circle) pretreatment strategies. Dashed 1:1 lines in each facet represent no response due to harvest. The total number of responses shown is $k = 746$.

5. Conclusions

We analyzed 945 studies from 112 publications to examine the effect of harvest on forest soil C around the globe. There is a significant overall reduction in forest soil C following harvest that occurs in both the O horizon and mineral soil. Significant variation in the response to harvesting was observed among different soil depths, among soil orders, between overstory forest types, and between different harvest intensities and pretreatment strategies. Broadcast burning, in particular, appears to exacerbate loss of soil C in both organic and mineral horizons following harvest. The recovery period of soil C following harvest depends upon soil type and takes at least 60 years in many production forests. One of the most important findings of this analysis is a significant loss (-17.7%) of soil C following harvest in very deep soil (60–100+ cm). Deep layers of the soil are greatly under-represented in the literature, and consequently, there is great uncertainty around this estimate. Examination of deep soil horizons in existing manipulative forest studies, in new studies, and in C inventory should be a clear objective for future research.

Supplementary Materials: The following are available online at www.mdpi.com/1999-4907/7/12/308/S1, Table S1: Harvest meta-analysis database (Excel file).

Acknowledgments: A huge thank you to Bryan James for assistance extracting data from publications and assembling the database. Thanks also to Lucas Nave and an anonymous reviewer for constructive recommendations on the manuscript. This manuscript was supported by funding from the University of Washington Stand Management Cooperative as well as a US Department of Agriculture McIntire-Stennis Grant.

Author Contributions: J.J. and R.H. conceived and designed the research questions for the meta-analysis; J.J. conducted the literature search; J.J. defined the inclusion criteria for the meta-analysis and extracted data from the publications; J.J. analyzed the data; and J.J. wrote the paper in consultation with R.H.

Conflicts of Interest: The authors declare no conflict of interest. The founding sponsors had no role in the design of the study; in the collection, analyses, or interpretation of data; in the writing of the manuscript, and in the decision to publish the results.

Appendix A. Publications Providing Response Ratios for This Analysis

Reference	Year	*k*	Max Depth (cm)	Time Since Harvest [a] (years)	Location
Alban and Perala [53]	1992	7	50	35	MN, USA
Bauhus et al. [54]	2004	6	40	9	Germany
Bisbing et al. [55]	2010	6	100	40	MT, USA
Black and Harden [56]	1995	15	20	23	CA, USA
Boerner et al. [57]	2006	4	10	2	SC, USA
Borchers and Perry [58]	1992	4	15	14	OR, USA
Bravo-Oviedo et al. [59]	2015	8	30	15	Spain
Cade-Menun et al. [60]	2000	12	26	5	BC, Canada
Carter et al. [61]	2002	8	15	2	LA, TX, USA
Chatterjee et al. [62]	2009	19	54	21	WY, USA
Chen et al. [63]	2013	24	100	29	China
Chiti et al. [64]	2016	24	100	24	Ghana, Cameroon, Gabon
Christophel et al. [65]	2013	6	30	15	Germany
Christophel et al. [66]	2015	18	30	33	Germany
Cromack et al. [67]	1999	1	100	10	OR, USA
Dai et al. [68]	2001	3	70	14	NH, USA
DeByle et al. [69]	1980	10	5	3	WY, USA
Deluca and Zouhar [52]	2000	6	8	5	MT, USA
Diochon et al. [70]	2009	28	50	35	NS, Canada
Edmonds and McColl [71]	1989	4	20	3	Australia
Edwards and Ross-Todd [72]	1983	6	45	1	TN, USA
Elliott and Knoepp [73]	2005	3	15	3	NC, USA
Ellis et al. [73]	1982	4	10	2	Tasmania
Ellis and Graley [74]	1983	2	10	1	Tasmania
Esquilin et al. [75]	2008	1	10	14	CO, USA
Falsone et al. [76]	2012	3	130	5	Russia
Fraterrigo et al. [77]	2005	1	15	30	NC, USA
Frazer et al. [78]	1990	4	14	12	CA, USA
Gartzia-Bengoetxea et al. [79]	2009	2	5	10	Spain
Gillon et al. [80]	1999	2	0	1	France
Goh and Phillips [81]	1991	4	60	2	New Zealand
Goodale and Aber [82]	2001	2	10	85	NH, USA
Gough et al. [83]	2007	15	80	41	MI, USA
Grady and Hart [84]	2006	2	15	12	AZ, USA
Grand and Lavkulich [85]	2012	6	80		BC, Canada
Gresham [86]	2002	6	30	10	SC, USA
Griffiths and Swanson [87]	2001	3	10	20	OR, USA
Gundale et al. [88]	2005	4	10	3	MT, USA
Gupta and DeLuca [89]	2012	12	50	5	Wales
Hart et al. [90]	2006	2	15	1	AZ, USA
Hendrickson and Chattarpaul [91]	1989	6	20	3	ON, Canada
Herman et al. [92]	2003	2	9	8	CA, USA
Holscher et al. [93]	2001	2	20	22	Germany
Hwang and Son [94]	2006	2	30	2	Korea
Jang and Page-Dumroese [95]	2015	8	30	38	MT, USA
Johnson [96]	1991	3	20	3	NH, USA
Johnson and Todd [97]	1998	6	45	15	TN, USA
Johnson [98]	1995	12		7	NH, USA
Johnson et al. [99]	1997	14	53	6	NH, USA
Johnson et al. [100]	2014	4	60	1	CA, USA
Jones et al. [101]	2011	12	30	15	New Zealand
Kaye and Hart [102]	1998	2	15	1	AZ, USA
Keenan et al. [103]	1994	1	20	4	BC, Canada
Kelliher et al. [104]	2004	4	50	22	OR, USA

Reference	Year	k	Max Depth (cm)	Time Since Harvest [a] (years)	Location
Kishchuk et al. [105]	2014	4	7	6	AB, Canada
Klockow et al. [106]	2013	9	20	1	MN, USA
Klopatek [107]	2002	6	20	30	WA, USA
Knoepp and Swank [108]	1997	4	30	33	NC, USA
Korb et al. [109]	2004	1	10	1	AZ, USA
Kraemer and Hermann [110]	1979	2	10	26	WA, USA
Kurth et al. [111]	2014	72	30	8	MI, MN, USA
Laiho et al. [112]	2003	5	22	5	NC, LA, USA
Latty et al. [113]	2004	2	15	90	NY, USA
Law et al. [114]	2001	3	100	21	OR, USA
Law et al. [115]	2003	9	100	62	OR, USA
Leduc and Rothstein [116]	2007	1	10	5	MI, USA
Maassen and Wirth [117]	2004	2	5		Germany
Mattson and Smith [118]	1993	30	10	11	WV, USA
Mattson and Swank [119]	1989	8	60	5	NC, USA
May and Attiwill [120]	2003	2	10	5	Australia
McLaughlin and Phillips [121]	2006	2	50	17	ME, USA
McKee et al. [122]	2013	8	60	24	AL, USA
McLaughlin [123]	1996	10	50	5	MI, USA
Merino and Edeso [124]	1999	6	15	1	Spain
Moreno-Fernandez et al. [125]	2015	54	50	60	Spain
Mu et al. [126]	2013	18	50	5	China
Murphy et al. [127]	2006	20	60	1	CA, USA
Neher et al. [128]	2003	3	20	2	NC, USA
Norris et al. [129]	2009	15	100	16	SK, Canada
O'Brien et al. [130]	2003	6	50	18	Australia
Powers et al. [131]	2011	20	30	13	MN, WI, USA
Prest et al. [132]	2014	5	50	35	NS, Canada
Prietzel et al. [133]	2004	4	0	1	WA, USA
Puhlick et al. [134]	2016	10	100		ME, USA
Rab [135]	1996	8	10	1	Australia
Riley and Jones [136]	2003	3	10	1	SC, USA
Roaldson et al. [137]	2014	16	20	5	CA, USA
Rothstein and Spaulding [138]	2010	6	30		MI, USA
Sanchez et al. [139]	2007	6	105	2	SC, USA
Sanscrainte et al. [140]	2003	4	70		WA, USA
Saynes et al. [141]	2012	8	5	11	Mexico
Selig et al. [142]	2008	3	30	14	VA, USA
Shelburne et al. [143]	2004	4	10	1	SC, USA
Sheng et al. [144]	2015	5	100	8	China
Skovsgaard et al. [145]	2006	12	30	0	Denmark
Slesak et al. [146]	2012	12	60	5	OR, WA, USA
Small and McCarthy [147]	2005	3	10	7	OH, USA
Stone et al. [148]	1999	1	15	1	AZ, USA
Stone and Elioff [149]	1998	4	30	5	MN, USA
Strong [150]	1997	8	40	18	MN, USA
Strukelj et al. [151]	2015	12	10	5	QC, Canada
Tang et al. [152]	2009	12	60	29	MI, WI, USA
Trettin et al. [153]	2011	6	150	11	MI, USA
Ussiri and Johnson [154]	2007	15	60	8	NH, USA
Vario et al. [155]	2014	6	60	49	NH, USA
Vesterdal et al. [156]	1995	9	0		Denmark
Waldrop et al. [157]	2003	3	0	1	CA, USA
Wu et al. [158]	2010	1	20	10	China
Xiang et al. [159]	2009	8	30	0	China
Yanai et al. [160]	2000	35	0	29	NH, USA
Zabowski et al. [161]	2008	2	20	25	OR, WA, USA
Zhong and Makeshin [162]	2003	2	10	16	Germany
Zummo and Friedland [163]	2011	15	60	3	NH, USA

[a] For chronosequence studies, time since harvest in this table is averaged across all response ratios for that study.

References

1. Dixon, R.K.; Brown, S.; Houghton, R.A.; Solomon, A.M.; Trexler, M.C.; Wisniewski, J. Carbon Pools and Flux of Global Forest Ecosystems. *Science* **1994**, *263*, 185–190. [CrossRef] [PubMed]
2. Prentice, I.C.; Farquhar, G.; Fasham, M.J.R.; Goulden, M.L.; Heimann, M.; Jarmillo, V.J.; Kheshgi, H.S.; Le Quere, C.; Scholes, R.J.; Wallace, D.W.R. *The Carbon Cycle and Atmospheric Carbon Dioxide*; Cambridge University Press: Cambridge, UK, 2001; pp. 183–237.
3. Batjes, N.H. Total carbon and nitrogen in the soils of the world. *Eur. J. Soil Sci.* **1996**, *47*, 151–163. [CrossRef]
4. Jobbagy, E.G.; Jackson, R.B. The vertical distribution of soil organic carbon and its relation to climate and vegetation. *Ecol. Appl.* **2000**, *10*, 423–436. [CrossRef]
5. Six, J.; Conant, R.T.; Paul, E.A.; Paustian, K. Stabilization mechanisms of soil organic matter: Implications for C-saturation of soils. *Plant Soil* **2002**, *241*, 155–176. [CrossRef]
6. Lal, R. Forest soils and carbon sequestration. *For. Ecol. Manag.* **2005**, *220*, 242–258. [CrossRef]
7. Schimel, D.S. Terrestrial ecosystems and the carbon cycle. *Glob. Chang. Biol.* **1995**, *1*, 77–91. [CrossRef]
8. Jandl, R.; Lindner, M.; Vesterdal, L.; Bauwens, B.; Baritz, R.; Hagedorn, F.; Johnson, D.W.; Minkkinen, K.; Byrne, K.A. How strongly can forest management influence soil carbon sequestration? *Geoderma* **2007**, *137*, 253–268. [CrossRef]
9. Nave, L.E.; Vance, E.D.; Swanston, C.W.; Curtis, P.S. Harvest impacts on soil carbon storage in temperate forests. *For. Ecol. Manag.* **2010**, *259*, 857–866. [CrossRef]
10. Schlesinger, W.H.; Bernhardt, E.S. *Biogeochemistry: An Analysis of Global Change*; Academic Press: Oxford, UK, 2013.
11. Brady, N.C.; Weil, R.R. *The Nature and Properties of Soils*; Macmillan Publ. Co.: New York, NY, USA, 2002; Volume 13, p. 960.
12. Krull, E.S.; Skjemstad, J.O.; Baldock, J.A. *Functions of Soil Organic Matter and the Effect on Soil Properties*; Final Rep. CSO00029; Grains Research and Development Corporation: Glen Osmond, South Australia, 2009; p. 128.
13. Six, J.; Paustian, K. Aggregate-associated soil organic matter as an ecosystem property and a measurement tool. *Soil Biol. Biochem.* **2014**, *68*, A4–A9. [CrossRef]
14. Buchholz, T.; Friedland, A.J.; Hornig, C.E.; Keeton, W.S.; Zanchi, G.; Nunery, J. Mineral soil carbon fluxes in forests and implications for carbon balance assessments. *Glob. Chang. Biol. Bioenergy* **2014**, *6*, 305–311. [CrossRef]
15. Ryan, M.G.; Harmon, M.E.; Birdsey, R.A.; Giardina, C.P.; Heath, L.S.; Houghton, R.A.; Jackson, R.B.; McKinley, D.C.; Morrison, J.F.; Murray, B.C.; et al. A synthesis of the science on forests and carbon for U.S. Forests. *Issues Ecol.* **2010**, *13*, 1–16.
16. McKinley, D.C.; Ryan, M.G.; Birdsey, R.A.; Giardina, C.P.; Harmon, M.E.; Heath, L.S.; Houghton, R.A.; Jackson, R.B.; Morrison, J.F.; Murray, B.C.; et al. A synthesis of current knowledge on forests and carbon storage in the United States. *Ecol. Appl.* **2011**, *21*, 1902–1924. [CrossRef] [PubMed]
17. Fahey, T.J.; Woodbury, P.B.; Battles, J.J.; Goodale, C.L.; Hamburg, S.P.; Ollinger, S.V.; Woodall, C.W. Forest carbon storage: Ecology, management, and policy. *Front. Ecol. Environ.* **2010**, *8*, 245–252. [CrossRef]
18. American Carbon Registry. *The ACR Forest Carbon Project Standard v 2.1*; Winrock International: Arlington, VA, USA, 2011.
19. Verified Carbon Standard. *Agriculture, Forestry and other Land Use (AFOLU) Requirements*; VCS Version 3 Requirements Doc., v3.4 2013; Verified Carbon Standard: Washington, DC, USA, 2013.
20. IPCC. 2006 IPCC Guidelines for national greenhouse gas inventories. In *Agriculture, Forestry and Other Land Use*; Institute for Global Environmental Strategies: Hayama, Japan, 2006; Volume 4, pp. 1–29.
21. Bechtold, W.A.; Patterson, P.L. The Enhanced Forest Inventory and Analysis Program—National Sampling Design and Estimation Procedures. In *USDA General Technical Report SRS-80*; U.S. Department of Agriculture, Forest Service, Southern Research Station: Asheville, NC, USA, 2005; p. 85.
22. Zanchi, G.; Pena, N.; Bird, N. Is woody bioenergy carbon neutral? A comparative assessment of emissions from consumption of woody bioenergy and fossil fuel. *GCB Bioenergy* **2012**, *4*, 761–772. [CrossRef]
23. Homann, P.S.; Bormann, B.T.; Boyle, J.R. Detecting treatment differences in soil carbon and nitrogen resulting from forest manipulations. *Soil Sci. Soc. Am. J.* **2001**, *65*, 463–469. [CrossRef]
24. Hedges, L.V.; Gurevitch, J.; Curtis, P.S. The meta-analysis of response ratios in experimental ecology. *Ecology* **1999**, *80*, 1150–1156. [CrossRef]
25. Johnson, D.W.; Curtis, P.S. Effects of forest management on soil C and N storage: Meta analysis. *For. Ecol. Manag.* **2001**, *140*, 227–238. [CrossRef]

26. Harrison, R.B.; Footen, P.W.; Strahm, B.D. Deep soil horizons: Contribution and importance to soil carbon pools and in assessing whole-ecosystem response to management and global change. *For. Sci.* **2011**, *57*, 67–76.

27. Schmidt, M.W.I.; Torn, M.S.; Abiven, S.; Dittmar, T.; Guggenberger, G.; Janssens, I.A.; Kleber, M.; Koegel-Knabner, I.; Lehmann, J.; Manning, D.A.C.; et al. Persistence of soil organic matter as an ecosystem property. *Nature* **2011**, *478*, 49–56. [CrossRef] [PubMed]

28. Deb Richter, D.; Yaalon, D.H. "The Changing Model of Soil" Revisited. *Soil Sci. Soc. Am. J.* **2012**, *76*, 76–778. [CrossRef]

29. Marin-Spiotta, E.; Gruley, K.E.; Crawford, J.; Atkinson, E.E.; Miesel, J.R.; Greene, S.; Cardona-Correa, C.; Spencer, R.G.M. Paradigm shifts in soil organic matter research affect interpretations of aquatic carbon cycling: Transcending disciplinary and ecosystem boundaries. *Biogeochemistry* **2014**, *117*, 279–297. [CrossRef]

30. Lehmann, J.; Kleber, M. The contentious nature of soil organic matter. *Nature* **2015**, *528*, 60–68. [CrossRef] [PubMed]

31. Fontaine, S.; Barot, S.; Barre, P.; Bdioui, N.; Mary, B.; Rumpel, C. Stability of organic carbon in deep soil layers controlled by fresh carbon supply. *Nature* **2007**, *450*, 277–280. [CrossRef] [PubMed]

32. Adams, D.C.; Gurevitch, J.; Rosenberg, M.S. Resampling tests for meta-analysis of ecological data. *Ecology* **1997**, *78*, 1277–1283. [CrossRef]

33. Gerlanc, D.; Kirby, K. *bootES: Bootstrap Effect Sizes*; R Package, Version 1.2; R Foundation for Statistical Computing: Vienna, Austria, 2015.

34. R Development Core Team. *A Language and Environment for Statistical Computing*; R Foundation for Statistical Computing: Vienna, Austria, 2016.

35. Currie, W.S. Responsive C and N biogeochemistryof the temperate forest floor. *Trends Ecol. Evol.* **1999**, *14*, 316–320. [CrossRef]

36. Silver, W.; Miya, R. Global patterns in root decomposition: Comparisons of climate and litter quality effects. *Oecologia* **2001**, *129*, 407–419. [CrossRef]

37. James, J.; Devine, W.; Harrison, R.; Terry, T. Deep soil carbon: quantification and modeling in subsurface layers. *Soil Sci. Soc. Am. J.* **2014**, *78*, S1–S9. [CrossRef]

38. Richter, D.D.; Markewitz, D. How Deep Is Soil? *Bioscience* **1995**, *45*, 600–609. [CrossRef]

39. Canadell, J.; Jackson, R.B.; Ehleringer, J.R.; Mooney, H.A.; Sala, O.E.; Schulze, E.D. Maximum rooting depth of vegetation types at the global scale. *Oecologia* **1996**, *108*, 583–595. [CrossRef]

40. Strahm, B.D.; Harrison, R.B.; Terry, T.A.; Flaming, B.L.; Licata, C.W.; Petersen, K.S. Soil solution nitrogen concentrations and leaching rates as influenced by organic matter retention on a highly productive Douglas-fir site. *For. Ecol. Manag.* **2005**, *218*, 74–88. [CrossRef]

41. Likens, G.E.; Bormann, F.H.; Johnson, N.M. Nitrification—Importance to nutrient losses from a cutover forested ecosystem. *Science* **1969**, *163*, 1205–1206. [CrossRef] [PubMed]

42. Kuzyakov, Y. Priming effects: Interactions between living and dead organic matter. *Soil Biol. Biochem.* **2010**, *42*, 1363–1371. [CrossRef]

43. Blagodatskaya, E.; Kuzyakov, Y. Mechanisms of real and apparent priming effects and their dependence on soil microbial biomass and community structure: Critical review. *Biol. Fertil. Soils* **2008**, *45*, 115–131. [CrossRef]

44. Mobley, M.L.; Lajtha, K.; Kramer, M.G.; Bacon, A.R.; Heine, P.R.; Richter, D.D. Surficial gains and subsoil losses of soil carbon and nitrogen during secondary forest development. *Glob. Chang. Biol.* **2015**, *21*, 986–996. [CrossRef] [PubMed]

45. Soil Survey Staff. *Keys to Soil Taxonomy*; Department of Agriculture, Natural Resources Conservation Service: Washington, DC, USA, 2014.

46. Wada, K. Minerals formed and mineral formation from volcanic ash by weathering. *Chem. Geol.* **1987**, *60*, 17–28. [CrossRef]

47. Nanzyo, M.; Shoji, S.; Dahlgren, R. *Volcanic Ash Soils—Genesis, Properties and Utilization*; Elsevier: Amsterdam, The Netherlands, 1993; Volume 21.

48. Blagodatskaya, E.V.; Blagodatsky, S.A.; Anderson, T.H.; Kuzyakov, Y. Priming effects in Chernozem induced by glucose and N in relation to microbial growth strategies. *Appl. Soil Ecol.* **2007**, *37*, 95–105. [CrossRef]

49. Kuzyakov, Y.; Friedel, J.K.; Stahr, K. Review of mechanisms and quantification of priming effects. *Soil Biol. Biochem.* **2000**, *32*, 1485–1498. [CrossRef]

50. Valette, J.; Gomendy, V.; Marechal, J.; Houssard, C.; Gillon, D. Heat Transfer in the Soil During Very Low-intensity Experimental Fires: The Role of Duff and Soil Moisture Content. *Int. J. Wildland Fire* **1994**, *4*, 225–237. [CrossRef]

51. Certini, G. Effects of fire on properties of forest soils: A review. *Oecologia* **2005**, *143*, 1–10. [CrossRef] [PubMed]
52. DeLuca, T.H.; Zouhar, K.L. Effects of selection harvest and prescribed fire on the soil nitrogen status of ponderosa pine forests. *For. Ecol. Manag.* **2000**, *138*, 263–271. [CrossRef]
53. Alban, D.H.; Perala, D.A. Carbon storage in lake states aspen ecosystems. *Can. J. For. Res.* **1992**, *22*, 1107–1110. [CrossRef]
54. Bauhus, J.; Vor, T.; Bartsch, N.; Cowling, A. The effects of gaps and liming on forest floor decomposition and soil C and N dynamics in a Fagus sylvatica forest. *Can. J. For. Res.* **2004**, *34*, 509–518. [CrossRef]
55. Bisbing, S.M.; Alaback, P.B.; DeLuca, T.H. Carbon storage in old-growth and second growth fire-dependent western larch (*Larix occidentalis* Nutt.) forests of the Inland Northwest, USA. *For. Ecol. Manag.* **2010**, *259*, 1041–1049. [CrossRef]
56. Black, T.A.; Harden, J.W. Effect of timber harvest on soil carbon storage at Blodgett Experimental Forest, California. *Can. J. For. Res.* **1995**, *25*, 1385–1396. [CrossRef]
57. Boerner, R.E.J.; Waldrop, T.A.; Shelburne, V.B. Wildfire mitigation strategies affect soil enzyme activity and soil organic carbon in loblolly pine (*Pinus taeda*) forests. *Can. J. For. Res.* **2006**, *36*, 3148–3154. [CrossRef]
58. Borchers, J.G.; Perry, D.A. The influence of soil texture and aggregation on carbon and nitrogen dynamics in southwest Oregon forest and clearcuts. *Can. J. For. Res.* **1992**, *22*, 298–305. [CrossRef]
59. Bravo-Oviedo, A.; Ruiz-Peinado, R.; Modrego, P.; Alonso, R.; Montero, G. Forest thinning impact on carbon stock and soil condition in Southern European populations of *P. sylvestris* L. *For. Ecol. Manag.* **2015**, *357*, 259–267. [CrossRef]
60. Cade-Menun, B.J.; Berch, S.M.; Preston, C.M.; Lavkulich, L.M. Phosphorus forms and related soil chemistry of Podzolic soils on northern Vancouver Island. II. The effects of clear-cutting and burning. *Can. J. For. Res.* **2000**, *30*, 1726–1741. [CrossRef]
61. Carter, M.C.; Dean, T.J.; Zhou, M.; Messina, M.G.; Wang, Z. Short-term changes in soil C, N, and biota following harvesting and regeneration of loblolly pine (*Pinus taeda* L.). *For. Ecol. Manag.* **2002**, *164*, 67–88. [CrossRef]
62. Chatterjee, A.; Vance, G.F.; Tinker, D.B. Carbon pools of managed and unmanaged stands of ponderosa and lodgepole pine forests in Wyoming. *Can. J. For. Res.* **2009**, *39*, 1893–1900. [CrossRef]
63. Chen, L.-C.; Liang, M.-J.; Wang, S.-L. Carbon stock density in planted versus natural *Pinus massoniana* forests in sub-tropical China. *Ann. For. Sci.* **2016**, *73*, 461–472. [CrossRef]
64. Chiti, T.; Perugini, L.; Vespertino, D.; Valentini, R. Effect of selective logging on soil organic carbon dynamics in tropical forests in central and western Africa. *Plant Soil* **2015**, *399*, 283–294. [CrossRef]
65. Christophel, D.; Spengler, S.; Schmidt, B.; Ewald, J.; Prietzel, J. Customary selective harvesting has considerably decreased organic carbon and nitrogen stocks in forest soils of the Bavarian Limestone Alps. *For. Ecol. Manag.* **2013**, *305*, 167–176. [CrossRef]
66. Christophel, D.; Höllerl, S.; Prietzel, J.; Steffens, M. Long-term development of soil organic carbon and nitrogen stocks after shelterwood- and clear-cutting in a mountain forest in the Bavarian Limestone Alps. *Eur. J. For. Res.* **2015**, *134*, 623–640. [CrossRef]
67. Cromack, K.; Miller, R.E.; Anderson, H.W.; Helgerson, O.T.; Smith, R.B. Soil Carbon and Nutrients in a Coastal Oregon Douglas-Fir Plantation with Red Alder. *Soil Sci. Soc. Am. J.* **1999**, *63*, 232–239. [CrossRef]
68. Dai, K.H.; Johnson, C.E.; Driscoll, C.T. Organic matter chemistry and dynamics in clear-cut and unmanaged hardwood forest ecosystems. *Biogeochemistry* **2001**, *54*, 51–83. [CrossRef]
69. DeByle, N.V. Harvesting and Site Treatment Influences on the Nutrient Status of Lodgepole Pine Forests in Western Wyoming. In *General Technical Report INT-90*; USDA Forest Service Intermountain Forest And Range Experimental Station: Ogden, UT, USA, 1979; pp. 137–156.
70. Diochon, A.; Kellman, L.; Beltrami, H. Looking deeper: An investigation of soil carbon losses following harvesting from a managed northeastern red spruce (*Picea rubens* Sarg.) forest chronosequence. *For. Ecol. Manag.* **2009**, *257*, 413–420. [CrossRef]
71. Edmonds, R.L.; McColl, G.J. Effects of forest management on soil nitrogen in Pinus radiata stands in the Australian Capital Territory. *For. Ecol. Manag.* **1989**, *29*, 199–212. [CrossRef]
72. Edwards, N.T.; Ross-Todd, B.M. Soil carbon dynamics in a mixed deciduous forest following clear-cutting with and without residue removal. *Soil Sci. Soc. Am. J.* **1983**, *47*, 1014–1021. [CrossRef]
73. Elliott, K.J.; Knoepp, J.D. The effects of three regeneration harvest methods on plant diversity and soil characteristics in the southern Appalachians. *For. Ecol. Manag.* **2005**, *211*, 296–317. [CrossRef]

74. Ellis, R.; Graley, A.M. Gains and losses in soil nutrients associated with harvesting and burning eucalypt rainforest. *Plant Soil* **1983**, *74*, 437–450. [CrossRef]

75. Esquilín, J.; Aida, E.; Stromberger, M.E.; Shepperd, W.D. Soil scarification and wildfire interactions and effects on microbial communities and carbon. *Soil Sci. Soc. Am. J.* **2008**, *72*, 111–118. [CrossRef]

76. Falsone, G.; Celi, L.; Caimi, A.; Simonov, G.; Bonifacio, E. The effect of clear cutting on podzolisation and soil carbon dynamics in boreal forests (Middle Taiga zone, Russia). *Geoderma* **2012**, *177–178*, 27–38. [CrossRef]

77. Fraterrigo, J.; Turner, M.; Pearson, S.M.; Dixon, P. Effects of past land use on spatial heterogeneity of soil nutrients in southern Appalachian forests. *Ecol. Monogr.* **2005**, *75*, 215–230. [CrossRef]

78. Frazer, D.W.; McColl, J.G.; Powers, R.F. Soil Nitrogen Mineralization in a Clearcutting Chronosequence in a Northern California Conifer Forest. *Soil Sci. Soc. Am. J.* **1990**, *54*, 1145. [CrossRef]

79. Gartzia-Bengoetxea, N.; González-Arias, A.; Merino, A.; De Arano, I.M. Soil organic matter in soil physical fractions in adjacent semi-natural and cultivated stands in temperate Atlantic forests. *Soil Biol. Biochem.* **2009**, *41*, 1674–1683. [CrossRef]

80. Gillon, D.; Houssard, C.; Valette, J.C.; Rigolot, E. Nitrogen and phosphorus cycling following prescribed burning in natural and managed Aleppo pine forests. *Can. J. For. Res.* **1999**, *29*, 1237–1247. [CrossRef]

81. Goh, K.M.; Phillips, M.J. Effects of clearfell logging and clearfell logging and burning of a Nothofagus forest on soil nutrient dynamics in South Island, New Zealand—Changes in forest floor organic matter and nutrient status. *N. Z. J. Bot.* **1991**, *29*, 367–384. [CrossRef]

82. Goodale, C.L.; Aber, J.D. The Long-term effects of land-use history on nitrogen cycling in northern hardwood forests. *Ecol. Appl.* **2001**, *11*, 253–267. [CrossRef]

83. Gough, C.M.; Vogel, C.S.; Harrold, K.H.; George, K.; Curtis, P.S. The legacy of harvest and fire on ecosystem carbon storage in a north temperate forest. *Glob. Chang. Biol.* **2007**, *13*, 1935–1949. [CrossRef]

84. Grady, K.C.; Hart, S.C. Influences of thinning, prescribed burning, and wildfire on soil processes and properties in southwestern ponderosa pine forests: A retrospective study. *For. Ecol. Manag.* **2006**, *234*, 123–135. [CrossRef]

85. Grand, S.; Lavkulich, L.M. Effects of forest harvest on soil carbon and related variables in Canadian spodosols. *Soil Sci. Soc. Am. J.* **2012**, *76*, 1816–1827. [CrossRef]

86. Gresham, C.A. Sustainability of intensive loblolly pine plantation management in the South Carolina Coastal Plain, USA. *For. Ecol. Manag.* **2002**, *155*, 69–80. [CrossRef]

87. Griffiths, R.P.; Swanson, A.K. Forest soil characteristics in a chronosequence of harvested Douglas-fir forests. *Can. J. For. Res.* **2001**, *31*, 1871–1879. [CrossRef]

88. Gundale, M.J.; DeLuca, T.H.; Fiedler, C.E.; Ramsey, P.W.; Harrington, M.G.; Gannon, J.E. Restoration treatments in a *Montana ponderosa* pine forest: Effects on soil physical, chemical and biological properties. *For. Ecol. Manag.* **2005**, *213*, 25–38. [CrossRef]

89. Das Gupta, S.; DeLuca, T.H. Short-term changes in belowground C, N stocks in recently clear felled Sitka spruce plantations on podzolic soils of North Wales. *For. Ecol. Manag.* **2012**, *281*, 48–58. [CrossRef]

90. Hart, S.C.; Selmants, P.C.; Boyle, S.I.; Overby, S.T. Carbon and nitrogen cycling in southwestern ponderosa pine forests. *For. Sci.* **2006**, *52*, 683–693.

91. Hendrickson, O.; Chatarpaul, L. Nutrient cycling following whole-tree and conventional harvest in a northern mixed forest. *Can. J. For. Res.* **2016**, *19*, 725–735. [CrossRef]

92. Herman, D.J.; Halverson, L.J.; Firestone, M.K. Nitrogen dynamics in an annual grassland: Oak canopy, climate, and microbial population effects. *Ecol. Appl.* **2003**, *13*, 593–604. [CrossRef]

93. Hölscher, D.; Schade, E.; Leuschner, C. Effects of coppicing in temperate deciduous forests on ecosystem nutrient pools and soil fertility. *Basic Appl. Ecol.* **2001**, *2*, 155–164. [CrossRef]

94. Hwang, J.; Son, Y. Short-term effects of thinning and liming on forest soils of pitch pine and Japanese larch plantations in central Korea. *Ecol. Res.* **2006**, *21*, 671–680. [CrossRef]

95. Jang, W.; Page-Dumroese, D.S.; Keyes, C.R. Long-term soil changes from forest harvesting and residue management in the Northern Rocky Mountains. *Soil Sci. Soc. Am. J.* **2016**. [CrossRef]

96. Johnson, C.E. Whole-tree clear-cutting effects on soil horizons and organic matter pools. *Soil Sci. Soc. Am. J.* **1991**, *55*, 497–502. [CrossRef]

97. Johnson, D.; Todd, D.E. Harvesting effects on long-term changes in nutrient pools of mixed oak forest. *Soil Sci. Soc. Am. J.* **1998**, *62*, 1725–1735. [CrossRef]

98. Johnson, C.E. Soil nitrogen status 8 years after whole-tree clear-cutting. *Can. J. For. Res.* **1995**, 1346–1355. [CrossRef]

99. Johnson, C.E.; Romanowicz, R.B.; Siccama, T.G. Conservation of exchangeable cations after clear-cutting of a northern hardwood forest. *Can. J. For. Res.* **1997**, *27*, 859–868. [CrossRef]

100. Johnson, D.W.; Walker, R.F.; Glass, D.W.; Stein, C.M.; Murphy, J.B.; Blank, R.R.; Miller, W.W. Effects of thinning, residue mastication, and prescribed fire on soil and nutrient budgets in a Sierra Nevada mixed-conifer forest. *For. Sci.* **2014**, *60*, 170–179. [CrossRef]

101. Jones, H.S.; Beets, P.N.; Kimberley, M.O.; Garrett, L.G. Harvest residue management and fertilisation effects on soil carbon and nitrogen in a 15-year-old Pinus radiata plantation forest. *For. Ecol. Manag.* **2011**, *262*, 339–347. [CrossRef]

102. Kaye, J.P.; Hart, S.C. Restoration and canopy-type effects on soil respiration in a ponderosa pine-bunchgrass ecosystem. *Soil Sci. Soc. Am. J.* **1998**, *62*, 1062–1072. [CrossRef]

103. Keenan, R.J.; Messier, C.; Kimmins, J.P. (Hamish) Effects of clearcutting and soil mixing on soil properties and understorey biomass in western red cedar and western hemlock forests on northern Vancouver Island, Canada. *For. Ecol. Manag.* **1994**, *68*, 251–261. [CrossRef]

104. Kelliher, F.M.; Ross, D.J.; Law, B.E.; Baldocchi, D.D.; Rodda, N.J. Limitations to carbon mineralization in litter and mineral soil of young and old ponderosa pine forests. *For. Ecol. Manag.* **2004**, *191*, 201–213. [CrossRef]

105. Kishchuk, B.E.; Thiffault, E.; Lorente, M.; Quideau, S.; Keddy, T.; Sidders, D. Decadal soil and stand response to fire, harvest, and salvage-logging disturbances in the western boreal mixedwood forest of Alberta, Canada. *Can. J. For. Res.* **2015**, *45*, 141–152. [CrossRef]

106. Klockow, P.A.; D'Amato, A.W.; Bradford, J.B. Impacts of post-harvest slash and live-tree retention on biomass and nutrient stocks in *Populus tremuloides* Michx.-dominated forests, northern Minnesota, USA. *For. Ecol. Manag.* **2013**, *291*, 278–288. [CrossRef]

107. Klopatek, J.M. Belowground carbon pools and processes in different age stands of Douglas-fir. *Tree Physiol.* **2002**, *22*, 197–204. [CrossRef] [PubMed]

108. Knoepp, J.D.; Swank, W.T. Forest management effects on surface soil carbon and nitrogen. *Soil Sci. Soc. Am. J.* **1997**, *61*, 928–935. [CrossRef]

109. Korb, J.E.; Johnson, N.C.; Covington, W.W. Slash pile burning effects on soil biotic and chemical properties and plant establishment: Recommendations for amelioration. *Restor. Ecol.* **2004**, *12*, 52–62. [CrossRef]

110. Kraemer, J.F.; Hermann, R.K. Broadcast burning: 25-year effects on forest soils in the western flanks of the Cascade Mountains. *For. Sci.* **1979**, *25*, 427–439.

111. Kurth, V.J.; D'Amato, A.W.; Palik, B.J.; Bradford, J.B. Fifteen-year patterns of soil carbon and nitrogen following biomass harvesting. *Soil Sci. Soc. Am. J.* **2014**, *78*, 624–633. [CrossRef]

112. Laiho, R.; Sanchez, F.; Tiarks, A.; Dougherty, P.M.; Trettin, C.C. Impacts of intensive forestry on early rotation trends in site carbon pools in the southeastern US. *For. Ecol. Manag.* **2003**, *174*, 177–189. [CrossRef]

113. Latty, E.F.; Canham, C.D.; Marks, P.L. The effects of land-use history on soil properties and nutrient dynamics in northern hardwood forests of the Adirondack Mountains. *Ecosystems* **2004**, *7*, 193–207. [CrossRef]

114. Law, B.E.; Thornton, P.E.; Irvine, J.; Anthoni, P.M.; Van Tuyl, S. Carbon storage and fluxes in *Ponderosa pine* forests at different developmental stages. *Glob. Chang. Biol.* **2001**, *7*, 755–777. [CrossRef]

115. Law, B.E.; Sun, O.J.; Campbell, J.; Van Tuyl, S.; Thornton, P.E. Changes in carbon storage and fluxes in a chronoseuence of *Ponderosa pine*. *Glob. Chang. Biol.* **2003**, *4*, 510–524. [CrossRef]

116. LeDuc, S.D.; Rothstein, D.E. Initial recovery of soil carbon and nitrogen pools and dynamics following disturbance in jack pine forests: A comparison of wildfire and clearcut harvesting. *Soil Biol. Biochem.* **2007**, *39*, 2865–2876. [CrossRef]

117. Maassen, S.; Wirth, S. Soil microbiological monitoring of a pine forest after partial thinning for stand regeneration with beech seedlings. *Soil Sci. Plant Nutr.* **2004**, *50*, 815–819. [CrossRef]

118. Mattson, K.G.; Smith, H.C. Detrital organic matter and soil CO_2 efflux in forests regenerating from cutting in West Virginia. *Soil Biol. Biochem.* **1993**, *25*, 1241–1248. [CrossRef]

119. Mattson, K.G.; Swank, W.T. Soil and detrital carbon dynamics following forest cutting in the Southern Appalachians. *Biol. Fertil. Soils* **1989**, *7*, 247–253. [CrossRef]

120. May, B.M.; Attiwill, P.M. Nitrogen-fixation by Acacia dealbata and changes in soil properties 5 years after mechanical disturbance or slash-burning following timber harvest. *For. Ecol. Manag.* **2003**, *181*, 339–355. [CrossRef]

121. McLaughlin, J.W.; Phillips, S.A. Soil carbon, nitrogen, and base cation cycling 17 years after whole-tree harvesting in a low-elevation red spruce (*Picea rubens*)-balsam fir (*Abies balsamea*) forested watershed in central Maine, USA. *For. Ecol. Manag.* **2006**, *222*, 234–253. [CrossRef]

122. McKee, S.E.; Seiler, J.R.; Aust, W.M.; Strahm, B.D.; Schilling, E.B.; Brooks, S. Carbon pools and fluxes in a tupelo (*Nyssa aquatica*)-baldcypress (*Taxodium distichum*) swamp 24-years after harvest disturbances. *Biomass Bioenergy* **2013**, *55*, 130–140. [CrossRef]

123. McLaughlin, J.W.; Liu, G.; Jurgensen, M.F.; Gale, M.R. Organic carbon characteristics in a spruce swamp five years after harvesting. *Soil Sci. Soc. Am. J.* **1996**, *60*, 1228–1236. [CrossRef]

124. Merino, A.; Edeso, J.M. Soil fertility rehabilitation in young *Pinus radiata* D. Don. plantations from northern Spain after intensive site preparation. *For. Ecol. Manag.* **1999**, *116*, 83–91. [CrossRef]

125. Moreno-Fernandez, D.; Diaz-Pines, E.; Barbeito, I.; Sanchez-Gonzalez, M.; Montes, F.; Rubio, A.; Canellas, I. Temporal carbon dynamics over the rotation period of two alternative management systems in Mediterranean mountain Scots pine forests. *For. Ecol. Manag.* **2015**, *348*, 186–195. [CrossRef]

126. Mu, C.C.; Lu, H.C.; Bao, X.; Wang, B.; Cui, W. Effects of selective cutting on vegetation carbon storage of boreal Larix gmelinii-Carex schmidtii forested wetlands in Daxing'anling, China. *Shengtai Xuebao/Acta Ecol. Sin.* **2013**, *33*, 5286–5298.

127. Murphy, J.D.; Johnson, D.W.; Miller, W.W.; Walker, R.F.; Blank, R.R. Prescribed fire effects on forest floor and soil nutrients in a Sierra Nevada Forest. *Soil Sci.* **2006**, *171*, 181–199. [CrossRef]

128. Neher, D.A.; Barbercheck, M.E.; El-Allaf, S.M.; Anas, O. Effects of disturbance and ecosystem on decomposition. *Appl. Soil Ecol.* **2003**, *23*, 165–179. [CrossRef]

129. Norris, C.E.; Quideau, S.A.; Bhatti, J.S.; Wasylishen, R.E.; MacKenzie, M.D. Influence of fire and harvest on soil organic carbon in jack pine sites. *Can. J. For. Res.* **2009**, *39*, 642–654. [CrossRef]

130. O'Brien, N.D.; Attiwill, P.M.; Weston, C.J. Stability of soil organic matter in *Eucalyptus regnans* forests and *Pinus radiata* plantations in south eastern Australia. *For. Ecol. Manag.* **2003**, *185*, 249–261. [CrossRef]

131. Powers, M.; Kolka, R.; Palik, B.; McDonald, R.; Jurgensen, M. Long-term management impacts on carbon storage in Lake States forests. *For. Ecol. Manag.* **2011**, *262*, 424–431. [CrossRef]

132. Prest, D.; Kellman, L.; Lavigne, M.B. Mineral soil carbon and nitrogen still low three decades following clearcut harvesting in a typical Acadian Forest stand. *Geoderma* **2014**, *214–215*, 62–69. [CrossRef]

133. Prietzel, J.; Wagoner, G.L.; Harrison, R.B. Long-term effects of repeated urea fertilization in Douglas-fir stands on forest floor nitrogen pools and nitrogen mineralization. *For. Ecol. Manag.* **2004**, *193*, 413–426. [CrossRef]

134. Puhlick, J.J.; Fernandez, I.J.; Weiskittel, A.R. Evaluation of forest management effects on the mineral soil carbon pool of a lowland, mixed-species forest in Maine, USA. *Can. J. Soil Sci.* **2016**, *96*, 207–218. [CrossRef]

135. Rab, M.A. Soil physical and hydrological properties following logging and slash burning in the *Eucalyptus regnans* forest of southeastern Australia. *For. Ecol. Manag.* **1996**, *84*, 159–176. [CrossRef]

136. Riley, J.M.; Jones, R.H. Factors limiting regeneration of *Quercus alba* and *Cornus florida* in formerly cultivated coastal plain sites, South Carolina. *For. Ecol. Manag.* **2003**, *177*, 571–586. [CrossRef]

137. Roaldson, L.M.; Johnson, D.W.; Miller, W.W.; Murphy, J.D.; Walker, R.F.; Stein, C.M.; Glass, D.W.; Blank, R.R. Prescribed Fire and Timber Harvesting Effects on Soil Carbon and Nitrogen in a Pine Forest. *Soil Sci. Soc. Am. J.* **2014**, *78*, S48. [CrossRef]

138. Rothstein, D.E.; Spaulding, S.E. Replacement of wildfire by whole-tree harvesting in jack pine forests: Effects on soil fertility and tree nutrition. *For. Ecol. Manag.* **2010**, *260*, 1164–1174. [CrossRef]

139. Sanchez, F.G.; Scott, D.A.; Ludovici, K.H. Negligible effects of severe organic matter removal and soil compaction on loblolly pine growth over 10 years. *For. Ecol. Manag.* **2006**, *227*, 145–154. [CrossRef]

140. Sanscrainte, C.L.; Peterson, D.L.; McKay, S. Carbon storage and soil properties in late-successional and second-growth subalpine forests in the North Cascade Range, Washington. *Northwest Sci.* **2003**, *77*, 297–307.

141. Saynes, V.; Etchevers, J.D.; Galicia, L.; Hidalgo, C.; Campo, J. Soil carbon dynamics in high-elevation temperate forests of Oaxaca (Mexico): Thinning and rainfall effects. *Bosque (Valdivia)* **2012**, *33*, 3–11. [CrossRef]

142. Selig, M.F.; Seiler, J.R.; Tyree, M.C. Soil carbon and CO_2 efflux as influenced by the thinning of loblolly pine (*Pinus taeda* L.) plantations on the Piedmont of Virginia. *For. Sci.* **2008**, *54*, 58–66.

143. Shelburne, V.B.; Boyle, M.F.; Lione, D.J.; Waldrop, T.A. Preliminary effects of prescribed burning and thinning as fuel reduction treatments on the piedmont soils of the clemson experimental forest. In *General Technical Report SRS-71*; U.S. Department of Agriculture, Forest Service, Southern Research Station: Asheville, NC, USA, 2004; pp. 35–38.

144. Sheng, H.; Zhou, P.; Zhang, Y.; Kuzyakov, Y.; Zhou, Q.; Ge, T.; Wang, C. Loss of labile organic carbon from subsoil due to land-use changes insubtropical China. *Soil Biol. Biochem.* **2015**, *88*, 148–157. [CrossRef]
145. Skovsgaard, J.P.; Stupak, I.; Vesterdal, L. Distribution of biomass and carbon in even-aged stands of Norway spruce (*Picea abies* (L.) Karst.): A case study on spacing and thinning effects in northern Denmark. *Scand. J. For. Res.* **2006**, *21*, 470–488. [CrossRef]
146. Slesak, R.A.; Schoenholtz, S.H.; Harrington, T.B. Soil carbon and nutrient pools in Douglas-fir plantations 5 years after manipulating biomass and competing vegetation in the Pacific Northwest. *For. Ecol. Manag.* **2011**, *262*, 1722–1728. [CrossRef]
147. Small, C.J.; McCarthy, B.C. Relationship of understory diversity to soil nitrogen, topographic variation, and stand age in an eastern oak forest, USA. *For. Ecol. Manag.* **2005**, *217*, 229–243. [CrossRef]
148. Stone, J.E.; Kolb, T.E.; Covington, W.W. Effects of restoration thinning on presettlement *Pinus ponderosa* in northern Arizona. *Restor. Ecol.* **1999**, *7*, 172–182. [CrossRef]
149. Stone, D.; Elioff, J. Soil properties and aspen development five years after compaction and forest floor removal. *Can. J. Soil Sci.* **1998**, *78*, 51–58. [CrossRef]
150. Strong, T.F. Harvesting intensity influences the carbon distribution in a northern hardwood ecosystem. In *Research Paper NC-329*; Department of Agriculture, Forest Service, North Central Research Station: St. Paul, MN, USA, 1997.
151. Strukelj, M.; Brais, S.; Paré, D. Nine-year changes in carbon dynamics following different intensities of harvesting in boreal aspen stands. *Eur. J. For. Res.* **2015**, *134*, 737–754. [CrossRef]
152. Tang, J.; Bolstad, P.V.; Martin, J.G. Soil carbon fluxes and stocks in a Great Lakes forest chronosequence. *Glob. Chang. Biol.* **2009**, *15*, 145–155. [CrossRef]
153. Trettin, C.C.; Jurgensen, M.F.; Gale, M.R.; McLaughlin, J.W. Recovery of carbon and nutrient pools in a northern forested wetland 11 years after harvesting and site preparation. *For. Ecol. Manag.* **2011**, *262*, 1826–1833. [CrossRef]
154. Ussiri, D.A.N.; Johnson, C.E. Organic matter composition and dynamics in a northern hardwood forest ecosystem 15 years after clear-cutting. *For. Ecol. Manag.* **2007**, *240*, 131–142. [CrossRef]
155. Vario, C.L.; Neurath, R.A.; Friedland, A.J. Response of mineral soil carbon to clear-cutting in a northern hardwood forest. *Soil Sci. Soc. Am. J.* **2014**, *78*, 309–318. [CrossRef]
156. Vesterdal, L.; Dalsgaard, M.; Felby, C.; Raulund-Rasmussen, K.; Jørgensen, B.B. Effects of thinning and soil properties on accumulation of carbon, nitrogen and phosphorus in the forest floor of Norway spruce stands. *For. Ecol. Manag.* **1995**, *77*, 1–10. [CrossRef]
157. Waldrop, M.P.; McColl, J.G.; Powers, R.F. Effects of forest postharvest management practices on enzyme activities in decomposing litter. *Soil Sci. Soc. Am. J.* **2003**, *67*, 1250–1256. [CrossRef]
158. Wu, J.-S.; Jiang, P.-K.; Chang, S.X.; Xu, Q.-F.; Lin, Y. Dissolved soil organic carbon and nitrogen were affected by conversion of native forests to plantations in subtropical China. *Can. J. Soil Sci.* **2010**, *90*, 27–36. [CrossRef]
159. Xiang, W.; Chai, H.; Tian, D.; Peng, C. Marginal effects of silvicultural treatments on soil nutrients following harvest in a Chinese fir plantation. *Soil Sci. Plant Nutr.* **2009**, *55*, 523–531. [CrossRef]
160. Yanai, R.D.; Arthur, M.A.; Siccama, T.G.; Federer, C.A. Challenges of measuring forest floor organic matter dynamics: Repeated measures from a chronosequence. *For. Ecol. Manag.* **2000**, *138*, 273–283. [CrossRef]
161. Zabowski, D.; Chambreau, D.; Rotramel, N.; Thies, W.G. Long-term effects of stump removal to control root rot on forest soil bulk density, soil carbon and nitrogen content. *For. Ecol. Manag.* **2008**, *255*, 720–727. [CrossRef]
162. Zhong, Z.; Makeschin, F. Soil biochemical and chemical changes in relation to mature spruce (*Picea abies*) forest conversion and regeneration. *J. Plant Nutr. Soil Sci.* **2003**, *166*, 291–299. [CrossRef]
163. Zummo, L.M.; Friedland, A.J. Soil carbon release along a gradient of physical disturbance in a harvested northern hardwood forest. *For. Ecol. Manag.* **2011**, *261*, 1016–1026. [CrossRef]

forests

MDPI

Article

Effect of Timber Harvest Intensities and Fertilizer Application on Stocks of Soil C, N, P, and S

Marcella L.C. Menegale [1,*], Jose Henrique T. Rocha [2], Robert Harrison [1], Jose Leonardo de M. Goncalves [2], Rodrigo F. Almeida [3], Marisa de C. Piccolo [3], Ayeska Hubner [2], Jose Carlos Arthur Junior [4], Alexandre de Vicente Ferraz [5], Jason N. James [1] and Stephani Michelsen-Correa [1]

[1] School of Environmental and Forest Sciences, University of Washington, Seattle, WA 98195-2100, USA; robh@uw.edu (R.H.); jajames@uw.edu (J.N.J.); smcorrea@uw.edu (S.M.-C.)

[2] Forest Science Department, "Luiz de Queiroz" College of Agriculture, University of São Paulo, Piracicaba SP 13418-900, Brazil; rocha.jht@gmail.com (J.H.T.R.); jlmgonca@usp.br (J.L.d.M.G.); ayeskahubner@yahoo.com.br (A.H.);

[3] Center of Nuclear Energy in Agriculture, University of São Paulo, Piracicaba, São Paulo 13416-000, Brazil; rodrigo.fa18@gmail.com (R.F.A.); mpiccolo@cena.usp.br (M.d.C.P.)

[4] Federal Rural University of Rio de Janeiro, Seropédica, Rio de Janeiro 23890-000, Brazil; jcarthur@ufrrj.br

[5] Forest Science and Research Institute (IPEF), Piracicaba, São Paulo 13400-970 Private Bag 530, Brazil; alexandre@ipef.br

* Correspondence: marcism@uw.edu; Tel.: +1-206-209-6421

Academic Editors: Scott X. Chang and Xiangyang Sun
Received: 12 September 2016; Accepted: 12 December 2016; Published: 21 December 2016

Abstract: The purpose of this study was to determine the stocks of available P and S, total N, and oxidizable C at depth in an Oxisol cultivated with Eucalyptus in Brazil following different timber harvest intensities and fertilizer application over 12 years. The harvest regimes considered were (i) conventional stem-only harvest (all forest residues were maintained on the soil); (ii) whole-tree harvest (only litter was maintained on the soil—all slash, stemwood, and bark were removed); and (iii) whole-tree harvest + litter layer removal. The site was planted in 2004 considering three timber harvest intensities, some with and some without N and P fertilization. In 2012 the experiment was reinstalled, and all the treatments were reapplied in the each plot. From 2004 to 2016, nutrient accumulation and soil N, P, and S stocks were assessed in the 0–20 cm layer. Also in 2016, soil N, P, S, and oxidizable C stocks were measured to 2 m depth. For each treatment, the net balance of N, P, and S were calculated from soil stocks and harvest outputs during two forest rotations. A reduction in all nutrient stocks was observed in the 0–20 cm layer for all treatments. For N, this reduction was 20% smaller in the stem-only harvest treatment and 40% higher when no N fertilizer was applied, when compared to other treatments. Stem-only harvest treatment was observed to reduce the loss of N, P, and S due to harvest by 300, 30, and 25 kg·ha^{-1}, respectively, when compared to the whole-tree harvest + litter layer removal treatment.

Keywords: nutrient cycling; forest residue management; Eucalyptus; nitrogen; carbon; phosphorus; sulfur; minimum tillage

1. Introduction

Forest soils form an important reservoir in ecosystem nutrient and carbon budgets, which are crucial for the sustained productivity of forests [1]. Nutrient fluxes and transformations in forest soils are a result of a complex interchange between the atmosphere, plants, and soil. Consequently, soils are a critical source of plant nutrition [2], and integral to the recovery of ecosystems following natural or human disturbances. As essential plant nutrients, the concentrations of plant available nitrogen (N), phosphorous (P), and sulfur (S) are crucial controls on the net primary productivity of a site. As the plants uptake these nutrients, their concentrations within the soil are reduced until they are replaced either by nutrient recycling or fresh inputs. In order to maintain the supply of these nutrients back into the soil, sustainable silviculture practices focus on high timber productivity while efficiently balancing the additional export of harvest residues with the cost of fertilizer to replace nutrients.

Highly weathered tropical soils are characteristically deficient in essential plant nutrients. This is especially true of the macronutrients essential for plant growth and development [3]. Organic matter (OM) plays a fundamental role in soil fertility, contributing to cation exchange capacity (CEC) and as a source of nutrients [4], while also stabilizing soil aggregates and increasing water holding capacity. Soil management practices can sustain adequate levels of OM while providing sufficient quantities of plant available P. The need for balancing high timber productivity while maintaining soil nutrient concentrations has led to the development of sustainable silvicultural practices [5,6].

Highly weathered soils may conserve more carbon (C) with increasing OM inputs because P frequently limits decomposition [7]. Such P limitations will also effect microbially driven processes, such as N mineralization. While most temperate ecosystems—which are usually N limited—may store added mineral N for long periods, a P limited tropical forest can rapidly loose substantial amounts of mineral N (NO_3^- and NH_4^+) following N fertilization [8]. The nutrient limitations of tropical soils not only decrease plant productivity but also impact decomposition rates and immobilization processes and therefore the soil's capacity to store and cycle C and N [9–11].

Many consider the proper management of forest residues to be a practice which maintains the harvest residues (i.e., slash, bark, and litter layer) on site as a source of nutrients. The practice of leaving forest residues on site can sustain soil quality with respect to the productivity of medium and long-term species such as Eucalyptus [12–16]. This practice helps to improve nutrient pools and OM content in the soil [3]. Harvest residues, when retained on site, deposit nutrients back to the soil following decomposition. Plants cultivated in the area will then readily absorb these nutrients. This reduces the need for synthetic fertilizer and other methods for soil preparation that can disrupt soil physicochemical properties such as aggregation [17,18]. In addition, the protective layer formed due to the presence of harvest residues on the soil helps to reduce extreme surface heating and water loss through evaporation, and decreases soil loss due to erosion.

Due to the coupling of soil nutrient status (i.e., C, N, P, and S) and forest productivity, elucidating the impacts of forest residue additions will aid in the sustainable management of high quality, rapidly growing timber crops. Under tropical conditions, it is still unclear how the presence of harvest residues in the soil can help to create a more efficient silviculture system with regards to productivity, improved preservation of soil characteristics, and reduced fertilizer applications. For this reason, the objective of this study was to determine the stocks of available P and S, total N, and oxidizable C at depth in an Oxisol which was cultivated with Eucalyptus in Brazil over 12 years, under different timber harvest intensities and fertilizer applications. During each forest rotation the net balance between soil N, P, and S stocks and harvest outputs was also calculated.

2. Materials and Methods

2.1. Study Site

The study was carried out at the Itatinga Forest Science Experimental Station of the University of São Paulo in Brazil (23°06′ S, 48°36′ W, and 857 m above sea level). The Köppen climate classification of the site is humid subtropical (Cfa) with an average annual temperature of 19.4 °C. In the coldest month (July), the temperatures average 15.6 °C, and in the hottest month (January), the temperatures average 22.3 °C. The mean annual rainfall is 1319 mm, with 75% of the rainfall concentrated between October and March [19].

The topography of the region is flat to undulating, and the soil is a very deep (>10 m) Ferralsol (International Union of Soil Sciences (IUSS) Working Group—World Reference Base for

Soil Resources (WRB), 2015; red-yellow Latosol—Brazilian Classification System, and Oxisols—United State Department of Agriculture (USDA) Soil Taxonomy) that developed on Cretaceous sandstone. The clay content ranges from 17% in the A1 horizon to 25% in deeper Bo horizons. The mineralogy was dominated by quartz, kaolinite, and oxyhydroxides of Al and Fe with a low pH and small amounts of exchangeable cations (Table 1).

Table 1. Physical and chemical attributes of experimental site *.

Depth	Sandy	Silt	Clay [1]	pH [2]	CEC$_7$ [3]	C [4]	N [5]	P [6]	Exchangeable Cations [4]			
									K	Ca	Mg	Al
cm		g·kg^{-1}			mmol$_c$ kg^{-1}	g·kg^{-1}		mg·kg^{-1}	mmol$_c$ kg^{-1}			
0–10	802	22	175	3.8	63.98	9.61	1.44	4	0.25	4.28	2.81	7.50
10–20	811	12	176	3.9	51.42	10.05	1.67	3	0.27	2.80	2.17	8.43
20–30	790	34	176	3.9	39.98	6.77	1.53	1	0.20	1.32	1.00	6.09
30–40	777	23	200	3.9	40.18	5.33	1.29	1	0.15	0.88	0.81	7.03
40–60	747	14	239	3.9	38.46	5.42	1.14	1	0.15	0.99	0.72	7.50
60–100	712	12	276	3.9	32.72	5.04	0.99	1	0.15	0.66	0.54	6.56
100–150	712	11	277	4.0	30.09	3.44	1.04	1	0.08	0.71	0.54	2.34
150–200	704	20	276	4.2	22.05	0.87	1.04	1	0.05	0.55	0.54	2.81

* Samples were taken in 2014. The total area of the experiment was considered (average of all plots). [1] Pipette method; [2] Determined in CaCl$_2$ 0.01 mol·L^{-1} in soil/solution reason of 1:5; [3] Cation exchange capacity with soil at pH 7; [4] wet oxidation; [5] Determined using the micro Kjeldahl method after sulphuric digestion; [6] Extracted with exchange ion resin [20].

The original vegetation of the site was Cerrado *stricto sensu* (Brazilian savannah) [21]. The site has been planted with *Eucalyptus* species since 1940. From 1940 to 1992, it was cropped with *Eucalyptus saligna* and managed by coppicing with clearcutting every seven or eight years. In 1992, the plantation was harvested and replanted with *Eucalyptus grandis*, which was harvested (via clear-cutting) in 2004 when the study site was installed.

2.2. Experimental Design and Treatments

The study site was installed in 2004 (R1) and reinstalled in 2012 (R2) with three replicates of five treatments in a randomized block design. The plot sizes were 27 m × 18 m, with 81 trees per plot. The assessments were carried out in an inner plot of 15 m × 10 m (25 trees per plot). Five treatments were implemented with different management levels of forest residue removal and fertilizer applications (Table 2). The forest residues manipulated in this experiment include all of the organic residues remaining on the soil after wood harvesting of *E. grandis* plantations after 12 years of growth: the leaves and branches less than 3 cm in diameter (canopy), bark, and litter layer. The treatments tested were:

1. ReM + F—Only stemwood was harvested; all of the forest residues (bark, canopy, and litter layer from the previous rotation) were maintained on the soil after the clear-cutting, and all nutrients were applied as fertilizer and the soil was dressed with limestone;

2. ReR + F—All of the forest residues (bark, canopy, and litter layer from the previous rotation) were removed from the plot after the clear-cutting, and all nutrients were applied as fertilizer and the soil was dressed with limestone;

3. ACR + F—The canopy (leaves and branches) and bark were removed after clear-cutting, but the litter layer was maintained; all nutrients were applied as fertilizer and the soil was dressed with limestone;

4. ACR − N—The canopy (leaves and branches) and bark were removed after the clear-cutting, but the litter layer was maintained; all nutrients except N fertilizer were applied and the soil dressed with limestone. However, a small quantity of N was applied to ensure tree survival;

5. ACR − P—The canopy (leaves and branches) and bark were removed after the clear-cutting, but the litter layer was maintained; all nutrients except P fertilizer were applied and the soil was dressed with limestone;

Table 2. Forest residue management and nutrients applied in treatments.

Treatment [1]	Forest Residue [2]			Nutrients [3]					
	Canopy	Bark	Litter Layer	N	P	K	Ca	Mg	S
				kg·ha^{-1}					
ReM + F	M	M	M	130	44	125	480	120	140
ReR + F	R	R	R	130	44	125	480	120	140
ACR + F	R	R	M	130	44	125	480	120	140
ACR − N	R	R	M	10	44	125	480	120	11
ACR − P	R	R	M	130	-	125	480	120	140

[1] Detailed description in Material and Methods; [2] M = maintained on the soil, R = removed from the area; [3] N, P, K, Ca and Mg sources were ammonium sulphate, triple superphosphate, potassium chloride, and limestone, respectively.

The ReR + F treatment was included to simulate a severe harvest condition where all forest residues were removed from the site, with fertilization being the only source of nutrients added to the soil. Although unusual, this harvest management system has already been implemented in some regions in Brazil, especially when biomass prices were high and all forest residues were removed with the objective of maximizing biomass production. Furthermore, by examining such an extreme treatment, this study brackets the range of harvest intensities actually utilized by different forest managers. Thus, the effects of more moderate treatments can be estimated by interpolating between two endmembers rather than extrapolating from a more limited range of harvest intensity.

2.3. Field Procedures

After clear-cutting the 12-year-old *Eucalyptus grandis* Hill Ex Maiden plantation, the treatments (Table 2) were applied and the soil was prepared by subsoiling to 0.4 m deep with a ripper. The plots were planted with a single progeny of *E. grandis* Hill ex Maiden seedlings in June 2004 (one month after harvesting the previous plantation). The fertilizer that was applied is shown in Table 2. Additionally, 3.4 kg·ha^{-1} of boron (B) and 30 kg·ha^{-1} of Fritted Trace Elements (FTE) (9% Zn, 1.8% B, 0.8% Cu, 2% Mn, 3.5% Fe, and 0.1% Mo) were applied in every treatment. The fertilizer was applied as one base fertilizer application and two topdressing applications. The base fertilizer application was made on the same day as the planting, the triple superphosphate, the FTE, and 10 kg·ha^{-1} of N and K were added in a small pit to the side of each seedling, whereas the lime (2 Mg·ha^{-1}) was applied to the whole area. The ACR − N treatment received a small amount of N with the base fertilizer to ensure the survival and the initial development of plants. The topdressing application of N, K, S and B was applied around the seedling within the ground area covered by the canopy at three and eight months after planting. The plantation was harvested at eight years of age and replanted one month later (November 2012). The treatments were repeated at each plot (Table 2).

During clear-cutting, the canopy and bark from the trees of each plot were retained on the same plots. The clear-cut trees were felled with a chainsaw; the bark and branches were then removed manually. The seedlings in R2 were planted between the stumps of R1 without ripping. R2 was managed in the same way as R1 to evaluate the long-term effects of the residue removal and fertilizer application practices. Each experimental unit was maintained weed free throughout the two rotations.

2.4. Soil Sampling and Analysis

Soil samples were collected before the application of the treatments in the first (2004) and second (2012) rotations at 0–10 and 10–20 cm. In 2016, samples were taken from the soil down to two meters (0–10, 10–20, 20–40, 40–60, 60–80, 80–100, 100–150, and 150–200 cm) across three replications for each treatment. Each sample was composed of six subsamples taken from six points in the inner plot arranged in a diagonal design. For each plot, soil bulk density was calculated from soil samples taken from the walls of trenches opened with known volume rings.

Samples were then air-dried at 45 °C for three days and passed through a 2 mm sieve for chemical analysis. Soil analyses were carried out for all treatments. Total N was determined by dry combustion using an Elemental Analyzer CHNS/2400 (Perkin Elmer, Waltham, MA USA) [22]. Available P was determined by displacement using ion-exchange resins. Sulfur was displaced with $Ca(H_2PO_4)_2$ 0.01 $mol \cdot L^{-1}$ solution and determined by turbidimetry using $BaSO_4$ [20]. Oxidizable C was determined by wet oxidation [23].

2.5. Nutrient Accumulation in the Biomass

In order to assess the accumulation of nutrients in the biomass (canopy, bark, wood, and coarse roots), 10 trees were felled in 2004, 10 trees per treatment were felled in 2012, and 3 trees per treatment were felled in 2016, with all removed trees coming from the inner border of each plot. Felled trees were separated into the following compartments: leaves, branches, stemwood (diameter > 3 cm at the thinner end), and stem bark. Coarse root (diameter > 1 cm) was removed by excavation. Sub-samples were collected from all of the compartments and dried (65 °C) until reaching a constant weight, and then the dry biomass of the compartments in each tree was proportionally calculated. To estimate the wood, bark biomass, and stem volume of the plantation from the sampled trees, diameter at breast high (DBH) and total high (H) were used as independent variables to adjust a model following the form of Schumacher and Hall [24].

Samples were analyzed for N, P, and S. Total N was determined using the micro Kjeldahl method after sulphuric digestion. P was determined through colourimetry after nitric perchloric digestion, and S was determined through turbidimetry [25]. Nutrient accumulation per hectare was determined by the sum of the product of biomass accumulation of each compartment and the nutrient concentrations.

2.6. Statistical Analysis

Prior to statistical analysis, the data were tested for normality (Shapiro-Wilk) and homoscedasticity (Box-Cox). A two-way Analysis of Variance (ANOVA) test was performed to identify the differences between the factors considered in the experiment (soil depth and treatment). Equal sample sizes were used in order to obtain maximum power and robustness of the test. In the case of overall significant differences in the group means, Fisher's Least Significant Difference (LSD) post hoc testing was performed to determine the differences between groups [26]. Statistical tests were considered significant at α = 0.05. Data was analyzed using SAS University Edition (University Edition, SAS Institute Inc., Cary, NC, USA).

3. Results

3.1. Nutrient Balance After Two Crop Rotations

Before the installation of the experiment in 2004, the site contained 370, 11, and 9.2 $kg \cdot ha^{-1}$ of N, P, and S, respectively, in the 0–20 cm layer (Figure 1). On that occasion, there was an accumulation of approximately 660, 90, and 100 $kg \cdot ha^{-1}$ of N, P, and S, respectively, in the biomass and litter layer (250 $t \cdot ha^{-1}$ in total). The nutrient losses following harvest within the ReR + F treatment were higher than that of the ReM + F treatment by 120%, 50%, and 40% of N, P, and S respectively. For the ACR treatments, the losses from harvest were 65%, 30%, and 20% higher, respectively, than the ReM + F treatment (Table 3).

In 2012 (the end of experiment's first rotation), the ReM + F treatment contained the largest stocks of N (206 kg·ha^{-1}) and P (13.8 kg·ha^{-1}) in the 0–20 cm layer, while the highest stocks of S (13.2 kg·ha^{-1}) for this layer were observed in the ACR − P treatment. The lowest stocks of N (151 kg·ha^{-1}) and P (9 kg·ha^{-1}) were observed in the ACR − N and ACR − P treatments, respectively (Figure 1).

At the end of the first rotation of the experiment (eight years) there were no observed differences in the volume of wood produced in the treatments ReM + F, ACR + F, ReR + F, and ACR − N, which were about 420 m^3·ha^{-1} on average. On the other hand, the absence of fertilization with P (ACR − P treatment) resulted in approximately a 10% reduction in the final wood volume [27]. Even with small differences in tree growth, the renewal of the experiment for the next crop rotation resulted in great differences in the quantity of nutrient losses among the treatments, due to the different parts of the plants removed at harvest (Table 3).

At 41 months into the second rotation, the ReM + F treatment contained 100 Mg·ha^{-1} of biomass (wood, bark, canopy, and coarse roots). Compared to the ReM + F treatment, the biomass accumulated in the ACR + F, ReR + F, ACR − N, and ACR − P treatments was lower by 14%, 15%, 16%, and 25%, respectively. The accumulated N, P, and S in the biomass of the ACR − P treatment was 33%, 40%, and 20% lower, respectively, compared to the ReM + F treatment.

Between 2004 and 2012, N contents in the soil from 0–20 cm depth were reduced by around 50% with larger losses in the ACR − N treatment and smaller losses in the ReM+F treatment. From 2012 to 2016, a smaller reduction in soil N was observed, with the exception of the ACR − P treatment. Only small losses of available P were observed between 2004 and 2012, but greater losses occurred after reinstallation of treatments, especially for repeated removal of harvest residues (ReR + F). In the ACR + F treatment the concentration of available P in surface soil was constant through the years. A larger reduction was observed for the ReR + F and ACR − P treatments with losses of approximately 30% from 2004 to 2016. In regards to available S, a reduction in the ReR + F treatment from 2004 to 2012 was observed, and an increase was observed in the ACR − P treatment. For the other treatments, S stocks were constant. From 2012 to 2016, available S stocks improved for all treatments (Figure 1).

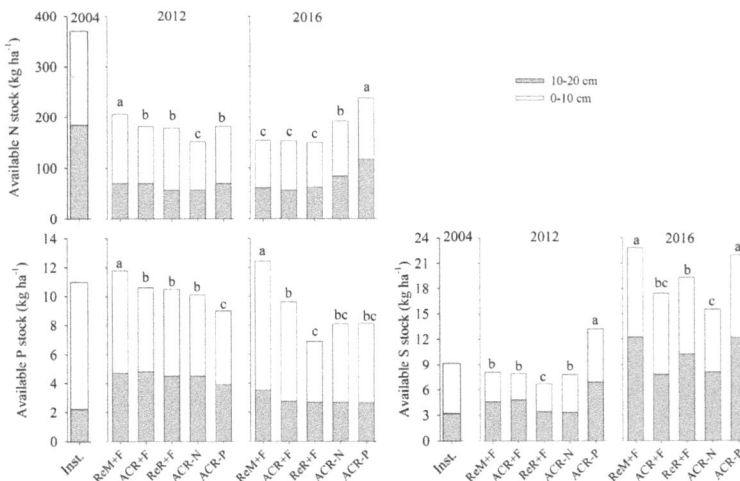

Figure 1. Available N, P, and S stocks in the 0–20 cm soil layer before the setup of treatments (2004), before the harvest of the first rotation (2012), and at four years into the second rotation (2016) in a Eucalyptus plantation under different harvest residue management and fertilizer application strategies. P was determined by resin extraction, S was determined by extraction with Ca(H$_2$PO$_4$)$_2$ 0.001 mol·L^{-1}, and available N was calculated assuming that 10% of total N is mineralizable [28]. Columns with the same letter are not significantly different from each other.

Table 3. Nutrient stocked in the forest biomass, inputs, and outputs of nutrients in two crop rotations.

Component	Biomass (t·ha^{-1})	N (kg·ha^{-1})	P (kg·ha^{-1})	S (kg·ha^{-1})
Nutrients Stocks Before 2004 Harvest				
Litter layer	24 (1) [1]	141 (17)	12 (1)	12 (1)
Stem wood	167 (15)	250 (20)	52 (3)	67 (5)
Stem bark	18 (1)	62 (3)	8 (1)	7 (1)
Canopy	9 (1)	100 (9)	7 (1)	6 (1)
Root	33 (5)	109 (18)	11 (2)	14 (2)
Total	251 (12)	662 (48)	90 (10)	106 (15)
Harvest Outputs in 2004				
Treatment ReM + F	167 (15)	250 (20)	52 (3)	67 (5)
Treatments ACR + F, ACR − N, and ACR − P	194 (18)	412 (46)	67 (4)	80 (7)
Treatment ReR + F	218 (18)	553 (49)	79 (15)	92 (8)
Inputs from 2004 to 2012				
Atmospheric deposition [2]		32	-	-
Fertilizer application [3]:				
Treatments ReM + F, ACR + F, and ReR + F		130	44	143
Treatment ACR − N		10	44	11
Treatment ACR − P		130	0	143
Harvest Outputs in 2012				
Treatment ReM + F	191 (10)	313 (17)	60 (3)	71 (4)
Treatment ACR + F	196 (20)	412 (37)	67 (7)	71 (7)
Treatment ReR + F	220 (16)	542 (39)	79 (6)	80 (6)
Treatment ACR − N	200 (32)	426 (60)	69 (11)	70 (11)
Treatment ACR − P	180 (23)	386 (48)	61 (8)	65 (8)
Inputs from 2012 to 2016				
Atmospheric deposition		16	-	-
Fertilizer application:				
Treatments ReM + F, ACR+F, and ReR + F		130	44	143
Treatment ACR − N		10	44	11
Treatment ACR − P		130	0	143
Accumulated in the Biomass in 2016				
Treatment ReM + F	100 (3)	255 (20)	25 (4)	40 (2)
Treatment ACR + F	86 (3)	207 (26)	21 (1)	37 (2)
Treatment ReR + F	85 (7)	202 (17)	18 (2)	37 (4)
Treatment ACR − N	84 (5)	195 (12)	18 (1)	36 (1)
Treatment ACR − P	75 (2)	172 (3)	15 (2)	32 (1)

[1] Standard deviation; [2] By Laclau et al. [29]; [3] It is assumed that 100% of fertilizer application was available.

3.2. Soil Nutrient Stocks in 2016

At four years into the second study rotation (2016), large differences between treatments were found in the stocks of Total N, Available P and S, and oxidizable C in the soil to 200 cm depth (Figure 2). The largest stock of oxidizable C in the soil was observed in the ReM + F treatment ($p > 0.001$), and this treatment contained the most oxidizable C at every individual sampling layer except between 60–100 cm. The ReR + F treatment contained the lowest stocks of oxidizable C ($p > 0.002$) up to 40 cm depth, and did not differ from the ACR treatments in the deeper layers. The supply of oxidizable carbon in the entire 0–200 cm profile of the ReM + F treatment was approximately 130 Mg·ha^{-1}, which is 10% greater than the ACR + F treatment and 20% higher than the ReR + F treatment.

The ACR-P treatment contained the highest stock of N ($p > 0.001$) in all the layers considered. ReM + F, ACR + F, and ReR + F treatments presented the lowest stocks of N, not differing statistically from each other. The ReR + F and ACR − N treatments presented the lowest stocks of P, and were not significantly different. The stocks of N in the 0–200 cm layer of the soil were approximately 10 Mg·ha^{-1} in the ReM + F, ACR + F, and ReR + F treatments. Stocks of available P in the soil were approximately 40 kg·ha^{-1} in the ReM + F and ACR + F treatments, which is 60% higher than the stock of P observed in the ReR + F treatment.

The ACR + F and ReR + F treatments contained the largest stocks of S in the soil ($p > 0.001$). The ACR − N and ACR − P treatments contained the lowest stocks of S ($p > 0.001$). The stock of S available in the 0–200 cm layer was 280 kg·ha^{-1} in the ReM + F, ACR + F, and ReR + F treatments. The ACR − N treatment had a stock 30% lower than the previous treatments.

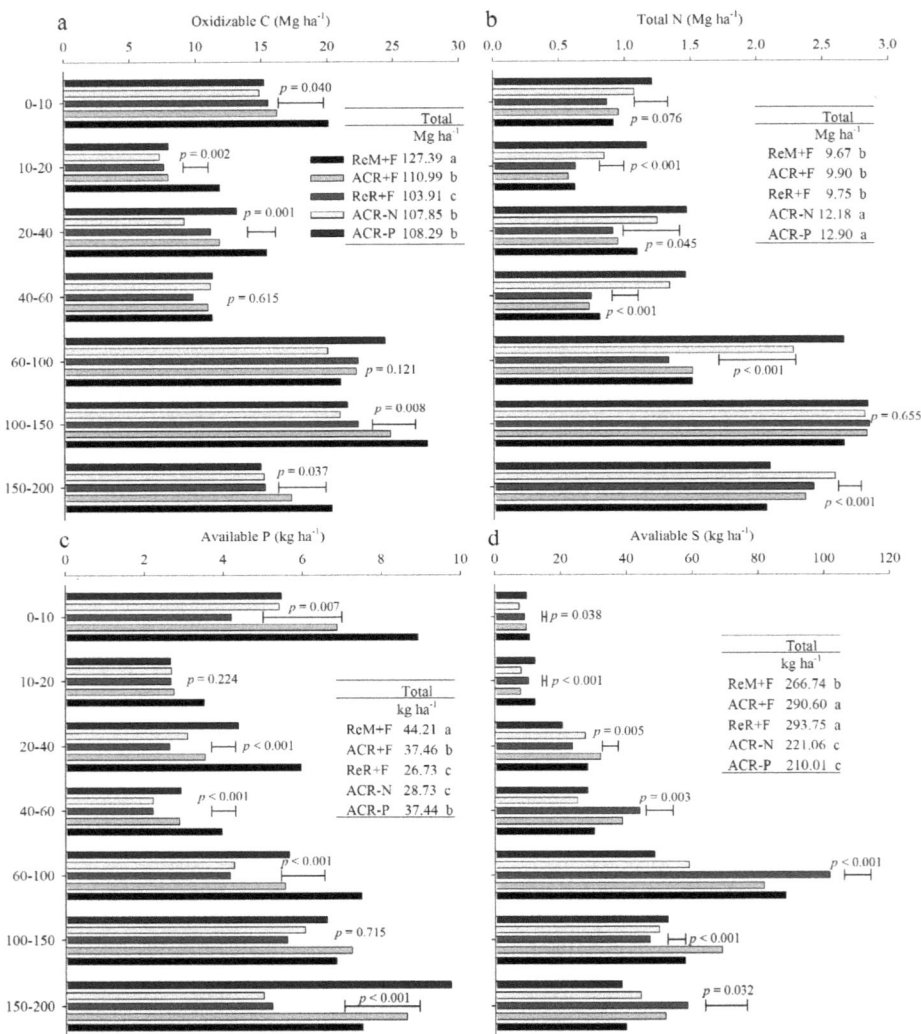

Figure 2. Stocks of oxidizable C (**a**), total N (**b**), available P (**c**) and S (**d**) to 2 m depth in a Eucalyptus plantation under different harvest residue management and fertilizer application strategies. The bars indicate the least significant difference based on the Fisher's Least Significant Difference (LSD) test at 5% probability, and the values next to the bars show the significance of the F test.

4. Discussion

4.1. Effects of Harvest Residue Strategies on Soil C and Available Nutrient Stocks

The effect of forest residue removal on soil nutrients stocks is inconsistent between different studies in the literature. In some cases, no losses in soil C or nutrient stocks have been reported, even with increases in nutrients exported during harvest [4,14,29–33]. This observation has been attributed to several mechanisms: high buffer capacity of the soil, slow decomposition of forest residues, a long harvesting return interval (more than seven years), and fast growth and litter deposition from new Eucalyptus plantations. However, reduction in soil C and nutrient stocks has been observed in wet tropical sites with sandy soils, high productivity forests, and successive harvest treatments [16,17,29,31,33]. Our study agrees with this latter set of studies and shows substantial changes in soil C and nutrients following removal of harvest residues.

In this study, the highest stocks of oxidizable C in the soil were observed in the 0–20 cm soil layer of the ReM + F treatment, followed by ACR + F and ReR + F treatments ($p > 0.001$). Similar results were observed by Moreno-Fernandez et al. [34] in Mediterranean mountain Scots pine forests, where the highest stocks of C were in the 0–20 cm layer of the soil under moderate intensity management.

Stocks of soil C decreased as soil depth increased, with the lowest stocks being observed in the ACR − P treatment ($p > 0.006$) at 150–200 cm depth. In a previous study developed in the same experimental site, Rocha [35] found that the removal of forest harvest residues from the soil reduces the oxidizable organic C from the surface layer of the soil by 50%, and 75% of this reduction happens in its labile fractions. Vanguelova et al. [36], on the other hand, concluded that there was no evidence that whole-tree harvest decreased soil organic C in a 28-year old second rotation stand of Sitka spruce in the UK. Slash removal has larger effects on soil C and nutrients when rotations are short, slash is removed repeatedly, clay content is low, temperatures are high, the site is relatively wet, and forest productivity is high [4,5]. It is well known that retaining harvest residues on the soil is extremely important to maintain high productivity of tropical forests [14–16]. However, more research is needed in order to reach a plausible conclusion about the subject due to the divergence in the data available, as observed by Nambiar and Harwood [14].

The forest residue removal for two rotations resulted in the cumulative loss of 23.5 t·ha^{-1} of C from 0–200 cm soil depth (ReM + F and ReR + F treatments). This reduction was more concentrated in the top 40 cm (Figure 2). This effect is a result of the additional removal of 80 t·ha^{-1} of biomass (slash and litter; 51 and 29 t·ha^{-1} in R1 and R2, respectively) in the ReR + F treatment. Furthermore, the reduced initial growth in the ReR + F treatment during both rotations resulted in less litter deposition [27]. In the second rotation, a 40% reduction in fine roots was observed in the 0–30 cm layer of the ReR + F treatment [37], which also contributed to low soil C content.

Forest residue removal resulted in a small reduction in the N stocks in the 0–20 cm layer in 2012 (Figure 1). However, no differences were observed between ReR + F and ReM + F treatments in 2016 (Figure 2) despite the higher harvest output in the ReR + F treatment. This can be attributed to the larger accumulation of N in the biomass in the ReM + F treatment, which was 33% higher when compared to the ReR + F treatment. This aligns with other long-term productivity studies that find that subsequent tree growth following stem only harvest tends generally to be higher than whole-tree harvest due to the maintenance of harvest residues on the soil [38].

The available P stock was larger in the ReM + F treatment when compared with ReR + F treatment. This result could be caused by two main factors: (i) less ReM + F nutrient outputs by harvest and (ii) larger quantities of organic matter in ReM + F which reduces the fixation of P on the soil colloids, thus improving its availability. No differences were observed between treatments after 100 cm in depth. In a long-term productivity experiment conducted in North Wales, UK, by Walmsley et al. [9], the alterations in soil characteristics and their effects were tested using different harvest treatments (whole tree harvest and bole-only harvest) on 23-year-old second rotation stands of Sitka spruce (*Picea sitchensis*). The authors found that the whole-tree harvest treatment is responsible for the depletion of three to four times greater quantities of N, P, and K than the conventional bole-only harvest in the first rotation.

ACR + F and ReR + F treatments contained the largest stocks of S in the 40–100 cm layer in the soil, being statistically different when compared to others. This can be attributed to the lower concentration of soil organic C (SOC) of the previously mentioned treatments. Low organic matter (OM) in highly weathered soils results in a higher quantity of positive charges, especially in the B horizon [39], with consequent retention of SO_4^{2-} in the soil.

Repeated removal of forest harvest residues is likely to reduce oxidizable C, available N, and total P stocks in the soil under *Eucalyptus* plantations. The use of harvest residues as soil coverage can protect the soil against erosion, improve or maintain the SOC, and improve both the quantity and availability of nutrients stored in the soil [5,11,15].

4.2. Soil Contribution to Nutrients Absorbed by Trees

The N and P losses as a result of the two harvest rotations (2004 and 2012) exceeded the inputs from fertilization and atmospheric deposition, resulting in a net loss of N and P from the site. In this period, the net N balance for ReM + F, ACR + F, ReR + F, and ACR − N treatments was −255, −516, −787, and −770 kg·ha^{-1}, respectively. With regards to P, there was also a negative net balance of −24, −48, −70, and −128 kg·ha^{-1} in the ReM + F, ACR + F, ReR + F, and ACR − P treatments, respectively. The net balance was positive only for S, with approximately +120 kg·ha^{-1} for all treatments with the exception of the ACR − N treatment, which contained a net balance of −128 kg·S·ha^{-1} (Table 4). N inputs through biological fixation are unlikely due to the absence of N-fixing weeds. In addition, one should not expect inputs of N from rock weathering as the soil was highly weathered (Ferralsol). The difference between nutrient losses due to harvesting and nutrients supplied via fertilization and atmospheric deposition must be made up for with nutrients absorbed from soil pools.

Table 4. Nutrient balance in two *Eucalyptus grandis* rotations under different timber harvest intensities and fertilizer applications.

Treatment	N (kg·ha^{-1})	P (kg·ha^{-1})	S (kg·ha^{-1})
Net Nutrient Balance [1]			
ReM + F	−255	−24	148
ACR + F	−516	−48	135
ReR + F	−787	−70	111
ACR − N	−770	−48	−128
ACR − P	−490	−128	141
Difference in Soil Nutrients Stocks from 2004 to 2012 (0–20 cm)			
ReM + F	−164	0.8	−1.1
ACR + F	−189	−0.4	−1.3
ReR + F	−192	−0.5	−2.5
ACR − N	−219	−0.9	−1.4
ACR − P	−189	−2.0	4.0

[1] Differences between the inputs by fertilizer application and atmospheric deposition with the harvest outputs (Table 3 and Figure 1).

From 2004 to 2012, soil N stocks declined in the 0–20 cm layer for all treatments (Figure 1). The ReR + F treatment was the most affected among the treatments due to the elevated nutrient outputs by harvest. For the ACR − N treatment, this was due to the small quantity of N applied via fertilization. In the ReM + F treatment, the 0–20 cm layer was responsible for providing 65% of the difference in the balance of soil N. The 0–20 cm layer was responsible for providing 35% of the difference in net soil N balance for the ACR treatments and 24% for the ReR + F treatment (Figure 1 and Table 4). The maintenance of harvest residues on the soil increases soil C and N pools—especially in labile fractions—thereby increasing the capacity of upper layers to provide N to plants [14–16,30].

The remaining difference possibly came from deeper layers of the soil profile and factors such as biological fixation by symbiotic associations which were not examined in this study [40]. Despite the large difference in the nutrient outputs between ReM + F and ReR + F treatments (more than 500 kg·ha^{-1} in the net balance), no differences in the N stocks between these treatments were found in the deeper layers of the soil (20–200 cm). More research is necessary in order to fully understand the additional sources of N in such highly weathered, tropical forests.

Little reduction was observed in the stocks of available P in the 0–20 layer (Figure 1). The ACR − P treatment presented the highest reduction (2 kg·ha^{-1}), while the ReM + F treatment presented a slight increase. This happened even with the negative balance of the nutrient for all treatments evaluated. No reduction in soil P below 40 cm depth was observed in 2016 (Figure 2). The small reduction in P content contradicts the highly negative net balance of this nutrient across treatments. P balance for all treatments was −60 kg·ha^{-1} on average. The ACR − P treatment had a deficit of −128 kg·ha^{-1} while the reduction was only 6 kg·ha^{-1} from 0–200 cm of soil depth. This discrepancy suggests that the Eucalyptus trees take up soil P hidden in fractions not extracted by the traditional resin method of analysis. Eucalyptus species can uptake organic and inorganic P fractions by phosphatase and exudation of low molecular mass acids [41]. More studies are necessary to clarify the complete origin of P absorbed by trees.

Even with a positive net balance of S stocks in the soil (Table 4), a small reduction was observed in the stocks of the nutrient when comparing the evaluations in 2004 and 2012. This is due to the elevated mobility of the element in the soil, especially in the upper layers [42], along with its consequent migration to and accumulation in the 40–100 cm layer (Figure 2) where it will adsorb to positive charges on soil surfaces exposed by losses of OM [39].

4.3. Management Considerations

Despite the small effect of whole-tree harvest on the soil nutrient stocks described by many authors [4,14,29–33], this and other studies emphasize the importance of the maintenance of forest residues on the soil, especially in tropical sites with low buffer capacity soils, high productivity, and short cycle plantations [14,16,29–31,33]. In these sites, removal of forest residues can result in loss of wood productivity of up to 40% due to the low nutrient pools remaining in the soil [16].

In the past several years, increasing demands for the use of forest residues for bioenergy purposes has grown in Brazil. Models for the utilization of these residues are based on research in temperate regions, where soils often have higher buffer capacities and higher organic C contents. This study shows that in tropical conditions the use of forest residues for bioenergy purposes should be carefully considered, taking into account the unique conditions of each site. On steep sites and/or those with low buffer capacity, all forest residues should be retained on the soil in order to avoid soil erosion [43] and depletion of soil nutrient pools [14,16,29,30,33,41]. In sites with favorable conditions for residue removal, preference should be given to the coarse residues due to their high caloric power and reduced nutrient concentration. With the removal of forest residues, the application of high rates of fertilizer is necessary in order to avoid productivity losses and to ensure the sustainability of the silviculture system.

The high negative net balance, especially for N and P, and the relative low reduction in the availability of these nutrients in the soil draws attention to two main points. First, more attention should be given to production sustainability, mainly in sites harvested in the whole-tree harvest system. Second, more studies should be implemented to better understand the contributions of organic and inorganic (mainly to P) fractions of low lability on the supply of nutrients to trees. Our results and others [29,41] suggest that Eucalyptus trees can access P fractions not identified by traditional soil analysis methods. Regarding N concentrations, even with high harvest outputs (more than 500 kg·ha^{-1} in each crop rotation) no response to N application by fertilization was found in terms of wood productivity in Brazilian conditions [28]. More research is essential to better understand the sources and the cycle of this nutrient within tropical soil-plant systems.

5. Conclusions

It was observed a reduction in organic C, total N, available P and S stocks in the whole tree harvest and the whole tree plus litter treatments, especially at 0–40cm. The balances of N and P in the two rotations of the experiment were negative in all treatments, with even lower values when whole tree and litter were removed or when no fertilizer was applied. The reduction in the total N and available P stocks in the soil explain around 50% of the negative balance observed for these nutrients; the source of the remaining 50% was not determined for this study. Our results indicate that N and P soil pools play an important role in forest nutrition; the reduction in these pools can possibly affect the sustainability of the production system in the long term. More research is necessary in order to better understand the balance of nutrients within the soil-plant system.

Acknowledgments: The authors would like to thank University of Washington/EUA, Experimental Station of Forest Sciences of Itatinga/Brazil, University of Sao Paulo/Brazil and FAPESP/Brazil (Processes: 2010/16623-9 and 2014/15876-1) for the financial support and for field collections.

Author Contributions: M.L.C.M. and J.H.T.R. wrote the manuscript as well as performed the soil sampling, and modeled and interpreted the results. A.R., J.C.A.J., A.V.F., J.N.J., and S.M.-C. helped on the interpretation of some of the results. R.R. and J.L.d.M.G. gave suggestions on the modeling and interpretation of the results. R.F.A. and M.d.C.P. performed the laboratory analysis.

Conflicts of Interest: The authors declare no conflict of interest.

References

1. Schlesinger, W.H.; Bernhardt, E.S. *Biogeochemistry: An Analysis of Global Change*, 3rd ed.; Elsevier: New York, NY, USA, 2013; pp. 1–664.
2. Jobbagy, E.; Jackson, R. The distribution of soil nutrients with depth: Global patterns and the imprint of plants. *Biogeochemistry* **2001**, *53*, 51–77. [CrossRef]
3. Vitousek, P.M.; Naylor, R.; Crews, T.; David, M.B.; Drinkwater, L.E.; Holland, E. Nutrient imbalances in agricultural development. *Science* **2009**, *324*, 1519–1520. [CrossRef] [PubMed]
4. Du Toit, B. Effects of site management on growth, biomass partitioning and light use efficiency in a young stand of *Eucalyptus grandis* in South Africa. *For. Ecol. Manag.* **2008**, *255*, 2324–2336. [CrossRef]
5. Gonçalves, J.L.N.; Stape, J.L.; Wichert, M.C.P.; Gava, J.L. *Conservação e Cultivo de Solos Para Plantações de Florestas (Soil Conservation and Cultivation for Forest Production)*; Instituto de Pesquisas e Estudos Florestais (IPEF): Piracicaba, (SP), Brazil, 2002; pp. 131–204.
6. Corbeels, M.; McMurtrie, R.E.; Pepper, D.A.; Mendham, D.S.; Grove, T.S.; O'Connell, A.M. Long-term changes in productivity of Eucalyptus plantations under different harvest residues and nitrogen and management practices: A modeling analysis. *For. Ecol. Manag.* **2005**, *217*, 1–18. [CrossRef]
7. Compton, J.E.; Cole, D.W. Phosphorus cycling and soil P fractions in Douglas-fir and red alder stands. *For. Ecol. Manag.* **1998**, *110*, 101–112. [CrossRef]
8. Hall, S.J.; Matson, P.A. Nutrient status of tropical rainforest influences soil N dynamics after N additions. *Ecol. Monogr.* **2003**, *73*, 107–129. [CrossRef]
9. Walmsley, J.D.; Jones, D.L.; Reynolds, B.; Price, M.H.; Healey, J.R. Whole tree harvesting can reduce second rotation forest productivity. *For. Ecol. Manag.* **2009**, *257*, 1104–1111. [CrossRef]
10. Jones, H.; Beets, P.; Kimberley, M.; Garrett, M. Harvest residue management and fertilization effect on soil carbon and nitrogen in a 15-year old *Pinus radiata* plantation forest. *For. Ecol. Manag.* **2011**, *262*, 339–347. [CrossRef]
11. Rocha, J.H.T.; Marques, E.R.G.; Goncalves, J.L.M.; Hubner, A.; Brandini, C.B.; Ferraz, A.V.; Moreira, R.M. Decomposition rates of forest residues and soil fertility after clear-cutting of *Eucalyptus grandis* stands in response to site management and fertilizer application. *Soil Use Manag.* **2016**. [CrossRef]
12. Xu, X.; Luo, Y.; Zhou, J. Carbon quality and the temperature sensitivity of soil organic carbon in a tallgrass prairie. *Soil Biol. Biochem.* **2012**, *50*, 142–148. [CrossRef]
13. Mendham, D.S.; White, D.A.; Battaglia, M.; McGrath, J.F.; Short, T.M.; Ogden, G.N.; Kinal, J. Soil water depletion and replenishment during first- and early second-rotation *Eucalyptus globulus* plantations with deep soil profiles. *Agric. For. Meteorol.* **2011**, *151*, 1568–1579. [CrossRef]

14. Nambiar, E.K.S.; Harwood, C.E. Productivity of acacia and eucalypt plantations in Southeast Asia 1. Biophysical determinants of production: Opportunities and challenges. *Intern. For. Rev.* **2014**, *16*, 225–248.
15. Achat, D.L.; Deleuze, C.; Landmann, G.; Pousse, N.; Ranger, J.; Augusto, L. Quantifying consequences of removing harvesting residues on forest soils and tree growth—A meta-analysis. *For. Ecol. Manag.* **2015**, *348*, 124–141. [CrossRef]
16. Rocha, J.H.T.; Goncalves, J.L.M.; Gava, J.L.; Godinho, T.O.; Melo, E.S.A.C.; Bazani, J.H.; Hubner, A.; Arthur Junior, J.C.; Wichert, M.P. Forest residue maintenance increased the wood productivity of a *Eucalyptus* plantation over two short rotations. *For. Ecol. Manag.* **2016**, *379*, 1–10. [CrossRef]
17. Peng, Y.; Thomas, S.C.; Tian, D. Forest management and soil respiration: Implications for carbon sequestration. *Environ. Rev.* **2008**, *16*, 93–111. [CrossRef]
18. Holub, S.M.; Terry, T.A.; Harrington, C.A.; Harrison, R.B.; Meade, R. Tree growth ten years after residual biomass removal, soil compaction, tillage, and competing vegetation control in a highly productive Douglas-fir plantation. *For. Ecol. Manag.* **2013**, *305*, 60–66. [CrossRef]
19. Alvares, C.A.; Stape, J.L.; Sentelhas, P.C.; Goncalves, J.L.M.; Sparovek, G. Koppen's climate classification map for Brazil. *Meteorol. Z.* **2013**, *22*, 711–728. [CrossRef]
20. Van Raij, B.; Andrade, J.C.; Cantarella, H.; Quaggio, J.A. *Análise Química Para Avaliação da Fertilidade de Solos Tropicais (Chemical Analysis for the Evaluation of Fertility in Tropical Soils)*; Instituto Agronomico de Campinas (IAC): Campinas, (SP), Brazil, 2001; pp. 1–58.
21. Ribeiro, J.F.; Walter, B.M.T. Fitofisionomias do Bioma Cerrado. Phytophisiognomy of Cerrado biome. In *Cerrado: Ambiente e Flora (Cerrado: Environment and Flora)*; Sano, S.M., Almeida, S.P., Eds.; Embrapa: Brasilia, Brazil, 1998; pp. 1–556.
22. Nelson, D.W.; Sommers, L.E. Total carbon, organic carbon, and organic matter. In *Methods of Soil Analysis. Part 3. Chemical Methods*; SSSA Book Series No. 5; Sparks, D.L, Ed.; Soil Science Society of America/SSSA and American Society of Agronomy (ASA): Madison, WI, USA, 1996; pp. 961–1010.
23. Walkley, A.; Black, I.A. An examination of the Degtjareff method for determining soil organic matter, and a proposed modification of the chromic acid titration method. *Soil Sci.* **1934**, *37*, 29–38. [CrossRef]
24. Schumacher, F.X.; Hall, F.D.S. Logarithmic expression of timber-tree volume. *J. Agric. Res.* **1933**, *47*, 719–734.
25. Malavolta, E.; Vitti, G.C.; Oliveira, A.S. *Avaliação do Estado Nutricional das Plantas: Princípios e Aplicações (Evaluation of Nutritional Condition of Plants: Principles and Application)*; Associação Brasileira para Pesquisa da Potassa e do Fosfato: Piracicaba, Brazil, 1989; pp. 1–201.
26. Zar, J.H. *Biostatistical Analysis*, 5th ed.; Pearson Education Ltd.: DeKalb, IL, USA, 2010.
27. Hubner, A. Ciclagem de Nutrientes e Produtividade de Madeira em Povoamento de Eucalyptus Grandis Sob Diferentes Manejos de Resíduos Florestais e Fertilização Mineral (Nutrient Cycling and Growth in *Eucalyptus Grandis* Plantation under Different Forest Residues Management and Fertilization). Ph.D. Thesis, ESALQ-USP, Piracicaba, Brazil, 2015.
28. Pulito, A.P.; Goncalves, J.L.M.; Smethurst, P.J.; Arthur Junior, J.C.; Alvares, C.A.; Rocha, J.H.T.; Hubner, A.; Moraes, L.F.; Miranda, A.C.; Kamogawa, M.Y.; et al. Available nitrogen and responses to nitrogen fertilizer in Brazilian Eucalypt plantations on soils of contrasting textures. *Forests* **2015**, *6*, 973–991. [CrossRef]
29. Laclau, J.-P.; Ranger, J.; Goncalves, J.L.M.; Maquere, V.; Krusche, A.V.; M'Bou, A.T.; Nouvellon, Y.; Saint-Andre, L.; Bouillet, J.-P.; Piccolo, M.D.C. Biogeochemical cycles of nutrients in tropical *Eucalyptus* plantations Main features shown by intensive monitoring in Congo and Brazil. *For. Ecol. Manag.* **2010**, *259*, 1771–1785. [CrossRef]
30. Mendham, D.S.; Ogden, G.N.; Short, T.; O'Connell, T.M.; Grove, T.S.; Rance, S.J. Repeated harvest residue removal reduces *E. globulus* productivity in the 3rd rotation in south-western Australia. *For. Ecol. Manag.* **2014**, *329*, 279–286. [CrossRef]
31. Mendham, D.S.; O'Connell, A.M.; Grove, T.S.; Rance, S.J. Residue management effects on soil carbon and nutrient contents and growth of second rotation eucalypts. *For. Ecol. Manag.* **2003**, *181*, 357–372. [CrossRef]
32. Huang, Z.; He, Z.; Wan, X.; Hu, Z.; Fan, S.; Yang, Y. Harvest residue management effects on tree growth and ecosystem carbon in a Chinese fir plantation in subtropical China. *Plant Soil* **2013**, *364*, 303–314. [CrossRef]
33. Kumaraswamy, S.; Mendham, D.S.; Grove, T.S.; O'Connell, A.M.; Sankaran, K.V.; Rance, S.J. Harvest residue effects on soil organic matter, nutrients and microbial biomass in eucalypt plantations in Kerala, India. *For. Ecol. Manag.* **2014**, *328*, 140–149. [CrossRef]

34. Moreno-Fernandez, D.; Dias-Pines, E.; Barbeito, I.; Sanchez-Gonzalez, M.; Montes, F.; Rubio, A.; Canellas, I. Temporal carbon dynamics over the rotation period of two alternative management systems in Mediterranean mountain Scots pine forests. *For. Ecol. Manag.* **2015**, *348*, 186–195. [CrossRef]

35. Rocha, J.H.T. Reflexos do Manejo de Resíduos Florestais na Produtividade, Nutrição e Fertilidade do Solo em Plantações de *Eucalyptus grandis* (Implications of Forest Harvest Residues Management on the Productivity, Nutrition and Soil Fertility in *Eucalyptus grandis* Plantation in Brazil). Master's Thesis, ESALQ-USP, Piracicaba, Brazil, 2014.

36. Vanguelova, E.; Pitman, R.; Luiro, J.; Helmisaari, H.-S. Long-term effects of whole tree harvesting on soil carbon and nutrient sustainability in the UK. *Biogeochemistry* **2010**, *101*, 43–59. [CrossRef]

37. Rocha, J.H.T.; Franci, A.F.; Gonçalves, J.L.M. Effect of forest residue management on fine roots density in an *Eucalyptus grandis* plantation. "Luiz de Queiroz" College of Agriculture, University of São Paulo: Piracicaba, Brazil, Unpublished data. 2016.

38. Smolander, A.; Kitunen, V.; Tamminen, P.; Kukkola, M. Removal of logging residue in Norway spruce thinning stands: Long-term changes in organic layer properties. *Soil Biol. Biochem.* **2010**, *42*, 1222–1228. [CrossRef]

39. Brady, N.C.; Weil, R.R. *The Nature and Properties of the Soil*, 14th ed.; Bookmam: DeKalb, IL, USA, 2013; pp. 1–716.

40. Silva, M.D.S.; Paula, T.D.; Moreira, B.C.; Carolino, M.; Cruz, C.; Bazzolli, D.M.S.; Silva, C.C.; Kasuya, M.C.M. Nitrogen-Fixing Bacteria in *Eucalyptus globulus* Plantations. *PLoS ONE* **2014**, *9*, e111313. [CrossRef] [PubMed]

41. Costa, M.G.; Gama-Rodrigues, A.C.; Gonçalves, J.L.M.; Gama-Rodrigues, E.F.; Sales, M.V.S.; Aleixo, S. Labile and non-labile fractions of phosphorus and its transformations in soil under *Eucalyptus* plantations, Brazil. *Forests* **2016**, *7*, 1–15. [CrossRef]

42. Bissani, C.A.; Tedesco, M.J. O enxofre no solo (Sulphur in the soil). In *Reunião Brasileira de Fertilidade do Solo*, 17nd ed.; Borkert, C.M., Lantmann, A.F., Eds.; EMBRAPA/IAPAR/SBCS: Londrina, Brazil, 1988; pp. 11–29.

43. Candido, B.M.; Silva, M.L.N.; Curi, N.; Batista, P.V.G. Water erosion post-planting in eucalyptus forests in the Parana river basin, in the eastern Mato Grosso do Sul. *Revista Brasileira De Ciencia Do Solo* **2014**, *38*, 1565–1575.

forests

MDPI

Article

Phosphorus in Preferential Flow Pathways of Forest Soils in Germany

Dorit Julich *, Stefan Julich and Karl-Heinz Feger

Institute of Soil Science and Site Ecology, TU Dresden, Pienner Str. 19, 01737 Tharandt, Germany;
Stefan.Julich@tu-dresden.de (S.J.); karl-heinz.feger@tu-dresden.de (K.-H.F.)
* Correspondence: dorit.julich@tu-dresden.de; Tel.: +49-351-463-31390

Academic Editors: Scott X. Chang and Xiangyang Sun
Received: 31 August 2016; Accepted: 23 December 2016; Published: 30 December 2016

Abstract: The transport of nutrients in forest soils predominantly occurs along preferential flow pathways (PFP). This study investigated the composition of phosphorus (P) forms in PFPs and soil matrix in several temperate beech forests with contrasting soil P contents in Germany. The PFPs were visualized using dye tracer experiments. Stained and unstained soil was sampled from three profile cuts per plot and analyzed for P fractions. The results show that labile P concentrations were highest in the O-layer and had the same range of values at all sites (240–320 mg·kg^{-1}), although total P (TP) differed considerably (530–2330 mg·kg^{-1}). The ratio of labile P to TP was significantly lower in the P-rich soil compared to the medium and P-poor soils. By contrast, the ratio of moderately labile P to TP was highest at the P-rich site. The shifts in P fractions with soil depth were generally gradual in the P-rich soil, but more abrupt at the others. The contents of labile and moderately labile P clearly differed in PFPs compared to soil matrix, but not statistically significant. The studied soils are characterized by high stone contents with low potential for P sorption. However, indications were found that labile organically bound P accumulates in PFPs such as biopores.

Keywords: phosphorus fractions; forest soil; preferential flow; PFP

1. Introduction

Phosphorus (P) in forest soils is distributed highly variably, both between and within soil types, with respect to contents, speciation, availability, and sources of P [1]. Soils contain various P forms, generally differentiated in organically bound (P_o) and inorganic (P_i) P. Organic P occurs as orthophosphate monoesters (e.g., inositols, mononucleotides, sugar phosphates), orthophosphate diesters (phospholipids, teichoic acids, nucleic acids, and their degradation products), and phosphonates. Inorganic P is bound in minerals (e.g., apatite) and is present in dissolved form as ortho-, pyro-, and/or polyphosphates [2,3]. The total content and stock of soil P varies from very low (<100 mg·kg^{-1} and <10 g·m^{-2}) to P-rich (>3000 mg·kg^{-1} and >1000 g·m^{-2}), and the mineral P supply and P speciation is strongly dependent on the parent material and pedogenesis [4,5]. Whereas P supply from the mineral phase is the main source of available P in P-rich soils, P availability for plants in P-poor soils is considerably influenced by P speciation [4]. These differences in P supply drive P fluxes in the systems and determine the dominating strategy of ecosystem nutrition [6]. Besides those internal factors, external environmental and anthropogenic impacts (as forest management, N deposition, etc.) substantially influence the distribution of P and nutrient cycling in temperate forest soils [7–10].

The predominant transport pathway for nutrients in forest soils is along preferential flow pathways (PFP), bypassing large parts of the soil matrix [11–13]. An established method to investigate PFPs and their role in water and nutrient transport are tracer experiments [14–17]. Previous tracer studies indicated that PFPs in forest soils can be stable for decades, resulting in differences in biological and chemical parameters between PFPs and the soil matrix, particularly when preferential flow along root channels dominates the flow system [12,18]. Accumulation in stable flow paths has been found in previous studies for radionuclides, SOC, N [15,18], and also for C, Ca, Mg, and Fe [12]. Backnäs et al. [19] investigated P fractions in PFPs as compared to the soil matrix in a Podzol under forest in Southern Lapland, Finland. The results indicated that the leaching of P through PFPs occurred primarily along stone surfaces, where PFPs are larger than PFPs related to coarse grains or roots. Thus, P sorption may be limited due to potentially faster water flow, weaker lateral mass exchange and less sorption interaction than in smaller PFPs [20]. By contrast, P accumulation might occur in small preferential pores with decelerated water flow and longer contact time between the solution and inner pore surfaces [19]. As a result, such PFPs with high potential to accumulate P increase the likelihood for elevated P concentrations in export fluxes [21]. However, there are very few data available regarding P accumulation and transport in forest soils. Therefore, this study investigated the composition of P forms in PFPs and the soil matrix in four temperate forest ecosystems in Germany. Furthermore, the influence of different initial conditions regarding the content of total and available P at the sites on the distribution of P forms in the soil profiles and in PFPs was examined. The correlation between P fractions and chemical parameters was tested to identify the most important factors influencing the occurrence of P fractions in the investigated soils.

2. Materials and Methods

2.1. Site Description

Field experiments and sampling were conducted at four beech dominated forests (*F. sylvatica* L.) across Germany with similar stand age (110–120 years). The four study sites Lüss (LUE), Vessertal (VES), Mitterfels (MIT), and Bad Brückenau (BBR) are core sites of the German DFG Priority Program SPP 1685 which aims at testing the overall hypothesis that the P-depletion of soils drives forest ecosystems from P acquiring systems (efficient mobilization of P from the mineral phase) to P recycling systems (highly efficient cycling of P) [6]. Further, all sites are long-term monitoring sites in Germany and part of the EU-ICP Forests Level II Program (International Co-operative Programme on the Assessment and Monitoring of Air Pollution Effects on Forests). The sites were selected with regard to their considerable different availability of mineral P following the sequence BBR > MIT > VES > LUE. The BBR site was defined as P-rich site, LUE as P-poor site, and MIT and VES as medium P-supplied sites. Three sites (VES, MIT, and BBR) are located in the low mountain ranges, whereas LUE is in the lowlands. The sites differ in parent material, soil type, and climate (Table 1). Soil properties are summarized in Table 2. For better comparisons, the soil horizons were aggregated for each site considering their physical and chemical properties to: O—organic layer, A—topsoil horizon, B—subsoil horizon, C—lower subsoil horizon containing weathered parent material.

Table 1. Overview over the study sites.

	LUE	VES	MIT	BBR
Location	52°50′32″ N 10°16′06″ E	50°36′26″ N 10°46′20″ E	48°58′35″ N 12°52′49″ E	50°21′38″ N 9°55′71″ E
Elevation (m a.s.l.)	115	810	1023	809
MAT (°C)	8.0	5.5	4.5	5.8
MAP (mm)	779	1200	1229	1031
Parent material	Pleistocene sands	Trachyandesite	Gneiss	Basalt
Soil type (WRB *)	Dystric skeletic Cambisol	Hyperdystric skeletic folic Cambisol	Hyperdystric skeletic chromic Cambisol	Hyperdystric folic Cambisol

LUE: Lüss (Lower Saxony); VES, Vessertal (Thuringian Forest); MIT, Mitterfels (Bavarian Forest); BBR, Bad Brückenau (Bavarian Rhön Mountains); MAT, Mean annual temperature; MAP, Mean annual precipitation; * World Reference Base for Soil Resources 2014, World Soil Resources Reports 106, FAO, Rome, Italy.

Table 2. Soil properties at four study sites in beech forests in Germany.

Study Site/ Horizon	Depth (cm)	Texture	pH (CaCl$_2$)	C (%)	N (%)	Al$_{ox}$	Al$_{di}$	Fe$_{ox}$ (g·kg^{-1})	Fe$_{di}$	P$_{ox}$	TP
Lüss (LUE)											
O	9 to 0		2.8	36.63	1.59	0.73	0.79	0.87	1.30	0.15	0.49
A	0 to −12	Loamy sand	3.0	1.69	0.07	0.18	0.37	0.29	2.17	<0.04	0.06
B	−12 to −52	Loamy sand	3.7	0.90	0.04	1.04	1.00	1.40	3.21	<0.04	0.08
C	−52 to −100	Sand	4.3	0.12	0.01	0.50	0.41	0.15	1.01	<0.04	0.04
Vessertal (VES)											
O	4 to 0		3.4	32.17	1.72	4.11	1.20	1.28	1.75	0.30	0.86
A	0 to −3	Sandy loam	3.0	7.49	0.45	2.75	1.68	3.14	7.46	0.53	0.70
B	−3 to −65	Sandy loam	4.1	3.03	0.20	9.44	7.40	5.13	9.23	0.83	0.98
C	−65 to −100	Loamy sand	4.1	0.37	0.03	4.64	2.83	0.59	5.55	0.16	0.32
Mitterfels (MIT)											
O	4 to 0		3.5	44.12	2.19	0.95	0.87	1.28	1.45	0.34	0.88
A	0 to −4	Silt loam	3.0	35.50	2.05	3.32	3.83	3.58	4.97	0.47	1.14
B	−4 to −80	Silt loam	3.9	5.45	0.34	6.26	5.62	8.05	12.17	0.53	0.91
C	−80 to −100	Loamy sand	4.3	1.09	0.09	5.73	4.63	3.61	10.89	0.35	0.66
Bad Brückenau (BBR)											
O	3 to 0		4.6	27.33	1.68	5.53	5.51	15.27	21.85	1.29	2.24
A	0 to −7	Silty clay	4.2	17.57	1.28	7.77	8.30	21.15	31.52	1.51	2.45
B	−7 to −30	Silt clay loam	4.3	7.45	0.54	10.58	10.83	24.46	35.41	1.84	2.62
C	−30 to −65	Loam	4.5	4.60	0.34	10.70	9.57	18.87	25.92	1.63	2.39

Horizon: O—organic layer, A—topsoil horizon, B—subsoil horizon, C—lower subsoil horizon containing weathered parent material. Values for C and N are total contents (for C this equals organic C); Al$_{ox}$, Fe$_{ox}$, and P$_{ox}$—ammonium oxalate-extractable contents; Al$_{di}$ and Fe$_{di}$—dithionite-extractable contents; TP—total P contents.

2.2. Field Experiments

At each study site a test plot of 1 m^2 was established at a minimum distance of 2 m from any trees. On the first day, the plots were irrigated with 30 L of tracer-water solution containing 3 g·L^{-1} of Brilliant Blue FCF (within 1 hour) to activate rapid flow paths (PFPs) in the soils. Brilliant Blue dye was used due to its good visibility, retardation behavior in soil, and its non-toxicity to the environment [16,17]. After the dye tracer irrigation, the plots were sprinkled with 2 L of water to wash off the dye residues from the organic layer [19], and then covered overnight with a tarpaulin to exclude disturbances (e.g., rainfall). The irrigated plots were excavated on the following day by successively preparing five horizontal profile cuts per plot (at 20, 35, 50, 65, and 80 cm from the plot border) down to 1 m depth and 1 m width (see Figure 1). Stained (flow region) and unstained (non-flow region) soil was sampled respectively from the three inner cuts at 35, 50, and 65 cm from all morphological horizons. The amounts of soil material per sample varied because the thickness of soil horizons and stained/unstained parts per horizon differed considerably. The samples were sieved (<2 mm), dried at 40 °C, and then stored in plastic bottles until analysis. For the analysis of C, N, and total P the samples were finely grounded.

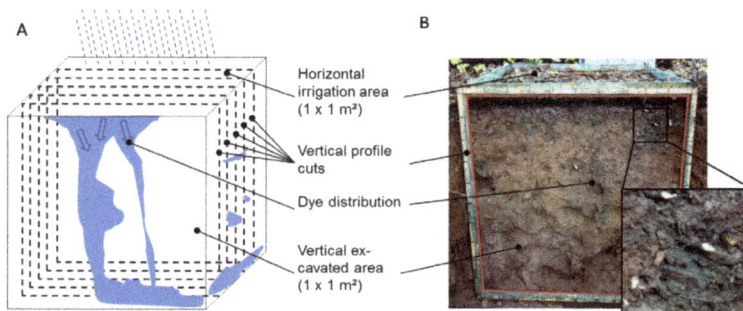

Figure 1. Scheme of tracer experiments (**A**): Dye application at the soil surface (1 m^2), covering overnight, and excavation of five vertical (1 m^2) profile cuts (example on the right (**B**)). The black framed enlargement shows staining of flow pathways along roots and stone surfaces.

2.3. Chemical Analyses

The P fractions in stained and unstained soil samples were analyzed using the Hedley fractionation method [22,23]. This method uses increasingly strong extractants to determine P forms with differing availability for plants. All samples were extracted with a replicate. Additionally, certified reference soil material from the Wageningen Evaluating Programs for Analytical Laboratories, International Soil-analytical Exchange (WEPAL-ISE 865 and 884) and blanks were treated equally to the samples for quality control of the extraction procedure.

In the fractionation procedure, a subsample of 0.5 g soil was first shaken in distilled water containing an anion exchange resin (Dowex 22, 16–50 mesh, Sigma-Aldrich, Taufkirchen, Germany) in netting bags (PP-405/230-41, Bückmann, Mönchengladbach, Germany) [24]. The following extractants applied were: 0.5 M sodium bicarbonate ($NaHCO_3$-P), 0.1 M sodium hydroxide (NaOH-$_i$), 1 M hydrochloric acid (HCl_{dil}-P), hot concentrated 3.5 M hydrochloric acid (HCl_{conc}-P), and finally nitric acid digestion of the residual P fraction (residual P). After each extraction step, the amount of inorganic P in the extract solution was measured using the method of Murphy and Riley [25] and O'Halloran and Cade-Menun [26]. For the determination of total P (TP) in $NaHCO_3$, NaOH, and HCl_{conc}, the extracts were oxidized in an autoclave using ammonium persulfate. Organically bound P (P_o) was calculated as the difference between TP and P_i [23]. The P concentrations in all extracts were measured colorimetrically with a spectrophotometer (UV-mini 1240, Shimadzu Deutschland GmbH, Duisburg, Germany). For a detailed description of the procedure applied here see Tiessen and Moir [23].

According to the concept of Hedley fractionation, the different P fractions represent (1) surface-adsorbed readily available P, (2) strongly-bonded or adsorbed moderate available P, and (3) very strongly-bonded or inaccessible mineral P with very low availability [22,23,27]. Therefore, the analyzed P fractions were classified for further data analyses as labile P (resin-P_i + $NaHCO_3$-$P_{i,o}$), moderately labile P (NaOH-$P_{i,o}$ + HCl_{dil}-P_i), and stable P (HCl_{conc}-$P_{i,o}$ + residual-P) (cf. [28]). The initial TP content of all samples was determined using the same method as the residual P fraction, by digestion with concentrated nitric acid and measurement with ICP-OES (CIROS, SPECTRO Analytical Instruments GmbH, Kleve, Germany). Total C and N contents were measured after dry combustion of grounded samples with a CN analyzer (Vario EL III Elementar GmbH, Hanau, Germany). The soils contained no inorganic C, thus soil organic C (SOC) equals total C. The concentrations of C in PFPs were not corrected for any C added as part of the Brilliant Blue dye. The C content of the dye is 56%, and the amount of total added C in the experiments was 50 g C m^{-2}. Bundt et al. [15] conducted comparable experiments with a dye-derived addition of 75 g C m^{-2} and found a resulting increase of C with a maximum of 2.5 % SOC in in PFPs in 0–9 cm depth, and a maximum of 3.5 % SOC in PFPs in 50–100 cm depth. In the current study, less dye was applied to the plots, and thus the change in C content in PFPs is expected to be very low. Amorphous metal sesquioxides (Al_{ox}, Fe_{ox}, Mn_{ox}) and P_{ox} were determined by extraction with acid 0.2 M ammonium oxalate (100 mL added to a subsample of 2 g). Dithionite-extractable (Fe_{di}, Al_{di}, Mn_{di}) was analyzed with 50 mL Na-citrate-dithionite in a subsample of 0.5 g (modified after Holmgren, 1967). The measurement of element contents in the extracts was conducted with ICP-OES.

2.4. Statistical Analysis

The statistical significance of differences between P contents of different chemical fractions in flow pathways and matrix were tested using the R software package (version R i386 3.1.0, R Foundation for Statistical Computing 2014, Vienna, Austria). Given that most of the data sets exhibited non-normal distributions (tested by Shapiro-Wilk test and q-q-plot), the non-parametric Wilcoxon rank-sum test was used to statistically compare mean values. The statistical tests were conducted pair-wise including all data of the four sites, and at horizon level due to the hierarchical structure of the data (profiles, cuts, horizons). For correlations between chemical parameters and P fractions, the Spearman's rho correlation coefficients were calculated and tested for significance.

3. Results and Discussion

The soil P fractions classified as labile, moderately labile, and stable P for PFP and soil matrix samples are shown in Figure 2. The values represent mean values for the three vertical cuts per profile. For the forest floor (O-layer), it was not possible to distinguish between stained and unstained material, due to the homogeneous distribution of the dye tracer from the irrigation. At the sites VES and MIT, the tracer did not reach the C horizon. Thus, the presented results contain no soil matrix data for the O-layer and no PFP data for the lower subsoil.

The TP values differed at the four study sites in their contents (45–2700 mg·kg^{-1}) and distribution within the soil profiles (Figure 2, TP = total bar length). Whereas TP at the P-poor site (LUE) was clearly highest in the O-layer and strongly reduced in the mineral soil, the P-rich soil (BBR) showed the lowest values in the O-layer and had a slight increase in the subsoil. However, the amount of TP in all horizons at the P-rich site was much higher compared to the other sites, and was more or less equally distributed through the profile. The summarized P fractions were compared to the initial TP contents in order to determine the recovery rate of the analytical method. Considering all samples, the TP recovery was very high, with a correlation coefficient of 0.98.

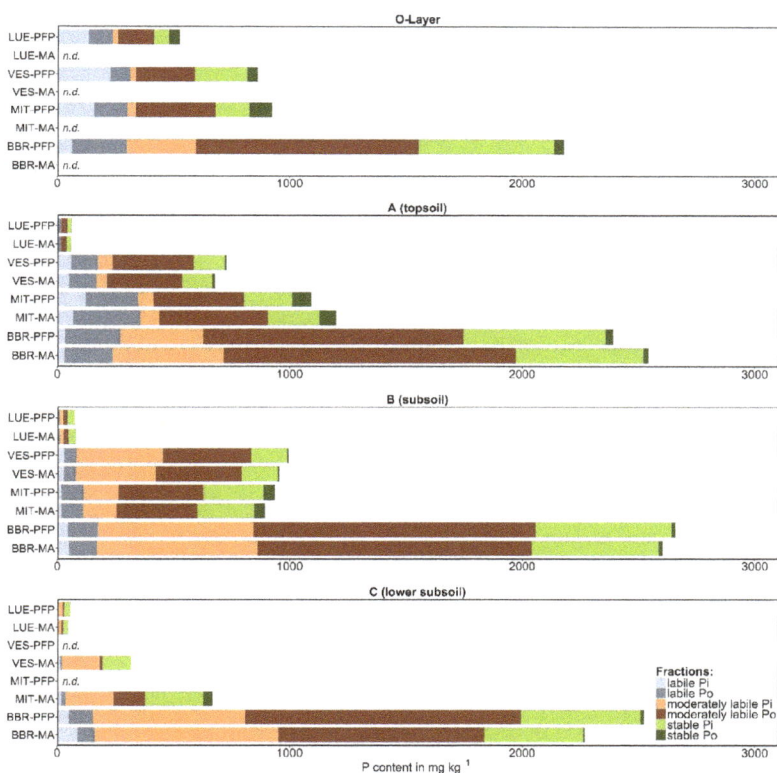

Figure 2. Phosphorus fractions summarized as labile (resin-P$_i$ + NaHCO$_3$-P$_{i,o}$), moderately labile (NaOH-P$_{i,o}$ + HCl$_{dil}$-P$_i$), and stable P (HCl$_{conc}$-P$_{i,o}$ + residual-P) for PFP (preferential flow pathways) and MA (soil matrix) samples; n.d., no data available; study sites in the sequence P-poor to P-rich are: LUE, Lüss, Lower Saxony; VES, Vessertal, Thuringian Forest; MIT, Mitterfels, Bavarian Forest; BBR, Bad Brückenau, Bavarian Rhön Mountains.

3.1. Labile P Fractions

Despite the large difference in TP between the sites, labile P (here defined as resin-P_i + NaHCO$_3$-$P_{i,o}$) was in the same range of values, at least for the medium P-supplied (VES, MIT), and the P-rich (BBR) sites. In the O-layer, the P-poor site (LUE) also had comparable values for labile P as the other sites. The share of labile P to TP in the O-layer was highest at the P-poor site (45 %), slightly lower at the medium sites (36 % and 33 %), and lowest in the P-rich site (14 %) (Figure 3a.1). There was a steep decrease in the ratio labile P to TP with soil depth at the P-poor to medium sites; and a more gradual decrease with depth at the P-rich site.

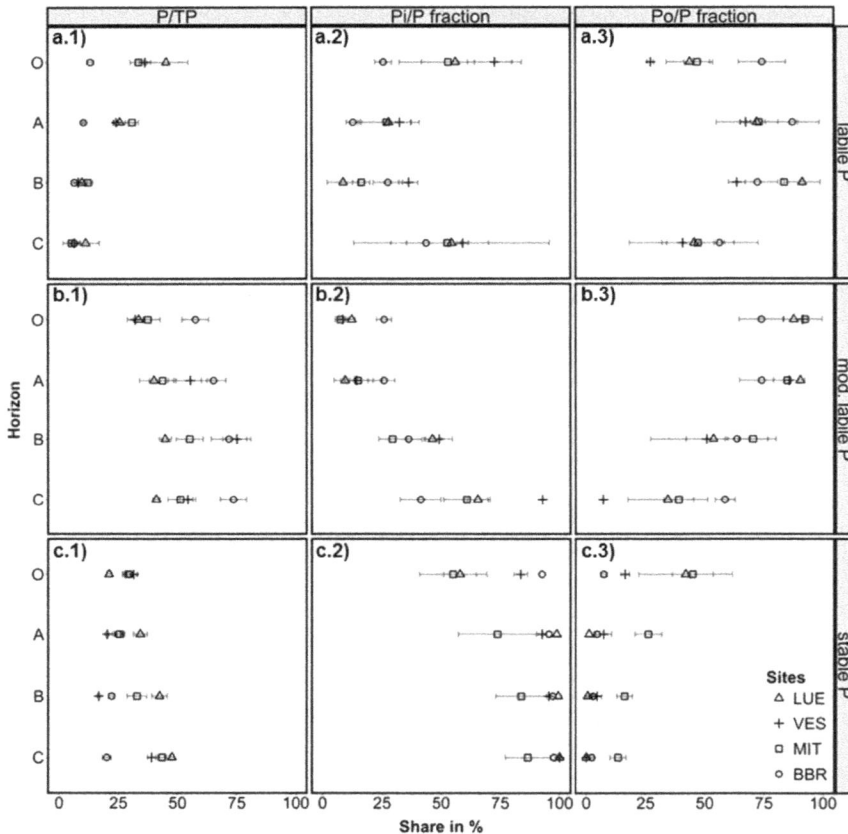

Figure 3. Share of **a.1)** the labile (resin-P_i + NaHCO$_3$-$P_{i,o}$), **b.1)** moderately labile (NaOH-$P_{i,o}$ + HCl$_{dil}$-P_i), and **c.1)** stable (HCl$_{conc}$-$P_{i,o}$ + residual-P) fractions on total P (TP) in organic layer (O) and soil (A, B, C horizons). Share of inorganic P (**a.2**: labile P_i, **b.2**: moderately labile P_i, **c.2**: stable P_i) and organically bound P (**a.3**: labile P_o, **b.3**: moderately labile P_o, **c.3**: stable P_o) on the particular pools. Error bars: standard deviation of three samples (profile cuts). Study sites in the sequence P-poor to P-rich are: LUE, Lüss, Lower Saxony; VES, Vessertal, Thuringian Forest; MIT, Mitterfels, Bavarian Forest; and BBR, Bad Brückenau, Bavarian Rhön Mountains.

Organically bound P dominated the labile P pool in the A and B horizon for all sites, and also in the O-layer for BBR. At VES, 27% of labile P in the O-layer consisted of P_o, while MIT and LUE had approximately equal amounts of P_i and P_o (Figure 3a.2,a.3). Thus, most of the easily plant-available P_i

was located in the O-layers at the medium and P-poor sites, whereas labile P was clearly dominated by P_o at the P-richest site. This finding was unexpected, as very low P supply from the mineral phase is assumed to enhance tight P recycling in forest ecosystems [6] (and therefore the fast uptake of P_i) which would result in a higher share of P_o at the P-poor site. In a recent study conducted at the same P-rich and P-poor study sites, Bünemann et al. [29] found that biological/biochemical processes were the dominant control of P_i availability in the low-P-sorbing sandy soil of the P-poor site. By contrast organic P mineralization was negligible when compared to the high P_i availability at the P-rich study site, which has a high-P-sorbing soil and very high TP content. Further, the high contents of amorphous and crystalline sesquioxides at the P-rich site (Table 2) may act as a sink for labile P_i, when it is adsorbed onto those surfaces and precipitated (depending on pH) (cf. [30,31]).

The differences in labile P between PFP and soil matrix were marginal considering the sum of P_i and P_o of this pool (Figure 2), but the ratio of P_i to P_o differed. Thus, the medium P-supplied site showed clearly higher contents of labile P_i and less labile P_o in PFPs compared to soil matrix. By contrast, labile P_o was significantly higher in PFPs of the P-rich site. However, the variation of results for labile P_i and P_o was quite high between the three profile cuts within the soil plot (see error bars in Figure 3), which indicates that conclusions from the results have to be drawn carefully.

3.2. Moderately Labile P Fractions

The sum of moderately labile P forms (NaOH-$P_{i,o}$ + HCl$_{dil}$-P_i) was the dominant fraction in mineral soil at the P-rich sites (Figures 2 and 3). At the P-poor site, this pool was equal or slightly lower than the stable P pool. At all sites, moderately labile P was predominantly organically bound in the O-layer and topsoil. In the subsoil, the ratio of P_i to P_o shifted towards P_i (Figure 3b.2,b.3). The shifts with depth were relatively gradual in the P-rich soil, but were more abrupt at the medium and P-poor sites.

The NaOH-extracted P represents the moderately available P forms related to Al, Fe, and humic substances. In acidic forest soils on silicate parent material, NaOH-P was observed to be involved in P_i adsorption (mainly to Al-oxyhydroxides) as well as P_o adsorption (to Fe-oxyhydroxides or Al-saturated soil organic matter) [4]. The share of HCl$_{dil}$-extractable P (apatite-bound) was very small for most samples, which indicates highly weathered soils [32]. In the P-rich soil, HCl$_{dil}$-P was found to be in the range of 150–300 mg·kg^{-1}, which confirms that the supply with mineral P is still high at this site. These results are in line with findings of Prietzel et al. [4] who identified apatite-P in this P-rich soil using P K-edge X-ray absorption near-edge structure (XANES) spectroscopy.

The sum of moderately labile P showed a variable distribution between PFP and soil matrix (Figure 2). In the topsoils, the P-poor to medium sites (LUE and VES) had slightly higher contents of this pool in PFPs, whereas clearly higher contents in soil matrix were observed at the medium to P-rich sites (MIT and BBR). Here, both higher P_i as well as higher P_o values contributed to the increased pool contents. For the subsoils, slightly higher contents of moderately labile P in PFPs as compared to soil matrix were detected at all sites which was mainly caused by higher P_o contents, especially at the P-rich site. For the P-poor site, the differences are hardly interpretable because the chemical analyses of the very low P concentrations led to high data uncertainty.

3.3. Stable P Fractions

Stable P in the soils (HCl$_{conc}$-$P_{i,o}$ + residual-P) followed the sequence of decreasing initial TP in the order of: BBR > MIT > VES > LUE, and was characterized by low contents of P_o compared to P_i in all soils (Figure 2). Elevated P_o portions in the O-layer were observed at the P-poor site and at the medium site MIT (Figure 3c.3), which indicates the presence of P associated with very stable organic compounds. However, the ratio of stable P to TP in the O-layer was comparable at all sites with around 30 % of TP at the P-rich to medium sites, and slightly lower at the P-poor site with 21 %. The depth distribution of this relation showed a slight decrease with depth in the P-rich soil, but an increase in the medium P-supplied soils and particularly increased in the P-poor soil. Here, almost 50 % of the

P in the subsoil is stable bound and not available for plants, which is in line with the assumption of decreasing availability of mineral-bound P from the P-rich site to the P-poor site.

The contents of stable P in PFPs and soil matrix were on the same level. Only for the P-rich soil, slightly higher values in PFPs were observed.

The distribution of P fractions revealed for the P-rich site comparable labile P concentrations as for the other sites, but a clearly lower ratio of labile P to TP, and much more P_o than P_i in this pool. Considering the assumption of negligible P_o mineralization [29], it can be assumed that the P-rich soil has enough labile P available from mineral sources and has no need for increased P_o mineralization. P_o therefore remains stable, but P_i is consumed and reduced thus relative to the amount of labile P_o. This may also explain the higher ratio of moderately labile P to TP at the P-rich site compared to the other sites. In this pool, P is moderately bound to inorganic and organic compound, and can be acquired if necessary.

3.4. Factors Affecting the Distribution of P Fractions in the Soil Profiles

The statistical test of correlations between parameters included in the first step all data from the four study sites and covered a wide range of values of chemical soil properties. The results confirmed a close relationship between labile P and the contents of soil organic C and total N, which is in accordance with previous studies [4,30,33,34] (for the correlation matrix and scatter plots see Table S1 of the supplementary material). This correlation is stronger for labile P_o (N: $R^2 = 0.98$, C: $R^2 = 0.97$) than for labile P_i (N: $R^2 = 0.94$, C: $R^2 = 0.93$). Also in agreement with previous studies [30,34], the easily available resin-P was strongly correlated with $NaHCO_3$-P_i ($R^2 = 0.79$) and $NaHCO_3$-P_o ($R^2 = 0.75$) indicating the contribution of inorganic as well as organic labile P fractions to the resin-P_i pool.

Moderately labile P was strongly related to sesquioxides (Al, Fe, Mn). In this fraction, P_i showed close correlations to Al_{ox}, Al_{di}, Mn_{ox}, Mn_{di}, and had slightly lower correlations with Fe_{ox} and Fe_{di}. Moderately labile P_i was also positively correlated with pH, but negatively correlated to the C/N ratio. By contrast, the strongest correlation of P_o in this fraction was found with Fe-oxyhydroxides, and to a lesser extent with Al- and Mn-oxyhydroxides. This strong contribution of Fe compounds to P_o adsorption, but the adsorption of mainly P_i to Al-oxyhydroxides, is in line with the study of Prietzel et al. [4] for acidic forest soils.

For stable P, a close correlation of P_i with Fe, Al, and Mn compounds was identified, whereas P_o was strongly related to C and N. This shows that P_o is occluded in stable organic compounds or particulate organic matter.

To indicate the dominant processes influencing the distribution of P_i and P_o in soils, the ratio C to P_o was calculated (Table 3). Low C/P_o values (<200) indicate the net mineralization of P, whereas high ratios (>300) are indicative for immobilization as result of microbial P_i utilization during decomposition of organic matter [35,36]. When comparing the test sites with considerable differences in soil P contents, the C to P_o ratio was highest at the P-poor site and lowest at the P-rich site. The strong reduction of C/P_o with depth at all sites can be attributed to the sharp decrease of organic C compared to P. In the O-layers, C/P_o was > 300 at the P-poor and medium sites, which indicates P immobilization (incorporation into biomass). C/P_o-values between 200 and 300, which occurred in the O-layer of the P-rich site, indicate a stable rate of cycling between P_i and P_o [35]. These findings support the hypothesis of intense P recycling at P-poor sites [6] leading to P depletion and high C to P_o ratios. The C/P_o ratio also differed considerably between PFP and soil matrix samples at all site with mostly clearly higher values in PFPs (Table 3). For the P-rich site, the difference between PFP and matrix was more pronounced in the topsoil and diminished with soil depth. In the P-poor soil, the contrasting behavior was observed: the difference increased with depth indicating stronger P_o depletion in the subsoil PFPs.

Table 3. Ratios of organic C to sum of organic P (Po) at four beech forest study sites in Germany for PFP (preferential flow pathways), Ma (soil matrix), and RC (root channel-PFP) samples.

C/P_o Horizon	LUE		VES		MIT		BBR	
	PFP	Ma	PFP	Ma	PFP	Ma	PFP	Ma
O	1143		840		910		223	
A	512	494	155	174	593	371	139	97
B	356	274	89	66	101	93	58	51
C	204	88		170		59	42	43

Horizon: O—organic layer; A—topsoil horizon; B—subsoil horizon; C—lower subsoil horizon containing weathered parent material. Study sites in the sequence P-poor to P-rich are: LUE—Lüss, Lower Saxony; VES—Vessertal, Thuringian Forest; MIT—Mitterfels, Bavarian Forest; and BBR—Bad Brückenau, Bavarian Rhön Mountains.

For all chemical parameters, the differences between PFP and the soil matrix were tested statistically (for test results see Table S2 of the supplementary material). The most statistical differences were found for the P-rich site. Significantly higher values in PFPs occurred here for C, labile P_o, and C to TP ratio with $p < 0.01$, and for N, stable P_o and stable P_i with $p < 0.05$. For the other sites, almost no statistically significant differences of P fractions and other parameters were identified. At the P-poor site, the ratio C to TP and the C content were statistically higher in the PFPs ($p < 0.05$). At the medium sites, the content of metals differed substantially, with high contents of Fe_{ox}, Mn_{ox}, Al_{di} and Mn_{di} in PFPs, but lower values of Fe_{di} compared to matrix samples. However, these statistical tests included all data per site and have to be considered carefully due to the hierarchical structure of the data. For the same tests at the horizon level per site, the data sets were much smaller (three samples for PFP and matrix respectively) and revealed no significant differences between PFP and matrix for all parameters (Table S3). In previous studies, Hagedorn and Bundt [18] suggested that PFPs in forest soils can be stable for decades, resulting in biological and chemical gradients between PFPs and soil matrix [15]. Bogner et al. [12] confirmed this hypothesis for PFPs along root channels and found an enrichment of C, N, Ca, Mg, and Fe in PFPs compared to the soil matrix. Such an accumulation in PFPs was also documented for different P fractions in flow pathways along stones by Backnäs et al. [19]. Our study sites are characterized by very high stone contents, resulting in very heterogeneous conditions for water and solute flow. PFPs were therefore located mostly along stone surfaces with low potential for P sorption [19], and preferential flow occurred less in root channels or other biopores where P accumulation would be expected to be more prevalent. Single biopores could be identified and sampled separately, which contained decayed root material. Indications were found in these biopores of an accumulation of labile organically bound P (P_o), but could not be statistically confirmed. With respect to the distribution of subsurface flow pathways at hillslopes (midrange mountain sites), it has been shown that a predominant surface-near lateral flow does not support chemical gradients between PFPs and soil matrix. Thus, non-significant results in the statistical test are a result of specific site conditions, spatial soil heterogeneity, and analytical uncertainties with regard to very low P concentrations at the P-poor site.

Nevertheless, several factors affecting the distribution of P within the tested soils became apparent. The results showed that the labile $P_{i,o}$ pool and stable P_o were strongly related to C and N, whereas moderately labile $P_{i,o}$ as well as stable P_i correlated with metal-oxides and -oxyhydroxides (Al, Fe, Mn). At the medium to P-rich sites, where higher contents of sesquioxide in topsoil matrix compared to PFPs were observed, moderately labile P_i and P_o also was clearly higher in soil matrix. The higher P_o contents of this moderately labile pool in subsoil PFPs could be related to Fe_{ox} accumulated in the flow pathways.

4. Conclusions

As expected, considerable differences were found for the four study sites in both the P contents and the distribution of P fractions within the soil profiles. Despite these large differences, the labile P contents in the O-layers were in the same range of values at all sites. At the site with very low supply of P from the mineral phase, the predominant portion of labile P was available in the O-layer, but was greatly reduced in the mineral soil. This indicates that for this site intense P recycling happens in the O-layer, which is also supported by a high C to P_o ratio in this layer. By contrast, labile P was available throughout the whole profile at the P-rich site. Here, the supply of plant-available P seems to be sufficient enough that there is no need for tight P recycling, and may therefore explain the very low share of labile P to TP, when compared to the P-poor site. The contents of labile and moderately labile P clearly differed in PFPs compared to soil matrix, and indicate that labile organically bound P (P_o) accumulates in PFPs with a high content of organic matter. However, these results could not be proved statistically at the tested sites. Therefore, more sites particularly for soils with a higher proportion of preferential flow through biopores should be studied to verify the indications.

Supplementary Materials: The following are available online at www.mdpi.com/1999-4907/8/1/19/s1, Table S1: Correlation matrix with Spearman's rho coefficients to test the correlation between chemical soil parameters (dataset: $n = 123$, includes data of four forest soils in Germany), Table S2: Results (p-values) of the non-parametric Wilcoxon rank-sum test for differences between PFPs and soil matrix samples (pair-wise test for all data per site), Table S3. Results (p-values) of the non-parametric Wilcoxon rank-sum test for differences between PFPs and soil matrix samples (pair-wise test at horizon level per sites).

Acknowledgments: We thank Gisela Ciesielski and Manuela Unger for technical support in the laboratory. Furthermore, our thanks go to Maximilian Kirsten, Jianhong Liang, Peer Appelfelder, and Nils Heide for their assistance in the field and laboratory works. Finally, we thank Dan Hawtree for helpful comments and linguistic help. This work was funded within the framework of Priority Program SPP 1685 "Ecosystem Nutrition: Forest Strategies for Limited Phosphorus Resources" by the German Research Foundation (DFG), grant No. JU 2940/1-1. We further acknowledge support by the Open Access Publication Funds of the TU Dresden.

Author Contributions: D.J. designed and performed the field experiments, and did the data analyses; all authors wrote the manuscript.

Conflicts of Interest: The authors declare no conflict of interest.

References

1. Bol, R.; Julich, D.; Brödlin, D.; Siemens, J.; Kaiser, K.; Dippold, M.A.; Spielvogel, S.; Zilla, T.; Mewes, D.; von Blanckenburg, F.; et al. Dissolved and colloidal phosphorus fluxes in forest ecosystems—An almost blind spot in ecosystem research. *J. Plant Nutr. Soil Sci.* **2016**, *179*, 425–438. [CrossRef]

2. Newman, R.H.; Tate, K.R. Soil phosphorus characterisation by ^{31}p nuclear magnetic resonance. *Commun. Soil Sci. Plant Anal.* **1980**, *11*, 835–842. [CrossRef]

3. Cade-Menun, B.; Liu, C.W. Solution phosphorus-31 nuclear magnetic resonance spectroscopy of soils from 2005 to 2013: A review of sample preparation and experimental parameters. *Soil Sci. Soc. Am. J.* **2014**, *78*, 19. [CrossRef]

4. Prietzel, J.; Klysubun, W.; Werner, F. Speciation of phosphorus in temperate zone forest soils as assessed by combined wet-chemical fractionation and XANES spectroscopy. *J. Plant Nutr. Soil Sci.* **2016**, *179*, 168–185. [CrossRef]

5. Walker, T.W.; Syers, J.K. The fate of phosphorus during pedogenesis. *Geoderma* **1976**, *15*, 1–19. [CrossRef]

6. Lang, F.; Bauhus, J.; Frossard, E.; George, E.; Kaiser, K.; Kaupenjohann, M.; Krüger, J.; Matzner, E.; Polle, A.; Prietzel, J.; et al. Phosphorus in forest ecosystems: New insights from an ecosystem nutrition perspective. *J. Plant Nutr. Soil Sci.* **2016**, *179*, 129–135. [CrossRef]

7. Prietzel, J.; Kaiser, K. De-eutrophication of a nitrogen-saturated Scots pine forest by prescribed litter-raking. *J. Plant Nutr. Soil Sci.* **2005**, *168*, 461–471. [CrossRef]

8. Neary, D.G.; Ice, G.G.; Jackson, C.R. Linkages between forest soils and water quality and quantity. *For. Ecol. Manag.* **2009**, *258*, 2269–2281. [CrossRef]

9. Van Miegrot, H.; Johnson, D.W. Feedbacks and synergism among biogeochemistry, basic ecology, and forest soil science. *Forest Ecol. Manag.* **2009**, *258*, 2214–2223. [CrossRef]

10. Santos, R.M.B.; Sanches Fernandes, L.F.; Pereira, M.G.; Cortes, R.M.V.; Pacheco, F.A.L. A framework model for investigating the export of phosphorus to surface waters in forested watersheds: Implications to management. *Sci. Total Environ.* **2015**, *536*, 295–305. [CrossRef] [PubMed]

11. Uchida, T.; Tromp-van Meerveld, I.; McDonnell, J.J. The role of lateral pipe flow in hillslope runoff response: An intercomparison of non-linear hillslope response. *J. Hydrol.* **2005**, *311*, 117–133. [CrossRef]
12. Bogner, C.; Borken, W.; Huwe, B. Impact of preferential flow on soil chemistry of a podzol. *Geoderma* **2012**, *175*, 37–46. [CrossRef]
13. Van der Heijden, G.; Legout, A.; Pollier, B.; Bréchet, C.; Ranger, J.; Dambrine, E. Tracing and modeling preferential flow in a forest soil—Potential impact on nutrient leaching. *Geoderma* **2013**, *195–196*, 12–22. [CrossRef]
14. Van Verseveld, W.J.; McDonnell, J.J.; Lajtha, K. The role of hillslope hydrology in controlling nutrient loss. *J. Hydrol.* **2009**, *367*, 177–187. [CrossRef]
15. Bundt, M.; Jäggi, M.; Blaser, P.; Siegwolf, R.; Hagedorn, F. Carbon and nitrogen dynamics in preferential flow paths and matrix of a forest soil. *Soil Sci. Soc. Am. J.* **2001**, *65*, 1529–1538. [CrossRef]
16. Flury, M.; Flühler, H. Brilliant Blue FCF as a dye tracer for solute transport studies—A toxicological overview. *J. Environ. Qual.* **1994**, *23*, 1108–1112. [CrossRef]
17. Weiler, M.; Flühler, H. Inferring flow types from dye patterns in macroporous soils. *Geoderma* **2004**, *120*, 137–153. [CrossRef]
18. Hagedorn, F.; Bundt, M. The age of preferential flow paths. *Geoderma* **2002**, *108*, 119–132. [CrossRef]
19. Backnäs, S.; Laine-Kaulio, H.; Kløve, B. Phosphorus forms and related soil chemistry in preferential flowpaths and the soil matrix of a forested podzolic till soil profile. *Geoderma* **2012**, *189*, 50–64. [CrossRef]
20. Jarvis, N.J. A review of non-equilibrium water flow and solute transport in soil macropores: principles, controlling factors and consequences for water quality. *Eur. J. Soil Sci.* **2007**, *58*, 523–546. [CrossRef]
21. Jensen, M.B. Subsurface transport of phosphorus in relation to its mobilization and immobilization in structured soil (steady state flow experiments). *Acta Agric. Scand. Sect. B-Soil Plant Sci.* **1998**, *48*, 11–17.
22. Hedley, M.J.; Stewart, J.W.B.; Chauhan, B.S. Changes in inorganic and organic soil phosphorus fractions induced by cultivation practices and by laboratory incubations. *Soil Sci. Soc. Am. J.* **1982**, *46*, 970. [CrossRef]
23. Tiessen, H.; Moir, J.O. Characterization of Available P by Sequential Extraction. In *Soil Sampling and Methods of Analysis*; Carter, M.R., Gregorich, E.G., Eds.; Canadian Society of Soil Science: Boca Raton, FL, USA, 2008; Chapter 25; pp. 293–306.
24. Sibbesen, E. A simple ion-exchange resin procedure for extracting plant-available elements from soil. *Plant Soil* **1977**, *46*, 665–669. [CrossRef]
25. Murphy, J.; Riley, J.P. A modified single solution method for the determination of phosphate in natural waters. *Anal. Chim. Acta* **1962**, *27*, 31–36. [CrossRef]
26. O'Halloran, I.P.; Cade-Menun, B.J. Total and Organic Phosphorus. In *Soil Sampling and Methods of Analysis*; Carter, M.R., Gregorich, E.G., Eds.; Canadian Society of Soil Science: Boca Raton, FL, USA, 2008; pp. 265–291.
27. Crews, T.E.; Brookes, P.C. Changes in soil phosphorus forms through time in perennial versus annual agroecosystems. *Agric. Ecosyst. Environ.* **2014**, *184*, 168–181. [CrossRef]
28. Negassa, W.; Leinweber, P. How does the Hedley sequential phosphorus fractionation reflect impacts of land use and management on soil phosphorus: A review. *J. Plant Nutr. Soil Sci.* **2009**, *172*, 305–325. [CrossRef]
29. Bünemann, E.K.; Augstburger, S.; Frossard, E. Dominance of either physicochemical or biological phosphorus cycling processes in temperate forest soils of contrasting phosphate availability. *Soil Biol. Biochem.* **2016**, *101*, 85–95. [CrossRef]
30. Tiessen, H.; Stewart, J.W.B.; Cole, C.V. Pathways of phosphorus transformations in soils of differing pedogenesis. *Soil Sci. Soc. Am. J.* **1984**, *48*, 853. [CrossRef]
31. Renneson, M.; Dufey, J.; Legrain, X.; Genot, V.; Bock, L.; Colinet, G. Relationships between the P status of surface and deep horizons of agricultural soils under various cropping systems and for different soil types: A case study in Belgium. *Soil Use Manag.* **2013**, *29*, 103–113. [CrossRef]
32. Hashimoto, Y.; Watanabe, Y. Combined applications of chemical fractionation, solution ^{31}P-NMR and P K-edge XANES to determine phosphorus speciation in soils formed on serpentine landscapes. *Geoderma* **2014**, *230*, 143–150. [CrossRef]
33. Frizano, J.; Johnson, A.H.; Vann, D.R.; Scatena, F.N. Soil phosphorus fractionation during forest development on landslide scars in the Luquillo Mountains, Puerto Rico. *Biotropica* **2002**, *34*, 17–26. [CrossRef]
34. Thomas, S.M.; Johnson, A.H.; Frizano, J.; Vann, D.R.; Zarin, D.J.; Joshi, A. Phosphorus fractions in montane forest soils of the Cordillera de Piuchué, Chile: Biogeochemical implications. *Plant Soil* **1999**, *211*, 139–148. [CrossRef]

35. McDowell, R.W.; Stewart, I. The phosphorus composition of contrasting soils in pastoral, native and forest management in Otago, New Zealand: Sequential extraction and ^{31}P NMR. *Geoderma* **2006**, *130*, 176–189. [CrossRef]

36. Stutter, M.I.; Shand, C.A.; George, T.S.; Blackwell, M.S.A.; Dixon, L.; Bol, R.; MacKay, R.L.; Richardson, A.E.; Condron, L.M.; Haygarth, P.M. Land use and soil factors affecting accumulation of phosphorus species in temperate soils. *Geoderma* **2015**, *257*, 29–39. [CrossRef]

forests

MDPI

Article

Sampling Method and Tree-Age Affect Soil Organic C and N Contents in Larch Plantations

Huimei Wang [1], Wenjie Wang [1,2,*] and Scott X. Chang [3]

1   Key Laboratory of Forest Plant Ecology, Northeast Forestry University, Harbin 150040, China; whm0709@hotmail.com
2   Northeast Institute of Geography and Agricultural Ecology, Chinese Academy of Sciences, Changchun 130102, China
3   Department of Renewable Resources, University of Alberta, 442 Earth Science Building, Edmonton, AB T6G 2E3, Canada; scott.chang@ualberta.ca
*   Correspondence: wjwang225@hotmail.com; Tel.: +86-431-8554-2336 or +86-451-8219-0092

Academic Editor: Timothy A. Martin
Received: 18 November 2016; Accepted: 11 January 2017; Published: 17 January 2017

Abstract: We currently have a poor understanding of how different soil sampling methods (pedogenetic horizon versus fixed-depth) influence the evaluation of soil properties. Here, 159 soil profiles were sampled from larch (*Larix gmelinii*) plantations in northeast China using both the pedogenetic horizon and fixed-depth sampling methods. Analysis of variance was used to determine how sampling method influences the assessment of the spatial variation in the concentration and storage of soil organic C (SOC) and N (SON), as well as how these properties are affected by tree age-group (<20, 20–40, and >40 years). In both the 20 cm (surface) and 80 cm (whole profile) sampling depths, pedogenetic sampling resulted in 1.2- to 1.4-fold higher SOC and SON concentrations than fixed-depth sampling. Surface soil nutrient storage between the two sampling methods was not significantly different, but was it was 1.2-fold higher ($p < 0.05$) with pedogenetic sampling than with fixed-depth sampling in the whole soil profile. For a given error limit in SOC and SON assessments, fixed-depth sampling had a 60%~90% minimum sampling intensity requirement compared with pedogenetic horizon sampling. Additionally, SOC was 1.1- to 1.3-fold greater in the >40 years age-group than in the <20 years age-group ($p < 0.05$), while SON was the highest in the 20–40 years age-group ($p < 0.05$). The total amount of SOC and nutrients in soil is fixed regardless how you sample, it is the different assumptions and different ways of extrapolation from samples to the population that cause sampling by horizon versus fixed depth to lead to different conclusions. Our findings highlight that soil sampling method and tree age-group affect the determination of the spatial variation of SOC and SON and future soil assessments should control for methodological differences.

Keywords: *Larix gmelinii*; tree-age effect; degraded farmland; Northeast China; soil nutrient variation; soil heterogeneity; pedogenetic horizon; fixed-depth sampling

1. Introduction

Accurate evaluation of soil properties is important for species-to-location matching during afforestation, precision agroforestry [1–3], and the investigation of forest ecological functions such as carbon (C) sequestration and nutrient dynamics [4]. Soil heterogeneity occurs horizontally and vertically [5]; thus, the lack of a standardized sampling procedure to assess soil fertility and C inventory could affect conclusions drawn from different studies. Therefore, research is needed to understand how sampling method influences the evaluation of the inter-site variations (i.e., interaction between sampling method and site variation) of soil nutrients. Accounting for the effect of sampling method could improve comparability among studies [6,7], particularly in terms of between-site soil fertility assessments [8].

Forest development with age can increase C storage and recover nutrient levels in soils degraded by long-term agricultural cultivation [9–12]. Soils can range from being a C sink (11 to 238 $g \cdot C \cdot m^{-2} \cdot year^{-1}$) [4,13–15] to a source of atmospheric CO_2 (0.1 to 14.1 $g \cdot C \cdot m^{-2} \cdot year^{-1}$) [11,16]. Additionally, studies examining post-afforestation conditions have reported both the depletion of soil nutrients [4,10,17–19] and the accumulation of soil organic N (SON) with increasing soil organic C (SOC) [9,12,20] over time (e.g., tree age effect). Previous land use, the tree species planted, and soil nutrient supply are thought to explain the contradictions in tree age effects on soil properties after afforestation [9–12,17–20]. Furthermore, forest age itself [21], soil depth, and other site-related variations [22–25] have been used to describe the spatial and temporal changes in nutrient accumulation. However, differences in sampling methods (fixed-depth [4,11,13] versus pedogenetic horizon [26–28] sampling) are a potentially powerful explanation for between-study differences in soil properties [29,30] that has not been fully studied. Indeed, comprehensive research including soil sampling method, sampling site, sampling depth, and tree age would provide a more holistic view of how different sampling methods alter evaluations of SOC and SON.

The objective of this study was to determine how soil sampling method, sampling site, sampling depth, and tree age independently and interactively influence SOC and SON, particularly variation across the soil profile in different soil depths and the effect of tree age. The investigation took advantage of widespread larch plantations (4.5 million hectares, or 70% of the total plantation area) available in northeastern China [11,31]. We hypothesized that soil sampling methods could greatly affect the evaluations of spatial changes of and tree-age effects on SOC and N, independent of soil depth inclusion and unit expression of storage or concentration. Our results may contribute to improving inter-study comparisons of SOC and SON measurements, leading to a broader study applicability and improved interpretation of conclusions drawn across disparate reports.

2. Materials and Methods

2.1. Study Site and Soil Sampling

All sampling was conducted in *Larix gmelinii* plantations located in northeast China. This region (45°20′–47°14′ N, 127°30′–128°55′ E) has a continental monsoon climate with mean annual temperatures ranging from −0.3 to 2.6 °C and mean annual precipitation ranging from 676 to 724 mm. The soil is generally a typical Dark-Brown Forest Soil, with occasional eluviation features. The soil in Chinese Soil Taxonomy is Mollic Bori-udic Cambosols or Albic Bori-udic Cambosols [32]; it is a Eutroboralf in the US Soil Taxonomy and a Halic Luvisol in the FAO (Food and Agriculture Organization) system [33]. Soil pH ranges from 4.9 to 6.5. The details of the study sites have been previously described [11].

A total of 159 plots were sampled from four sites: the Laoshan experimental station (LS; 57 plots), the Maoershan experimental farm (MES; 37 plots), the Dongshan forest farm (DS; 30 plots), and the Daqingchuan forest farm (DQC; 35 plots).

Fixed-depth and pedogenetic horizon sampling were employed for soil sampling. Soil cutting rings with a 100 cm^3 volume were used for sampling soil from a soil pit. A soil pit was excavated in each of four subplots established in each of the 159 20 m× 20 m plots (for a total of 636 soil pits), and the samples from the four pits from the same horizon/layer of the same sampling method were mixed to form a composite sample for that layer and plot. Soil samples were collected from one or two sides of the pit for sampling the same layer or depth. Those composite samples from a plot make up the samples for a soil profile for each plot. Such composite samples better represent the soil from a plot. Under fixed-depth sampling, the soil was collected from four depths: 0–20, 20–40, 40–60, and 60–80 cm. Under pedogenetic horizon sampling, soil samples were collected from the A, B, and C horizons. The first horizon contains partially humified organic matter, giving the soil a darker color than soils from lower horizons. The second horizon has undergone sufficient changes during soil genesis, with maximum accumulation of materials such as silicate clay, metal (e.g., Fe and Al) oxides, and organic material. The third horizon comprises unconsolidated material below the solum (A and B horizons) that is little affected by soil-forming processes and lacks pedogenic development.

A total of 1067 composite soil samples were collected (pedogenetic: 113 plots × 3 samples with one each from horizons A, B, and C, plus 46 plots with only A and B horizons, for a total of 431 samples; fixed-depth: 159 plots × 4 samples from the four depths = 636 samples). At the time of sampling, the soil samples were placed in cloth soil-sampling bags for transportation back to the laboratory. To compare the effect of sampling method, the surface soil layer (0–20 cm) and the whole profile soil (0–80 cm) as two sampling depths were examined in this study. In other words, within the 0–20 cm and 0–80 cm sampling depths, we assessed the impact of sampling method on the assessment of SOC and SON.

Tree age was determined using an increment borer (Zhonglinweiye, Beijing, China) on at least five trees in each plantation plot. Fifty-six plots contained trees younger than 20 years (<20 years age-group), with an average tree age of 8.1 ± 7.1 years (SD, standard deviation). The second age-group (20–40 years) had 73 plots, with an average tree age of 25.5 ± 5.3 years. The third age-group (>40 years) had 30 plots and an average tree age of 44.8 ± 2.3 years.

2.2. Bulk Density, SOC, and SON Measurements

Once the soil samples were brought back to the laboratory, they were air-dried in a ventilated room until they reached a constant weight. Cutting rings with a 100 cm^3 volume were used for intact bulk density measurement, and bulk density was determined based on the oven-dried mass and volume (400 cm^3 of soil from four cutting rings as a composite sample) of each site. In the fixed-depth sampling, we measured one soil bulk density at each soil depth (0–20, 20–40, 40–60, and 60–80 cm) of each site. In the pedogenetic sampling, we measured one soil bulk density in each horizon (A, B, and C) of each site, and in some places, the soil bulk density at some transitional horizons (e.g., AB horizon) was also measured. Soil was ground to <0.25 mm for measuring SOC and SON concentrations with a heated dichromate/titration and the semimicro-Kjeldahl method, respectively. Details about those methods are available in a previous publication of our group [11].

2.3. Storage Calculation

The storage of SOC and SON was calculated with the following equation:

$$\text{Storage } (g \cdot m^{-2}) = a \times \rho_b \times depth \times (1 - V_{\text{gravel}}), \qquad (1)$$

where a is SOC or SON concentration (g·kg^{-1}), ρ_b is soil bulk density (g·cm^{-3} = Mg·m^{-3}, Mg: million gram), and V_{gravel} is the fraction of gravel volume (determined via water displacement) in the soil.

2.4. Assessing the Effect of Sampling Method on Tree Age Effect and Spatial Variation in SOC and SON

Analysis of variance (ANOVA) was used to determine whether soil sampling method affected the spatial variation (among study sites) of SOC and SON and differences among tree age-groups. Specifically, we examined whether study site and age-group interacted with sampling method (all were included in the model as fixed factors). The concentration and storage of SOC and SON at 0–20 and 0–80 cm soil were analyzed. Duncan post-hoc tests were then conducted to identify the specific parameter (SOC concentration, SOC storage, SON concentration, and SON storage) that was affected by site, age-group, and sampling method. Marginal means were estimated and pairwise comparisons (LSD test) were used to separate the means across predictor variables, and their interactions with sampling method (site × method and age-group × method). All of these analyses were performed in SPSS 17.0 (SPSS Inc., Armonk, NY, USA).

2.5. Number of Samples Required for Estimating SOC and SON at Specific Error Limits

As a way of quantifying differences between the two soil sampling methods, statistical analyses [34,35] were conducted to compare the number of sampling points (n) required for estimating mean SOC and SON at a site within 5%, 10%, 20%, and 30% of its actual value, at the 95% probability level for both datasets (pedogenetic and fixed-depth). The following equation was used:

$$n = \frac{t_\alpha^2 s^2}{D^2},\tag{2}$$

where t_α is the Student's t statistic with degrees of freedom at the α probability level, s is standard deviation, and D is the specified error limit. Only the SOC and SON concentration data were used as patterns in the storage data were similar.

3. Results

3.1. Differences between Sites, Sampling Methods, and Age-Groups

The SOC and SON concentrations in both the surface and whole profile soil differed significantly among sites ($p < 0.001$) and age-groups ($p < 0.05$), and between sampling methods ($p < 0.001$). However, SOC and SON storage did not follow the same pattern. For example, no difference between sampling methods was observed in surface soil SOC storage ($p > 0.05$; Table 1).

Table 1. ANOVA result on the influence of sampling method, age-group, and study site on soil organic C (SOC) and N (SON) concentrations, storage, in surface (0–20 cm), and the whole profile soil (0–80 cm). Bold fonts indicate statistical significance ($p < 0.05$).

	Concentration (g·kg^{-1})				Storage (kg·m^{-2})			
	SOC		SON		SOC		SON	
	F	p-Value	F	p-Value	F	p-Value	F	p-Value
Surface soil (0–20 cm)								
Site	104.31	**<0.001**	119.5	**<0.001**	131.26	**<0.001**	118.77	**<0.001**
Age-group	8.50	**<0.001**	5.51	**0.004**	11.17	**<0.001**	4.13	**0.017**
Sampling method	15.29	**<0.001**	15.50	**<0.001**	1.65	0.200	1.83	0.177
Site × age-group	0.51	0.771	1.76	0.122	0.62	0.682	2.38	**0.039**
Site × method	8.08	**<0.001**	6.18	**<0.001**	4.71	**0.003**	2.12	0.098
Age-group × method	2.68	0.070	0.35	0.708	2.46	0.087	0.06	0.946
Site × age-group × method	0.75	0.586	0.66	0.656	0.67	0.647	0.41	0.839
Whole profile soil (0–80 cm)								
Site	87.56	**<0.001**	95.16	**<0.001**	135.95	**<0.001**	122.43	**<0.001**
Age-group	9.74	**<0.001**	11.32	**<0.001**	9.39	**<0.001**	9.25	**<0.001**
Sampling method	37.45	**<0.001**	25.57	**<0.001**	28.60	**<0.001**	10.42	**0.001**
Site × age-group	1.30	0.262	1.64	0.149	1.46	0.203	1.94	0.088
Site × method	6.37	**<0.001**	3.25	**0.022**	13.76	**<0.001**	4.89	**0.002**
Age-group × method	0.18	0.839	0.11	0.892	0.18	0.832	0.47	0.625
Site × age-group × method	0.33	0.897	0.49	0.782	0.21	0.958	0.48	0.790

The highest SOC and SON concentrations were usually observed at DQC or DS, while the lowest values were found in LS (Table 2). Inter-site variation (the highest/lowest ratio) was 2.8 and 3.5 for surface soil SOC and SON, respectively, whereas inter-site variation was 2.9 and 3.2 for SOC and SON, respectively, in the whole profile soil. The highest SOC and SON storage in surface and whole profile soils were found at DS and DQC, respectively, with the lowest values at LS. The ratios of the highest to the lowest values for SOC and SON storage in the surface soil were 2.3 and 2.6, respectively, while those for the whole profile soil were 2.5 and 2.7, respectively (Table 2).

Pedogenetic sampling generally yielded higher SOC and SON concentrations (Table 2). Surface and whole profile soil SOC and SON concentrations were 1.2- to 1.4-fold higher in pedogenetic than in fixed-depth sampling (Table 2). Different patterns were found in their storage. Whole profile soil storage of both SOC and SON was 1.2-fold higher under pedogenetic than under fixed-depth sampling ($p < 0.05$), but the difference was only 1.05- to 1.07-fold ($p > 0.05$) for the surface soil (Table 2).

Table 2. Differences in soil organic C (SOC) and N (SON) concentration and storage across four sites, three age-groups, and two soil sampling methods.

Item	Site	Concentration (g·kg⁻¹)	Storage (kg·m⁻²)	Age-Group (Year)	Concentration (g·kg⁻¹)	Storage (kg·m⁻²)	Method	Concentration (g·kg⁻¹)	Storage (kg·m⁻²)
SOC 0–20 cm	DQC	50.3 c	8.96 c	<20	35.5 a	6.83 a	FD	36.3 a	7.35 a
	DS	54.5 c	9.95 c	20–40	40.7 b	7.65 b	PE	43.8 b	7.71 a
	LS	19.4 a	4.29 a	>40	45.2 b	8.30 b			
	MES	39.9 b	6.61 b						
	Highest/lowest	2.8	2.3	>40/<20	1.3	1.2	PE/FD	1.2	1.05
SOC 0–80 cm	DQC	23.8 c	21.1 c	<20	18.5 a	16.6 a	FD	17.5 a	16.2 a
	DS	28.6 d	24.1 d	20–40	22.02 b	18.7 b	PE	23.6 b	19.7 b
	LS	10.0 a	9.71 a	>40	21.4 ab	18.8 b			
	MES	19.4 b	16.4 b						
	Highest/lowest	2.9	2.5	>40/<20	1.2	1.1	PE/FD	1.4	1.2
SON 0–20 cm	DQC	3.80 c	0.68 c	<20	2.58 a	0.51 a	FD	2.6 a	0.52 a
	DS	3.70 c	0.67 c	20–40	3.02 b	0.57 b	PE	3.2 b	0.56 a
	LS	1.10 a	0.26 a	>40	3.05 b	0.55 ab			
	MES	2.80 b	0.55 b						
	Highest/lowest	3.5	2.6	20–40/<20	1.2	1.1	PE/FD	1.2	1.08
SON 0–80 cm	DQC	1.80 c	1.66 c	<20	1.35 a	1.21 a	FD	1.30 a	1.19 a
	DS	1.90 c	1.51 bc	20–40	1.69 b	1.42 b	PE	1.70 b	1.37 b
	LS	0.60 a	0.61 a	>40	1.37 a	1.19 a			
	MES	1.60 b	1.39 b						
	Highest/lowest	3.2	2.7	20–40/<20	1.3	1.2	PE/FD	1.3	1.2

Note: DQC, Daqingchuan site; DS, Dongshan site; LS, Laoshan site; MES, Maoershan site; FD, fixed-depth sampling; PE, pedogenetic horizon sampling. Different letters in the same column and sampling depth indicate significant differences. Means that are significantly different among the treatment under comparison are highlighted.

239

Age-group differences increased linearly from <20, to 20–40, and >40 years groups in SOC concentration and storage in the surface soil, SON concentration in the surface soil, and SOC storage in the whole profile soil (Table 2). Forests in the 20–40 years age-group exhibited peak whole profile soil.

SOC and SON concentrations, as well as whole profile soil and surface soil SON storage (Table 2). In general, the >40 years group had 1.1- to 1.3-fold higher SOC than the <20 years age-group, while the 20–40 years group tended to have higher SON (Table 2).

3.2. SOC and SON Concentrations Across Study Sites: Interaction with Sampling Method

Marked site by sampling method interaction was found for SOC and SON concentrations in the surface and whole profile soil (Table 1; Figure 1). Both surface and whole profile soil SOC concentrations were higher with pedogenetic than with fixed-depth sampling. Inter-site differences were also found (Figure 1). The highest increases from one sampling method to another in surface soil SOC concentrations generally occurred in DQC (52%), while moderate increases (7%–10%) were observed in DS, LS, and MES (Figure 1). The whole profile soil (0–80 cm) typically accompanied higher SOC in pedogenetic sampling than in fixed-depth sampling. For example, SOC concentration at DS was 8% higher with pedogenetic sampling than with fixed-depth sampling in the surface soil, whereas it was 38% higher when the whole profile soil was evaluated. Furthermore, at LS, SOC in the surface soil was 10% higher, but that in the whole profile soil was 20% higher in pedogenetic than in fixed-depth sampling (Figure 1).

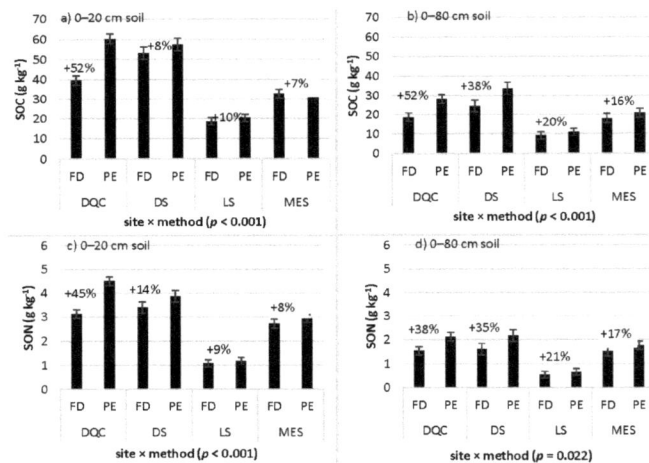

Figure 1. Sampling method influences soil organic C (SOC) and N (SON) concentrations depending on the sampling site (site × sampling method interaction). Data are presented for the two different sampling depths (surface soil: 0–20 cm and whole profile soil: 0–80 cm). A plus sign (+) indicates that the pedogenetic method resulted in higher values than the fixed-depth method, while a minus sign (−) indicates the opposite result. DQC, Daqingchuan site; DS, Dongshan site; LS, Laoshan site; MES, Maoershan site; FD, fixed-depth sampling; PE, pedogenetic horizon sampling. The error bars represent standard error of the mean. 0–20 cm soil for SOC (**a**); 0–80 cm soil for SOC (**b**); 0–20 cm soil for SON (**c**); 0–80 cm soil for SON (**d**).

Sampling influence on SON concentration was similar to that on SOC. Pedogenetic sampling led to higher SON in both surface and whole profile soil as compared with fixed-depth sampling, again with differences among sites (Figure 1). The largest increases (38%–45%) of pedogenetic over fixed-depth sampling occurred at DQC, whereas the other sites had relatively smaller sampling-induced changes

(14%–35% at DS, 9%–21% at LS, and 8%–17% at MES; Figure 1). In the surface soil, pedogenetic sampling resulted in 14% (DS), 9% (LS), and 8% (MES) increases in SON concentration as compared with fixed-depth sampling, whereas the increases were 35% (DS), 21% (LS), and 17% (MES) when the whole profile soil was evaluated (Figure 1).

3.3. SOC and SON Storage across Study Sites: Interaction with Sampling Method

The interaction between sampling method and study site significantly affected SOC and SON storage in the two sampling depths, except SON storage in the surface soil (Table 1; Figure 2). At DQC, pedogenetic sampling resulted in 16%–22% higher SOC and SON storage compared with fixed-depth sampling in the surface soil; and 30%–44% higher SOC and SON storage when the whole profile soil was considered. At DS, there was a <6% difference in SOC and SON storage in the surface soil, and a 14%–27% increase in the whole profile soil from pedogenetic to fixed-depth sampling. Additionally, SOC and SON storage values were similar under both methods at LS and MES (Figure 2).

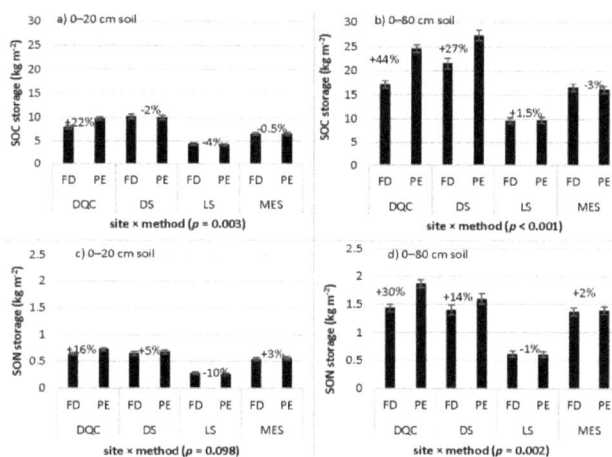

Figure 2. The influence of sampling method on soil organic C (SOC) and N (SON) storage across study sites (site×method interaction) and sampling depth (surface soil: 0–20 cm, whole profile soil: 0–80 cm). A plus sign (+) indicates that the pedogenetic method resulted in higher values than the fixed-depth method, while a minus sign (−) indicates the opposite result. DQC, Daqingchuan site; DS, Dongshan site; LS, Laoshan site; MES, Maoershan site; FD, fixed-depth sampling; PE, pedogenetic horizon sampling. The error bars represent standard error of the data. 0–20 cm soil for SOC (**a**); 0–80 cm soil for SOC (**b**); 0–20 cm soil for SON (**c**); 0–80 cm soil for SON (**d**).

3.4. Age-Group Differences in SOC and SON Concentrations: Interaction with Sampling Method

Age-group effects were similar between the two sampling methods for both SOC and SON concentrations (age-group* method interaction: $p > 0.05$; Table 1 and Figure 3). Moreover, age-group related differences were apparent in surface versus whole profile soil (Figure 3).

Both pedogenetic and fixed-depth sampling found sharp increases (6%–34%) in surface soil SOC concentrations from the <20 years to the 20–40 years age-group, which then moderately increased (9%–10%) from the latter to the >40 years age-group. Similarly, both fixed-depth and pedogenetic sampling found sharp increases (11%–22%) in SON concentration in the surface soil from the <20 years to the 20–40 years age-group, also followed by a small increase (2%) to the >40 years age-group (Figure 3).

Unlike in the surface soil, SOC and SON concentrations in the whole profile soil were lower in the >40 years age-group compared with the 20–40 years age-group (Figure 3). Both sampling methods found sharp increases (19%–28%) in SOC concentration in the 20–40 years age-group from the <20 years age-group, but 4%–5% decreases from the 20–40 years age-group to the >40 years group. Both sampling methods also found sharp increases (25%–27%) in SON concentration in the 20–40 years age-group from the <20 years age-group, followed by 17%–21% decreases in the >40 years age-group from the 20–40 years age-group (Figure 3).

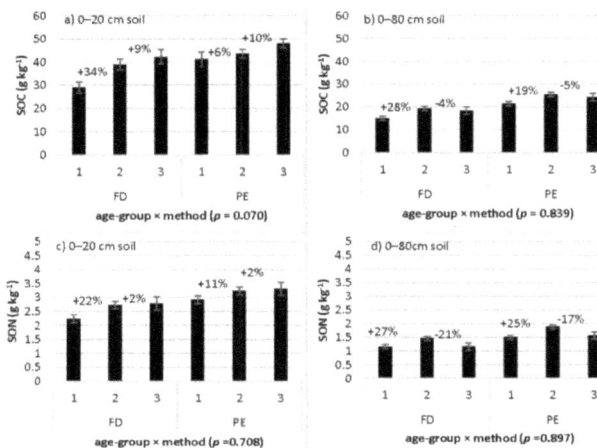

Figure 3. The interaction of age-group and sampling method affected soil organic C (SOC) and N (SON) concentrations, as well as their variation between surface (0–20 cm) and whole profile soil (0–80 cm). A plus sign (+) indicates that the pedogenetic method resulted in higher values than the fixed-depth method, while a minus sign (−) indicates the opposite result. DQC, Daqingchuan site; DS, Dongshan site; LS, Laoshan site; MES, Maoershan site; FD, fixed-depth sampling; PE, pedogenetic horizon sampling. The error bars represent standard error of the data. 0–20 cm soil for SOC (**a**);. 0–80 cm soil for SOC (**b**); 0–20 cm soil for SON (**c**); 0–80 cm soil for SON (**d**).

Owing to non-significant interactions between the age-group and sampling method (Figure 3), the age-group effect could be expressed as group means of the two sampling methods (Table 2). For the surface soil, marked increases in SOC and SON concentrations were found in the 20–40 years age-group from the <20 years age-group, while no marked changes were found between the 20–40 years and >40 years age-groups. However, different patterns were found in the 0–80 cm soil, with marked decreases of SON concentrtion from the 20–40 years age-group to the >40 years age-group (Table 2).

3.5. Age-Group Differences in SOC and SON Storage: Interaction with Sampling Method

Similar to SOC and SON concentrations, the sampling method did not alter age-group effects on SOC and SON storage (non-significant age-group × method interaction; Table 1 and Figure 4). Both fixed-depth and pedogenetic sampling revealed increases (11%–13%) in SON storage from the <20 years to the 20–40 years age-group, then small decreases (2%–3%) from the 20–40 years to the >40 years age-group. In addition, both sampling methods found sharp increases (7%–24%) in SOC storage from the <20 years to the 20–40 years age-group, then small increases (7%) from the 20–40 years to the >40 years age-group. Age-group differences in the whole profile soil SOC and SON storage clearly contrasted those in the surface soil (Figure 4). For example, there were sharp increases (15%–17%) in SOC storage from the <20 years to the 20–40 years age-group, but then no obvious changes (0%–2% decreases) between the 20–40 years and the >40 years age-groups regardless of the sampling method used. Furthermore, there were sharp increases (15%–20%) in SON storage from the <20 years to the 20–40 years age-group, and 13%–20% decreases from the 20–40 years to the >40 years age-group regardless of the sampling method used (Figure 4).

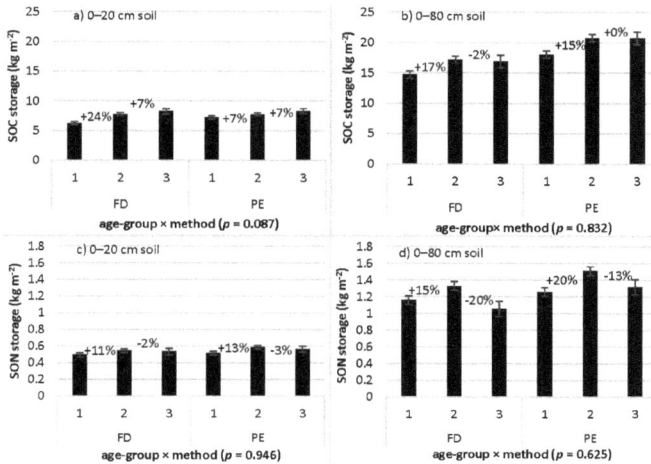

Figure 4. The interaction of age-group and sampling method on variation in soil organic C (SOC; upper two panels) and N (SON; lower two panels) storage, and differences between surface (0–20 cm; **a** and **c**) and whole profile soil (0–80 cm; **b** and **d**). A plus sign (+) indicates higher values in an older group than in the immediately younger one (e.g., >40 years versus 20–40 years), while a minus sign (−) indicates the opposite. 1 = <20 years age-group; 2 = 20–40 years age-group; 3 = >40 years age-group. FD, fixed-depth sampling; PE, pedogenetic horizon sampling. The error bars represent standard error of the data. 0–20 cm soil for SOC (**a**); 0–80 cm soil for SOC (**b**); 0–20 cm soil for SON (**c**); 0–80 cm soil for SON (**d**).

There were moderate increases in SOC storage and sharper decreases in SON storage from the 20–40 years to the >40 years age-group when the whole profile soil was considered (Figure 4). For example, surface soil SOC storage in the >40 years age-group was 7% higher than that in the 20–40 years age-group, whereas the percentage difference was a decrease of 0%–2% in the whole profile soil. Moreover, surface soil SON storage in the >40 years age-group was 2%–3% lower than that in the 20–40 years age-group, but the percentage difference increases to 13%–20% in the whole profile soil (Figure 4).

Owing to non-significant interactions (Figure 4), the age-group effect could be expressed as pooled means of two sampling methods (Table 2). For the surface soil, marked increases in SOC and SON storage were usualy found from the <20 years to the 20–40 years age-group, while no marked changes were found between the 20–40 years and >40 years age-groups. However, different patterns were found in the 0–80 cm soil, with marked decreases of SON storage in the >40 years age-group (Table 2).

3.6. Minimum Sample Number Required for Estimating SOC and SON at Specific Error Limits: Effects of Sampling Method

Fewer sampling points are required with fixed-depth sampling than with pedogenetic sampling, across multiple error limits (Figure 5). At least 538 samples should be collected for analyzing whole profile soil SOC concentration under pedogenetic sampling at the 5% error limit, whereas 37% fewer points are required with fixed-depth sampling to secure the same precision. This pattern of requiring more data points for pedogenetic sampling holds as the error limit increases (Figure 5). Similarly, under the same error limit, at least 543 pedogenetic samples are needed to analyze whole profile soil SON concentrations, compared with 480 samples for in fixed-depth sampling (12% fewer sampling points required) (Figure 5).

Similar results were observed for whole profile soil SOC and SON storage. At the 5% limit, pedogenetic sampling and fixed-depth required 387 and 233 samples, respectively, to estimate SOC storage, as well as 395 and 364 samples for SON storage estimation. At other error limits, these general patterns again held (Figure 5). Outcomes for surface-soil SOC and SON storage and concentration data were similar to the whole profile soil results (data not shown).

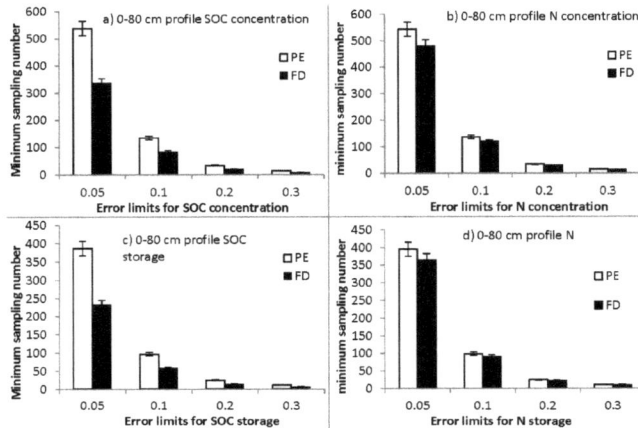

Figure 5. Effects of sampling method on minimum sampling number at different error limits to estimate differences in soil organic C (SOC) and soil organic N (SON) concentration and storage. PE, pedogenetic sampling; FD, fixed-depth sampling. Vertical bars represent a 5% error. 0–80 cm soil profile SOC concentration (**a**); 0–80 cm soil profile SON concentration (**b**); 0–80 cm soil profile SOC storage (**c**); 0–80 cm soil profile SON storage (**d**).

4. Discussion

4.1. Age-Group Effect on SOC and SON Is Dependent on Sampling Depth but Not Sampling Method

Compared with younger forests, older forests can have both higher and lower SOC and SON [11,14–16], but few have examined how sampling methods may affect such evaluations. Here, we demonstrated that sampling methods did not alter the estimate of the age-group effect on SOC and SON in larch plantations (Table 1, Figures 3 and 4). Pedogenetic horizon sampling in previous studies has revealed large increases (31.3 to 48.2 $g \cdot kg^{-1}$) in soil organic matter (SOM) concentration due to thinning [36], as well as large SOM variations (184 to 510 $g \cdot kg^{-1}$) within different larch forests [37]. Using fixed-depth sampling, SOM was found to increase from 47.7 to 51.6 $g \cdot kg^{-1}$ after a second rotation of a 36-years old larch forest at the same site [38]. Therefore, differences in sampling methods do not explain contradictory patterns of SOC and SON variation over time in the studied plantations. As seen in previous reports [9–12,17,18,20], prior land use, tree species planted, and soil nutrient supply may all affect SOC and SON dynamics after afforestation.

The patterns of tree age effect on SOC and SON was strongly dependent on soil sampling depth under both sampling methods. Under a deeper soil sampling depth, the SOC and SON estimates (storage and concentration) were accompanied by moderate SOC increases or sharper SON decreases from the 20–40 years to the >40 years age-group. Previous soil studies in larch forests were focused on the more dynamic surface soil (0–20 cm), which also contains the highest root density [39,40]. However, this focus may have overlooked the importance of deep soil, which is significant given that SOM and nutrient storage in deep soils are important for forest vegetation with deep root systems [41,42], such as the larch trees studied in this paper. A recent, 12 years chronosequence study in larch plantations showed that SOC initially decreased before increasing with stand age [43]. Larch plantations in northeast China could accumulate SOC at a rate that ranges from 57.9 to 139.4 $g \cdot m^{-2} \cdot year^{-1}$ in the top 20 cm of soil [11]. Moreover, researchers also noted a slight increase in surface soil N storage ($\sim$0.33 $kg \cdot m^{-2}$), which decreased from 0.11–0.16 to 0.06–0.11 $kg \cdot m^{-2}$ in the 20–60 cm soil during larch forest development [11]. Larch tree growth could cause divergent changes in SIC (soil inorganic carbon) and SOC levels, particularly in terms of their vertical distribution, and therefore, these effects should be fully considered in SIC-rich calcareous soils [44]. Larch reforestation could markedly affect the temporal dynamics in the concentration, storage, and vertical distribution of most soil nutrients, as indicated by findings showing eight SOC, SON, P, and K related parameters with differences between surface and subsurface soils [21]. These studies suggest that deeper soil C and nutrient depletion was possible. In contrast, this study suggests that larch plantation establishment in northeast China could sequester C in the mineral soil without depleting SON, at least in the surface soil of stands in the >40 years age-group. However, the inclusion of deeper soil layers could strongly modify these patterns to indicate possible depletion of SOC and SON with stand age.

4.2. Sampling Method Effects on SOC and SON Differ between Their Concentration and Storage

Understanding SOC and SON spatial variation is important in order to improve intensive forest management [2], but the influence of pedogenetic versus fixed-depth sampling on the assessment of SOC and SON spatial variation has not been statistically assessed [29,30,45,46]. This study demonstrated that sampling method significantly affected ($p < 0.05$) the identified size of inter-site variations of SOC and SON. Moreover, such sampling effects differed between concentration and storage of SOC and SON, as well as across the two soil sampling depths. These effects suggest that the effect of sampling method and its interaction with soil sampling depth on nutrient storage or concentration should be considered in future research, particularly when comparing results across different studies.

The A and B horizons in this study were generally thicker than 20 cm, with the A horizon at LS, MES, DS, and DQC averaging 17.8, 27.0, 40.5, and 25.0 cm, respectively, whereas the corresponding B horizon averaged 55.2, 66.1, 69.2, and 49.6 cm. Thus, fixed-depth sampling with a 20-cm increment may collect mixed A and B horizon samples. This corroborates with the literature that if soil horizons with different SOC concentrations are mixed during fixed-depth sampling, SOC and SON concentrations could be much higher in fixed-depth sampling than in pedogenetic sampling [47]. Others have also reported that SOC stock variability decreases under fixed-depth sampling as compared with pedogenetic horizon sampling [29]. Therefore, fixed-depth sampling could result in higher SOC and SON estimates than pedogenetic sampling depending on the depth of the soil sampled and also sampling intensities at different horizons [29,47].

Effects of sampling method on SOC and SON have been reported previously [29,30,45], but quantitative assessment on the effect of sampling method on inter-site variation and differences between sampling depths is lacking. In the upper 30 cm of ploughed Gleysols, VandenBygaart et al. found that pedogenetic horizon sampling reduced the variability in SOC stock [45]. Sampling by pedogenetic horizon is also recommended over fixed-depth when monitoring SOC stock variation in hydromorphic soils of the agricultural landscape [46], as well as when studying pedogenetic processes controlling SOC stocks. In contrast, fixed-depth sampling is preferred for determining regional SOC stock [29]. Yet another report [30] indicated that fixed-depth sampling tends to overestimate cultivation-induced SOC depletion, and thus they recommended the use of pedogenetic soil information (e.g., soil classes) in SOC stock calculation and C sequestration when assessing the impact of land-use change. Our findings differed from these reports; we showed that pedogenetic sampling typically resulted in much higher SOC and SON than fixed-depth sampling (Table 2), which was 1.05- to 1.2-fold higher for SOC and SON storage and 1.2- to 1.4-fold higher for SOC and SON concentration. Pedogenetic sampling was also used in several major datasets in China (e.g., the 1979–1982 national survey) and has since become the main sampling method for nationwide evaluation of SOC and SON budgets [48–50]. Currently, however, more studies use the fixed-depth method for estimating SOC and SON storage in China [11,21,51]. This study provides a statistical assessment of differences in SOC and SON between two major sampling methods, which should be useful for meta-analysis and soil fertility diagnoses across multiple locations.

4.3. Lower SOC and SON Spatial Variations Requiring Lower Sampling Intensity in Fixed-Depth Sampling

Many studies have addressed the effect of forest development on SOC and SON dynamics [11,21], inter-site and vertical soil variations [29,52], as well as the minimum sample number required for accurate assessment [20–23]. The minimum number of samples required should be determined for detecting a specified change in soil properties [34,53]. For designing efficient soil sampling methods, Conant et al. found that differences in the order of 2.0 $Mg \cdot C \cdot ha^{-1}$ could be detected through collecting and analyzing samples from at least five (tilled) or two (forest) microplots in Tennessee [5]. Compared with the large number required as was identified in this paper (300–600 samples), a proper sampling procedure, such as the microplot averaging method, could strongly decrease the number of samples required for accurate evaluation [5]. Moreover, the smaller variation with the fixed-depth sampling indicates that this method requires a lower sampling intensity (10%–40% lower) for a given precision as compared with pedogenetic horizon sampling in assessing changes in SOC and SON over time. Average SOC and SON changing rates are smaller relative to their total amount in the soil, resulting in a small signal to noise ratio [5]. The methodological effect on the minimum sample number required, together with previous findings on proper soil sampling design [5], may increase the precision of SON and SOC evaluations for a given error limit.

5. Conclusions

Larch plantation establishment on degraded farmland increased SOC and SON storage in the 20–40 years age-group, and sampling method (fixed-depth versus pedogenetic horizon sampling) did not alter this pattern ($p > 0.05$). Furthermore, pedogenetic horizon sampling resulted in higher SOC and SON storage and inter-site variations than fixed-depth sampling. At a given error limit for SOC and SON estimation, the minimum number of sampling points required for fixed-depth sampling was 60%~90% of that required for pedogenetic horizon sampling. These sampling method effects can help interpret differences in data collected in disparate studies (e.g., for meta-analysis), and improve our understanding of the assessment of spatial variations of SOC and SON.

Acknowledgments: This study was supported financially by the National Key Research and Development Program (2016YFA0600802), Fundamental Research Funds for the Central Universities (2572016EAJ1&DL13EA03-03), China's National Foundation of Natural Sciences (41373075, 31100457), and the "111" project (B16010). Thanks are due to De'an Xia from Forestry College, Northeast Forestry University for his kind support with statistical analysis and data explanation.

Author Contributions: W.W. conceived and designed the experiment; H.W. performed the experiment and analyzed the data; and H.W., W.W. and SC wrote the manuscript.

Conflicts of Interest: The authors declare no conflict of interest.

References

1. Rushton, B.T. Matching tree species to site conditions in reclamation. In *Evaluation of Alternatives for Restoration of Soil and Vegetation on Phophatic Clay Settling Ponds*; Odum, H.T., Rushton, B.T., Paulic, M., Everett, S., McClanahan, T.R., Munroe, M., Wolfe, R.W., Eds.; Florida Institute of Phosphate Research: Bartow, FL, USA, 1991.
2. Kravchenko, A.N. Influence of spatial structure on accuracy of interpolation methods. *Soil Sci. Soc. Am. J.* **2003**, *67*, 1564–1571. [CrossRef]

3. Franzen, D.W.; Hopkins, D.H.; Sweeney, M.D.; Ulmer, M.K.; Halvorson, A.D. Evaluation of soil survey scale for zone development of site-specific nitrogen management. *Agron. J.* **2002**, *94*, 381–389. [CrossRef]
4. Wang, H.-M.; Wang, W.-J.; Chen, H.; Zhang, Z.; Mao, Z.; Zu, Y.-G. Temporal changes of soil physic-chemical properties at different soil depths during larch afforestation by multivariate analysis of covariance. *Ecol. Evol.* **2014**, *4*, 1039–1048. [CrossRef] [PubMed]
5. Conant, R.T.; Smith, G.R.; Paustian, K. Spatial variability of soil carbon in forested and cultivated sites: Implications for change detection. *J. Environ. Qual.* **2003**, *32*, 278–286. [CrossRef] [PubMed]
6. Guo, L.B.; Gifford, R.M. Soil carbon stocks and land use change: A meta analysis. *Glob. Chang. Biol.* **2002**, *8*, 345–360. [CrossRef]
7. Laganiere, J.; Angers, D.A.; Pare, D. Carbon accumulation in agricultural soils after afforestation: A meta-analysis. *Glob. Chang. Biol.* **2010**, *16*, 439–453. [CrossRef]
8. Zinkevičius, R. Influence of soil sampling for precision fertilizing. *Agron. Res.* **2008**, *6*, 423–429.
9. Morris, S.J.; Bohm, S.; Haile-Mariam, S.; Paul, E.A. Evaluation of carbon accrual in afforested agricultural soils. *Glob. Chang. Biol.* **2007**, *13*, 1145–1156. [CrossRef]
10. Berthrong, S.T.; Jobbagy, E.G.; Jackson, R.B. A global meta-analysis of soil exchangeable cations, pH, carbon, and nitrogen with afforestation. *Ecol. Appl.* **2009**, *19*, 2228–2241. [CrossRef] [PubMed]
11. Wang, W.J.; Qiu, L.; Zu, Y.G.; Su, D.X.; An, J.; Wang, H.Y.; Zheng, G.Y.; Sun, W.; Chen, X.Q. Changes in soil organic carbon, nitrogen, pH and bulk density with the development of larch (*Larix gmelinii*) plantations in china. *Glob. Chang. Biol.* **2011**, *17*, 2657–2676.
12. Li, D.; Niu, S.; Luo, Y. Global patterns of the dynamics of soil carbon and nitrogen stocks following afforestation: A meta-analysis. *New Phytol.* **2012**, *195*, 172–181. [CrossRef] [PubMed]
13. Springsteen, A.; Loya, W.; Liebig, M.; Hendrickson, J. Soil carbon and nitrogen across a chronosequence of woody plant expansion in North Dakota. *Plant Soil* **2010**, *328*, 369–379. [CrossRef]
14. Covington, W. Changes in forest floor organic matter and nutrient content following clear cutting in northern hardwoods. *Ecology* **1981**, *62*, 41–48. [CrossRef]
15. Garten, J. Soil carbon storage beneath recently established tree plantations in Tennessee and South Carolina, USA. *Biomass Bioenergy* **2002**, *23*, 93–102. [CrossRef]
16. Wirth, C.; Czimczik, C.J.; Schulze, E.D. Beyond annual budgets: Carbon flux at different temporal scales in fire-prone siberian scots pine forests. *Tellus B* **2002**, *54*, 611–630. [CrossRef]
17. Kueffer, C.; Klingler, G.; Zirfass, K.; Schumacher, E.; Edwards, P.J.; Gusewell, S. Invasive trees show only weak potential to impact nutrient dynamics in phosphorus-poor tropical forests in the seychelles. *Funct. Ecol.* **2008**, *22*, 359–366. [CrossRef]
18. Cavard, X.; Bergeron, Y.; Chen, H.Y.H.; Pare, D. Effect of forest canopy composition on soil nutrients and dynamics of the understorey: Mixed canopies serve neither vascular nor bryophyte strata. *J. Veg. Sci.* **2011**, *22*, 1105–1119. [CrossRef]
19. Wei, X.; Shao, M.; Gale, W.; Li, L. Global pattern of soil carbon losses due to the conversion of forests to agricultural land. *Sci. Rep.* **2014**, *4*, 4062. [CrossRef] [PubMed]
20. Sauer, T.J.; James, D.E.; Cambardella, C.A.; Hernandez-Ramirez, G. Soil properties following reforestation or afforestation of marginal cropland. *Plant Soil* **2012**, *360*, 375–390. [CrossRef]
21. Wang, W.; Wang, H.; Zu, Y. Temporal changes in SOM, N, P, K, and their stoichiometric ratios during reforestation in China and interactions with soil depths: Importance of deep-layer soil and management implications. *For. Ecol. Manag.* **2014**, *325*, 8–17. [CrossRef]
22. Yim, M.H.; Joo, S.J.; Shutou, K.; Nakane, K. Spatial variability of soil respiration in a larch plantation: Estimation of the number of sampling points required. *For. Ecol. Manag.* **2003**, *175*, 585–588. [CrossRef]
23. Adachi, M.; Bekku, Y.S.; Konuma, A.; Kadir, W.R.; Okuda, T.; Koizumi, H. Required sample size for estimating soil respiration rates in large areas of two tropical forests and of two types of plantation in malaysia. *For. Ecol. Manag.* **2005**, *210*, 455–459. [CrossRef]
24. Laiho, R.; Penttilä, T.; Laine, J. Variation in soil nutrient concentrations and bulk density within Pearland forest sites. *Silva Fenn.* **2004**, *38*, 29–41. [CrossRef]
25. Kulmatiski, A.; Vogt, D.J.; Siccama, T.G.; Beard, K.H. Detecting nutrient pool changes in rocky forest soils. *Soil Sci. Soc. Am. J.* **2003**, *67*, 1282–1286. [CrossRef]
26. Verboom, W.H.; Pate, J.S. Evidence of active biotic influences in pedogenetic processes. Case studies from semiarid ecosystems of south-west Western Australia. *Plant Soil* **2006**, *289*, 103–121. [CrossRef]

27. Funakawa, S.; Hirooka, K.; Yonebayashi, K. Temporary storage of soil organic matter and acid neutralizing capacity during the process of pedogenetic acidification of forest soils in Kinki District, Japan. *Soil Sci. Plant Nutr.* **2008**, *54*, 434–448. [CrossRef]

28. Eger, A.; Almond, P.C.; Condron, L.M. Pedogenesis, soil mass balance, phosphorus dynamics and vegetation communities across a Holocene soil chronosequence in a super-humid climate, South Westland, New Zealand. *Geoderma* **2011**, *163*, 185–196. [CrossRef]

29. Grueneberg, E.; Schoening, I.; Kalko, E.K.V.; Weisser, W.W. Regional organic carbon stock variability: A comparison between depth increments and soil horizons. *Geoderma* **2010**, *155*, 426–433. [CrossRef]

30. Wiesmeier, M.; Sporlein, P.; Geuss, U.; Hangen, E.; Haug, S.; Reischl, A.; Schilling, B.; von Lutzow, M.; Kogel-Knabner, I. Soil organic carbon stocks in Southeast Germany (Bavaria) as affected by land use, soil type and sampling depth. *Glob. Chang. Biol.* **2012**, *18*, 2233–2245. [CrossRef]

31. Sun, Z.H.; Jin, G.Z.; Mu, C.C. *Study on the Keeping Long-Term Productivity of Larix Olgensis Plantation*; Science Press: Beijing, China, 2009.

32. Chen, Z.; Gong, Z.; Zhang, G.; Zhao, W. Correlation of soil taxa between Chinese Soil Genetic Classification and Chinese Soil Taxonomy on various scales. *Soils* **2004**, *36*, 584–595.

33. Gong, Z.; Chen, Z.; Luo, G.; Zhang, G.; Zhao, W. China Soil Classification Reference System. *Soils* **1999**, 57–63.

34. Petersen, R.; Calvin, L. Sampling. In *Methods of Soil Analysis. Part 1. Physical and Mineralogical Methods. Agronomy Monograph 9*, 2nd ed.; Klute, A., Ed.; American Society of Agronomy, Soil Science Society of America: Madison, WI, USA, 1986; pp. 33–51.

35. Robertson, G.P.; Coleman, D.C.; Bledsoe, C.S.; Sollins, P. *Standard Soil Methods for Long-Term Ecological Research*; Oxford University Press: Oxford, UK, 1999; p. 480.

36. Chen, X.Q. The effect of community structure of artificial larch wood on physic-chemical properties of Baijing soil. *J. Northeast For. Univ.* **1986**, *14*, 113–116.

37. Niu, X.; Wei, J.S.; Zhou, M.; Liu, B. Study on the soil organic matter under *Larix gmelinii* forest. In *Soil Fertilizer Society of Inner Mongolia 2007 Conference on Soil Fertilizer and Sustainable Development*; Inner Mogolica Publisher: Hohhot, China, 2007; Volume 2007, pp. 1145–1149.

38. Yan, D.; Wang, J.; Yang, M. Tendency of Soil Degradation in the Pure Larch Plantations. *Chin. J. Ecol.* **1997**, *16*, 62–66.

39. Kajimoto, T.; Osawa, A.; Matsuura, Y.; Abaimov, A.P.; Zyryanova, O.A.; Kondo, K.; Tokuchi, N.; Hirobe, M. Individual-based measurement and analysis of root system development: Case studies for *Larix gmelinii* trees growing on the permafrost region in Siberia. *J. For. Res.* **2007**, *12*, 103–112. [CrossRef]

40. Liu, S.R.; Wang, W.Z.; Wang, M.Q. The characteristics of energy in the formative process of net primary productivity of larch artificial forest ecosystem. *Acta Phytoecol. Geobot. Sin.* **1992**, *16*, 209–219.

41. Jobbágy, E.G.; Jackson, R.B. The vertical distribution of soil organic carbon and its relation to climate and vegetation. *Ecol. Appl.* **2000**, *10*, 423–436. [CrossRef]

42. Rumpel, C.; Kögel-Knabner, I. Deep soil organic matter—A key but poorly understood component of terrestrial c cycle. *Plant Soil* **2011**, *338*, 143–158. [CrossRef]

43. Wang, C.M.; Shao, B.; Wang, R. Carbon sequestration potential of ecosystem of two main tree species in Northeast China. *Acta Ecol. Sin.* **2010**, *30*, 1764–1772.

44. Wang, W.; Su, D.; Qiu, L.; Wang, H.; An, J.; Zheng, G.; Zu, Y. Concurrent changes in soil inorganic and organic carbon during the development of larch, *Larix gmelinii*, plantations and their effects on soil physicochemical properties. *Environ. Earth Sci.* **2012**, *69*, 1559–1570. [CrossRef]

45. VandenBygaart, A.J.; Gregorich, E.G.; Angers, D.A.; McConkey, B.G. Assessment of the lateral and vertical variability of soil organic carbon. *Can. J. Soil Sci.* **2007**, *87*, 433–444. [CrossRef]

46. VandenBygaart, A.J. Monitoring soil organic carbon stock changes in agricultural landscapes: Issues and a proposed approach. *Can. J. Soil Sci.* **2006**, *86*, 451–463. [CrossRef]

47. Palmer, C.J.; Smith, W.D.; Conkling, B.L. Development of a protocol for monitoring status and trends in forest soil carbon at a national level. *Environ. Pollut.* **2002**, *116* (Suppl. 1), S209–S219. [CrossRef]

48. Wu, H.B.; Guo, Z.T.; Peng, C.H. Land use induced changes of organic carbon storage in soils of China. *Glob. Chang. Biol.* **2003**, *9*, 305–315. [CrossRef]

49. Yan, X.Y.; Cai, Z.C.; Wang, S.W.; Smith, P. Direct measurement of soil organic carbon content change in the croplands of China. *Glob. Chang. Biol.* **2011**, *17*, 1487–1496. [CrossRef]

50. Yang, Y.H.; Fang, J.Y.; Ji, C.J.; Ma, W.H.; Mohammat, A.; Wang, S.F.; Wang, S.P.; Datta, A.; Robinson, D.; Smith, P. Widespread decreases in topsoil inorganic carbon stocks across China's grasslands during 1980s-2000s. *Glob. Chang. Biol.* **2012**, *18*, 3672–3680. [CrossRef]

51. Lv, H.; Wang, W.; He, X.; Xiao, L.; Zhou, W.; Zhang, B. Quantifying tree and soil carbon stocks in a temperate urban forest in Northeast China. *Forests* **2016**, *7*, 200. [CrossRef]

52. Wang, Q.; Wang, W. Grsp amount and compositions: Importance for soil functional regulation. In *Fulvic and Humic Acids: Chemical Composition, Soil Applications and Ecological Effects*; Barrett, K.D., Ed.; Nova Science Publishers, Inc.: Hauppauge, NY, USA, 2015.

53. Johnson, C.E.; Johnson, A.H.; Huntington, T.G. Sample size requirements for the determination of changes in soil nutrient pools. *Soil Sci.* **1990**, *150*, 637–644. [CrossRef]

MDPI AG

St. Alban-Anlage 66

4052 Basel, Switzerland

Tel. +41 61 683 77 34

Fax +41 61 302 89 18

http://www.mdpi.com

Forests Editorial Office

E-mail: forests@mdpi.com

http://www.mdpi.com/journal/forests

www.ingramcontent.com/pod-product-compliance
Lightning Source LLC
Chambersburg PA
CBHW051724210326

41597CB00032B/5599